ALIEN INTELLIGENCE
AND THE
PATHWAY TO MARS

ALIEN INTELLIGENCE

AND THE

PATHWAY TO MARS

The Hidden Connections between the Red Planet and Earth

MARY BENNETT

WITH

DAVID S. PERCY

Bear & Company
Rochester, Vermont

Bear & Company
One Park Street
Rochester, Vermont 05767
www.BearandCompanyBooks.com

Bear & Company is a division of Inner Traditions International

Cataloging-in-Publication Data for this title is available from the Library of Congress

ISBN 978-1-59143-400-9 (print)
ISBN 978-1-59143-401-6 (ebook)

Printed and bound in China by Reliance Printing Co., Ltd.

10 9 8 7 6 5 4 3 2 1

Text design and layout by Debbie Glogover
This book was typeset in Garamond Premier Pro with Arquitecta, Noyh Geometric, Gill Sans MT Pro and American Typewriter used as display fonts

To send correspondence to the author of this book, mail a first-class letter to the author c/o Inner Traditions • Bear & Company, One Park Street, Rochester, VT 05767, and we will forward the communication, or contact the authors directly at **www.aulis.com/pathway**.

For Kalina, Philippe, and Dylan,
Vivien and Tom.

Contents

List of Acronyms

ARE: Association for Research and Enlightenment Inc.

ASKAP: Australian Square Kilometre Array Pathfinder

CERN: Conseil Européen pour la Recherche Nucléaire (Geneva, Switzerland). In English: European Organization for Nuclear Research

CNES: Centre National d'Études Spatiales, the French space agency. In English: the National Center for Space Studies

CPS: cycles per second

CSIRO: Commonwealth Scientific and Industrial Research Organisation

DARPA: Defense Advanced Research Projects Agency

DLR: Deutsches Zentrum für Luft- und Raumfahrt. In English: German Aerospace Center

DOD: Department of Defense

DSP: David S. Percy, image source identity

EDI: Extra Dimensional Intelligence, also considered as paranormal phenomena

ESA: European Space Agency

ESP: extrasensory perception

ETI: extraterrestrial intelligence

FU Berlin: Free University of Berlin

GAI: German Archaeological Institute of Cairo

GPS: Global Positioning System

IAU: International Astronomical Union

ISS: International Space Station

JPL: Jet Propulsion Laboratory (Pasadena, California)

LANL: Los Alamos National Laboratories

LEO: low-Earth orbit

LML: Lookout Mountain Laboratory (Los Angeles, California)

MOLA: Mars Orbiter Laser Altimeter

NEA: National Endowment for the Arts

NOAA: National Oceanic and Atmospheric Administration

NGS: National Geographic Society

NRAO: National Radio Astronomy Observatory

OSIRIS-Rex: Origins, Spectral Interpretation, Resource Identification, Security-Regolith Explorer (space probe)

PSI: psychic faculties

PSI-Ops: intelligence gathering programs using human psychics

R&D: research and development

RSD: rapidly spinning disk

RV: remote viewing (protocol-led psychic intelligence gathering)

SAIC: Science Applications International Corporation

SCA: Supreme Council of Antiquities

SETI: Search for Extraterrestrial Intelligence (Institute)

SRI: Stanford Research Institute

UAP: unidentified aerial phenomena

UFO: unidentified flying object

USAF: US Air Force

USGS: United States Geological Survey

UTC: universal time coordinated

VLA: Very Large Array (Radio Telescope facility)

WHOI: Woods Hole Oceanographic Institution

Foreword

David Hatcher Childress

I have known the work of Mary Bennett and David S. Percy since the mid-1990s and have been constantly amazed by the depth of their research as well as their startling conclusions. Here, as in their previous works, they ask all the right questions and shine a light on even the most distant objects, such as Mars.

This book, written by Bennett, is foremost a captivating account of the search for extraterrestrial intelligence and the quest to travel into deep space. It looks closely at the anomalous Cydonia region of Mars, which is shown to be laid out as a special right triangle that in turn is enclosed within a phi spiral. While revealing the geologically and mathematically verifiable evidence pointing to the intelligent design of this area on Mars, it also shows, with Percy's excellent illustrations, the geometrical and cultural links between this design and Earth's ancient sites and monuments.

The book goes on to detail how exploratory probes sent to Mars in the 1970s triggered a plethora of anomalous events in various countries around the world over a thirty-year period. It reveals that the special right triangle found at Cydonia has been unwittingly mirrored in the development of Washington, DC, and indicates that various elements of the Cydonian geometry, already replicated in the ancient Avebury landscape in Wiltshire, England, have been more recently deliberately and discreetly replicated elsewhere on the planet.

I find the relationships between megalithic monuments on Earth and those on Mars extremely compelling, and this book takes a fascinating look at the relationship between the complex on Mars, the Nazca plateau in Peru, the Great Pyramid, and quantum computing.

Alien Intelligence and the Pathway to Mars demonstrates that the data and information held within the anomalous events of recent times can be decoded and, indeed much has been decoded, producing knowledge that can assist us humans individually in our personal growth and also help us collectively in overcoming present environmental and technological difficulties both on Earth and in space.

One thing that I have always appreciated in Bennett and Percy's research, including this book, is their insightful examinations of mainstream science and into whether what we are being told on a daily basis may be just plain wrong. When the average citizen is confronted with "facts and statistics" from rocket scientists and astrophysicists at NASA and elsewhere that are confusing and complicated, it is difficult for that person to seriously doubt the statements from respected scientists. But Bennett and Percy break down these facts and statistics so readers can make up their own minds. Whether it concerns the extreme dangers of space radiation, the bizarre inability to duplicate the Apollo missions of fifty years ago, or the diagraming of structures on Mars, Bennett and Percy discuss and elucidate all these topics in a credible, scientific way.

Here, with great clarity in their arguments and conclusions, they demonstrate how the geometric data present at Cydonia interact with the decoded knowledge to provide the blueprints of a viable spacecraft able to swiftly and safely take humans to Mars without being exposed to space radiation.

I have always liked books that get into the technical specifications of advanced aerospace designs, and in *Alien Intelligence and the Pathway to Mars,* thanks to Percy's illustrations, future concepts for a spacecraft using propulsion without propellant are brought into a present reality. Elaborating on some of the subject matter presented in the 1993 book by David P. Myers and Percy, *Two-Thirds: A History of Our Galaxy*, this spacecraft concept is based partly on Tesla's theories of electromagnetism and gravity and is described as using thrusting energy drawn from space for propulsion. At the time an outlandish suggestion. But by 2011, photon-based propulsion was being actively explored by some thirty-five teams under the aegis of NASA and DARPA (Defense Advanced Research Projects Agency). These teams have not adopted the concept of the rapidly spinning disk drive technology described by Myers and Percy, and little progress has been made toward a viable crewed craft for deep-space travel.

The proposed spacecraft design will deliver a totally different experience whereby one day it will be possible for astronauts to travel directly from the surface of Earth to the surface of Mars in the same craft. They would be able to transit between destinations rapidly while having total control of the craft within the vicinity of planets.

Bennett and Percy cover a great deal of ground in this extensively researched work and literally take us to another world! It is a world of ancient structures, astonishing science, and the promise of a future in space for mankind. They have deftly combined astronomy, math, art, music, philosophy, mythology, and history while imparting a wealth of information and fresh ideas. All the while they challenge cultural biases about who we are and where we come from. Take the leap into this book, and you will never think the same way about the worlds we live in.

DAVID HATCHER CHILDRESS

DAVID HATCHER CHILDRESS is a world explorer, an authority on ancient monuments, and an investigator into advanced technology and free energy. The owner of Adventures Unlimited Press, he is the author of *Technology of the Gods* and *The Anti-Gravity Handbook.*

Introduction

We human beings are a curious bunch. We want to know all about our world and how we got to be here on this planet, in this solar system, in this galaxy. Are we alone, or are there other beings like us out there? These age-old questions have only intensified as space engineers and scientists analyze data received from probes sent out into deep space.

The people of Earth have long been fascinated by Mars, and the notion that there might be life on the red planet has absorbed astronomers ever since Giovanni Schiaparelli observed channels on its surface over 150 years ago.

In the 1970s, we began physically interacting with Mars, first launching probes to fly past and orbit Mars, then sending small craft to touch down and, in more recent times, sending robots and roving vehicles to explore its surface. As a result of all this activity, Mars has been photographed and imaged extensively, but the next step —getting humans all the way there and landing them alive and in a healthy state—is a totally different matter.

In the early 1960s, NASA commissioned the Brookings Institution to study the ramifications of space exploration. This included looking at the question of how contact with any extraterrestrial intelligence (ETI) might occur. Given the perceived limitations of the speed of light and the state of rocket technology in the 1960s, no one in the space communities or at Brookings expected ETI to be anywhere close enough to warrant genuine concern about actual physical contact. However, the events described in the following pages reveal that the idea that ETI is far enough away to be irrelevant has not prevented the space agencies from employing their own strategic signaling to any intelligent life that may be out there, although whether they are asking for help or signaling "stay away" is a moot point.

The *Brookings Report* indicated that should there ever be any form of exchange, it was unlikely that the means of communication adopted by ETI would be similar to that used by us. In that respect, as it turns out, they would be confounded.

Whether ETI would use the same technology as ours for either contact or travel purposes is another debatable point. Naturally, the *Brookings Report* was founded on the technological understanding of human space travel capabilities as they were back in the 1960s. And this takes us to the nub of the problem we face today.

Imagine a conversation between a NASA representative and a member of the general public who is keenly following the human space program:

NASA spokesperson: *"We are going be sending astronauts to Mars, and we're planning to go soon."*

Everyman: *"How soon?"*

"Just as soon as we can perfect a rocket large enough to launch a spacecraft that can fully protect the crew from cosmic and solar radiation and sustain them all the way to Mars so that they arrive well enough and strong enough to walk unaided."

"I understand that a crewed launch is scheduled for sometime around 2033, so what's the problem?"

"Actually, we are finding that even in the International Space Station (ISS), orbiting 250 miles above Earth and mostly below the protective radiation belts, human bodies don't function too well. We don't know how to improve that situation right now. Nor do we know how to launch an adequately protected Mars-bound spacecraft with a suitable lander. Such a craft would be too heavy even for our big launchers."

"But wasn't that sorted in the 1960s with Apollo? Surely you just scale it all up as you are doing with the Space Launch System."

"Scaling up isn't that simple; there are several major challenges to overcome. The crew module/lander combo is still too heavy, we still haven't mastered the essential technique of a skip re-entry, and a Mars return is even faster than a lunar return."

The *Brookings Report* and any possible contact with ETI now come into focus. It is the premise of this book that the achievement of successfully landing a probe on the surface of Mars demonstrated to ETI that as a species we are now on the verge of addressing the challenge of sending human astronauts to land on another planet. But in reality, however good automated probes might be, the lack of progress in crewed deep-space flight has demonstrated that our existing rockets and spacecraft technology are woefully inadequate when it comes to protecting human beings anywhere beyond low-Earth orbit (LEO).

During the 1970s, the United States and the Soviet Union managed to send probes to Mars equipped with cameras to make planetary observations. And as NASA's *Mariner 9* was traveling toward the red planet, space engineers, having decided that mathematics would be understood by any civilization advanced enough to interface with the probes, were preparing messages for ETI. Greetings plaques were attached to Pioneer craft numbers 10 and 11. Launching in 1972 and 1973, respectively, both of these deep-space probes bore an aluminum plaque engraved with diagrams encoding mathematical, biometric, and locating data.

Meanwhile, in June 1972, *Mariner 9,* having had its mission brief extended, was taking a second round of photos of the Cydonia region, and here on Earth anomalous events occurred indicating interactions from ETI that, as the following chapters will show, have continued in various forms.

Four years after *Mariner 9,* the Viking probes were orbiting Mars and once again photographing the Cydonia region. In July 1976, an image was spotted by Tobias Owen of the Jet Propulsion Laboratory (JPL). It looked so much like an upturned face it was specifically mentioned at the July 26 press briefing by the Viking project scientist and astrobiologist Gerry Soffen, and immediately dismissed by him: "When we took another picture a few hours later, it all went away; it was just a trick, just the way light fell on it." Coincidentally, a committee led by astrophysicist and author Carl Sagan was in the process of selecting elements to be incorporated into another ETI communication. This time, gold plated copper disks were to be attached to the two *Voyager* probes launching in 1977. These disks included a message from then United States President Jimmy Carter:

> This is a present from a small, distant world, a token of our sounds, our science, our images, our music, our thoughts and our feelings. We are attempting to survive our time so we may live into yours. We hope some day; having solved the problems we face, to join a community of galactic civilizations.[1]

After collating the available evidence during research for this book, it has become abundantly clear that communications and messages from Earth have not gone unheeded. Just as humans have sent out these encoded and engraved diagrams and disks so have messages been returned to us here on Earth, inserted into our material world as symbols and shapes containing layers of decipherable meaning.

We have evolved our means of transportation considerably over the years. From the early harnessing of animals leading to horse-drawn carts and carriages came the invention of steam, in turn leading to the internal combustion engine, progressing to the age of electric power, which is ideally suited to vehicles of all kinds, especially trains. In the air, propeller-driven aircraft were greatly improved with the advent of the jet engine. This is progress through incremental steps.

However, despite the brilliant scientific minds of rocket engineers, the space industry is the exception—it has not progressed in a comparable way. The principle of deploying a rocket for ascending flight has not changed substantially since its invention by the Chinese in the tenth century, and the very basic idea of burning or combusting fuel in one form or another to power these rockets continues to this day.

To put this into context, one might ask what it would be like if the telephone, instead of developing from the basic, rather chunky and limited instruments of the 1960s into the compact smart pocket computer used today by billions, had remained static for fifty years. Unimaginable!

Yet, today humans can only travel to the ISS orbiting just 250 miles off the surface of the Earth—just one thousandth of the way to the Moon. NASA, the agency that built the lunar rockets and modules of the 1960s, having retired the inadequate Space Shuttle, has relied completely on Russian spacecraft to transport their astronauts up to the ISS. A situation that only began changing in the summer of 2020, when the American-built Space X had its first test run of a human rated spacecraft.

During the 1960s, when the term "rocket science" became shorthand for doing something really brainy and clever, the public was encouraged to leave the difficult stuff to the scientists. But with the hindsight of 20/20 vision, maybe we should all start asking the awkward questions since, clearly, there are profound challenges to be overcome when designing, building, and launching a craft for journeying into deep space with a full crew aboard and a fit-for-purpose lander.

Not least among the challenges is the fact that every rocket launch is a major polluter of the atmosphere. The Earth is ringed with ever increasing numbers of orbiting satellites and potentially dangerous space junk, while the oceans are the repository of much of the hardware, fuel, nuclear power units, and any other items that don't make it into space or "deorbit" and survive atmospheric re-entry.

The stagnating technology underscores the necessity for a big rethink. A totally new approach to human space travel is seriously overdue, and a paradigm shift is needed to make a genuine breakthrough.

In February 2020, the SETI (Search for Extraterrestrial Intelligence) Institute announced that the search for ETI had finally "gone mainstream" when it was granted access to real-time data coming from the Radio Telescope facility in New Mexico known as the Very Large Array (VLA).

Should this new SETI-VLA collaboration succeed, then a change of perspective is on the horizon. Tony Beasley, the director of the National Radio Astronomy Observatory (NRAO), which runs the VLA, made this statement:

> Determining whether we are alone in the universe as technologically capable life is among the most compelling questions in science, and [our] telescopes can play a major role in answering it.[2]

This 2020 initiative, along with the assertion that it is primarily the NRAO and VLA telescopes in charge of this search for ETI, indicates a profound change of attitude. One that is only partly due to the sheer quantity of planets that have now been found to conform to the criteria defined by scientists as "potentially life-supporting."

This book suggests that energetic interactions with our world (some via the VLA), mostly considered anomalous events by scientists, are actually encoded aspects of physics that, to date, mainstream science and the space agencies appear not to have fully grasped.

The messages sent from Earth encode factual information for ETI about the people of Earth, with a sprinkling of culture, such as the first two bars of Beethoven's String Quartet no. 13 in B-flat Major, added to the mix. The messages received on Earth also contain extra content to stimulate creative thinking, for the apparent high strangeness of the events described ultimately offer each of us the opportunity to explore our own response to other realities, to become more in tune with our environment and each other, and to become capable of embracing change harmoniously, should we so choose.

This book contains the story of how we got to where we are today in the matter of ETI and the quest to travel into space, and it also suggests how we might proceed in decoding more of the data and information held within those anomalous events described. It further proposes that the knowledge transmitted can assist us here on Earth. This is because we are living on our very own spaceship and so the blueprints that can provide us with the means to travel safely to Mars also offer solutions for overcoming present technological difficulties here on our home planet.

The research that has led to this book has involved a journey of some twenty-five years for my colleague David Percy and me, making connections with moments of insight and realization along the way. It is still very much a work in progress, and we shall continue developing themes from this book online. In the meantime we hope that the information we have amassed will enable the next step along the pathway to Mars, bringing successful deep-space travel to future generations.

MARY BENNETT

The Quest for Answers

Who Are We and Where
Do We Really Come From?

The quest for meaningful answers to the major questions of all time has preoccupied humanity since our ancestors first gazed upward and contemplated the starry skies.

Of what stuff are we made? Physicists, biologists, and chemists agree we are of the same elements as the stars.[1] This conclusion implies a deep and meaningful physical connection to the star that is the very genesis of the energy on which we depend for life—the Sun. It also implies a direct connection to all those other stars at their various stages of development—at the very least those contained within our own galaxy.

When did some of this star stuff become a self-aware creature, a human being? Anthropologists, archaeologists, and historians dig down through the earth looking for traces of earlier cultures in the search of answers to this complex question, and as they do so, they find their boundaries pushed ever further back into the mists of time. The assumptions made as to the capacity and intelligence of ancient humans are in constant need of revision. The focus of attention for ancient man and for modern humans may have been and is very different, but ancient peoples were able to resolve problems that today we would not know how to address, or even know why we should bother to do so.

Where do we actually come from? In attempting to address the emergence and spread of humankind, the "out of Africa theory" is now in question. So it is legitimate to ask if the traditional ladder or path of evolution is worth re-evaluating. After all, if essentially we are composed of star stuff and have an innate connection with "out there," then, digging down into our own psyche, we could ask ourselves some really big questions.

Who are we, Homo sapiens sapiens, *relative to any other putative civilizations?* Are there other sentient beings in other star systems who have their own experiences of living and are perhaps at other stages of development—physical, intellectual, and spiritual? And is our geocentric, solar-centric opinion of ourselves necessarily the same as any external observer's opinion of ourselves?

TOO CLOSE FOR COMFORT

It was humankind's early exploration into space that set us on the path to seeking real answers to questions formerly the preserve of philosophers and fans of science fiction. And although we humans are finding space travel challenging, our probes are wandering the solar system, and when apparent evidence of intelligent interaction with the landscape on our neighboring planet Mars emerged into the public domain in 1976, something in our collective consciousness was woken, and it became impossible to keep the subject of our ancestry a fiction, but that has not been for want of trying.

Human beings are a curious species, and when exploring our environment we have variously achieved the ability to travel, for the most part successfully, on water, on land, and in the air. For most of our energy needs, we have harnessed aspects of another basic element—fire. We then turned our eyes back to the starry skies we first contemplated millennia ago. Consequently, the most recent frontier for us to consider conquering is how to travel to those stars. Over the last fifty years or so, we have had some small successes with launching satellites, robotic probes, telescopes, shuttles, and space stations. But today, there are more unanswered questions than those that have been solved when it comes to human space travel. Even within LEO, astronauts orbiting Earth on the ISS are subject to various unfortunate biological effects—not all clearly understood or fully manageable. Yet despite this lack of a comprehensive understanding as to how to overcome these fundamental difficulties, the world's space agencies intend to fast-forward with human exploration/exploitation of both the Moon and Mars.

Clearly, it is vital to have goals and to continue this desire to explore, which is innate within us all precisely because of the very stuff of which we are made. And that makes these fundamental questions ever more pertinent. To those already posed, we might then add: *Why are we here, and what are we doing on this gem of a blue-green planet?* Some say that we are still "at school" or that we are currently Earth-bound "in quarantine" for our own good and for the safety of everyone else "out there." Others suggest that our predecessors trashed our original home planet in the distant past and that we settled and evolved here instead. These ideas might have some validity, but then again, shouldn't we have matured over time? Is it inevitable that we trash the oceans and the environment of our own home planet all over again? As a result of which we will need to exploit and eventually colonize the perceived resources of our Moon or Mars. For it appears we are doing just that, out of greed, perceived energy needs, and continuing disagreements between each other. Has such immature behavior sent a clear signal to anyone observing that we are not yet mature enough to proceed in a responsible manner? Or worse? From the point of view of an outsider observing our space exploration programs, to be doing the same thing over and over again while expecting to achieve different results would appear insane.[2]

THOSE OTHERS

Considering the general behavior between nations, cultures, and professions, it is obvious that, for the most part, we are somewhat belligerent toward those considered to be "others," mostly those whom we do not know or who do not look like us or speak like us or behave like us or hold the same ideas as we do. Finding common ground both on and off the planet might be the best way forward. Because if we are made of the same stuff as the stars, then surely each one of us is composed of the same stardust as all these "others"—both here on Earth and elsewhere too. Even if we, sentient beings, may look different, with differing languages and varying viewpoints and ideas, we are all of the same essence.

When it comes to exploring the so-called High Frontier, it's not necessarily the case that the raw fire we deployed in the past is the answer to the power source needed to get us humans from here into deep space and beyond.

Although that message does not seem to be getting through. The principles set out by Princeton physicist and professor Gerard O'Neill in his influential 1976 book *The High Frontier: Human Colonies in Space* retain their grip. The appendix to the updated 1988 edition still had this advice for start-ups in the space business: "Avoid, if at all possible, developing new technologies or stretching old ones. Instead, assemble building blocks of existing technology in such a way as to build a new capability that serves a real need." That book is much respected by NASA staff; however, it will become clear over the course of this book that if we are going to succeed in our inquiry into our origins and our desire to get out into deep space, we are going to have to develop new technologies and factor in "the fifth element."

This fifth element has gone under different names in the past. In relation to humans, it has been described as the etheric: some have called it the invisible energy of space—the ether—and have given as its material symbol the fifth Platonic solid, the dodecahedron. Physicists today refer to dark matter and dark energy; ancient and modern philosophers refer to it as the filaments of connection between (a) so-called nonliving systems, (b) all living things, and (c) all self-aware beings. Traditionally and metaphorically, the nonliving systems are represented by the number 6, all living things by the number 5, and the self-aware beings' physical body by the number 7. It is suggested that multiples of these three different levels of consciousness connect self-aware beings to everything in the universe, no doubt at very subtle levels. As we stand on the threshold of space, if we truly intend to travel up and away from the planet to work and live in harmony elsewhere without repeating the errors of the past, then it's likely that we shall have to take these subtle effects of the fifth element into account before we can turn our dreams of space travel into reality. Otherwise, like ancient man, we will still look up at the stars and continue to feel and wonder. However, unlike ancient man, we will not be able to translate any of those feelings and wonderment into a useful technology.

The study of Earth's ancient monuments combined with the use of modern technology has revealed that our ancestors built many of our most ancient monuments as practical tools for living on the planet. In the site locations and layouts of their buildings, it is clear that they also had the ability for considerable wise thinking, which means having the ability to operate their intuitive senses far more efficiently than the majority of us can do today. This planet's ancient monuments have revealed considerable amounts of information passed on through the ages. Many researchers suspect that the works laid down by ancient stargazing ancestors contain resources hitherto unrecognized by modern humans. Those interested in cosmic connections and future space travel suspect that any such data will only be considered of practical use when our species has sufficiently matured mentally—first to understand the data and then to use them wisely and peacefully. As Jacquetta Hawkes famously remarked, "Every age has the Stonehenge it deserves—or desires."[3]

Despite the considerable achievements of robotic space exploration thus far, and even though research on the ISS has demonstrated that the human body in space can benefit from specific harmonic resonances, NASA appears to be using an entirely inadequate technology for future human spacecraft and space stations. For the most part, the technologies employed to date have been developed as a result of wars, tension, and dissonance.

Thanks to the perceived necessities of deterring attacks from the air, radar was developed rapidly during World War II. And once again that age-old question resurfaced: *Are we alone?* Experiences by both allies and axis pilots of seeing lights (subsequently dubbed "foo fighters") flying near to their planes, along with other unexplained aerial phenomena never accounted for either during or after the conflict, had already convinced the authorities of all combatting nations that there was a need to search for any possible ETI. Post–World War II, rocket-powered craft, devolved directly from Germany's World War II rocket program, were later adapted to missile warfare and then to satellites, probes, and spacecraft for people. Post–World War II, thanks to the surplus of both technicians and equipment, radio astronomy progressed equally rapidly, setting us all on the road to the stars.

Parallel to these developments, the exploration and domination of the human mind through the use of drugs had been a subject of research for many governments. Warfare used as the excuse for developing and deploying drugs to produce specific physical conditions and mental states in a human being. Unsurprisingly, drugs are now being proposed as the means by which an astronaut might survive long spells traveling in space.

Indeed, the stress of conflict has often been credited with forcing nations into new discoveries by focusing the financing of new technologies. Although the rocketry that fueled the so-called Cold War was literally one of pushing and shoving a vehicle up into space, paradoxically, exploring far beyond the atmosphere is more than likely going to require a peaceful cooperation between the technological components we invent and the environment in which they, and ourselves, must function. And for this to be achieved, instead of scrapping like difficult teenagers, we might even have to grow up and become responsible adults. And invent a new way of traveling in space.

Down through the successive generations there have been those who know in their bones that form also generates function. What if the ideas we need to pursue have already been encoded not only by nature itself in the design of the solar system but also in the designs and layouts of Earth's ancient monuments and structures, which are also waiting to be decoded in conjunction with the various motifs used to create their unique architectural features? This might seem ridiculous, yet many of these structures, while spread out across the planet, have a demonstrable coherence of design and methodology. How so? These cultures are not recorded as having intermingled at the time, but the inspiration for these similar architectural forms came from somewhere. What if ancient man's use of intuition and openness to creativity resulted in constructions encoding layers of data and information, some useful for their times and some waiting for the time when we would need them? What if a set of blueprints showing us alternative

Photo: DSP.

Figure 1.1. The high Nazca plateau on the Pampas de Jumana, Peru.
This side view from the air shows the principle plateau in the background
and a separate flattened area at a lower level in the foreground. Attributing
the geoglyphs on this UNESCO World Heritage Site to the indigenous population
is to ignore these facts: that several of the stylized depictions of life-forms originate
far from Peru and that considerable engineering was required prior to placing miles
of trapezoid forms across the plateau and the surrounding mountains.

processes for living in harmony with a planet, and traveling off planet, has been waiting for us to come of age?

Some of this thinking is not news; there are many who consider that there must be an underlying pattern to the seemingly random placement of each of the great monuments across the world. Various arrangements of geometric patterns are discussed in books on this subject, but with each author proposing a different geometric design underlying their "world grid."[4] While these works are strong on pattern and placement within their selected sites, they fail to address the fundamental issue: What coherence of thought produced various ancient archaeological sites based on the same design principles while separated by vast distances and, to our eyes at least, built by different cultures? Pyramids are a good example of this perplexing problem. Whether stepped or smooth-sided, adorned or plain, these square-based monuments were constructed across the globe at different periods of history by different cultures, yet they are instantly recognizable as belonging to the same architectural family.

Let's look at some examples of two pyramid complexes and a city built on the same principles. The pyramid city of Teotihuacán, the pyramid complex at Giza, and the Forbidden City of Beijing all have connections and correspondences, but at the time of their building, according to the available records, there was little if any trading of importance occurring between Mexico, Egypt, and China. It is, however, a fact that all three complexes are based on the golden spiral, the phi spiral. The city of Teotihuacán also echoes the layout of the Forbidden City in Beijing in that both have waterways manipulated to intersect the site at the same level. At Teotihuacán, the natural waterway flowing east of the site has been deviated to cross the site a quarter of the way up, separating the sunken Quetzalcoatl complex from the rest of the site.[5]

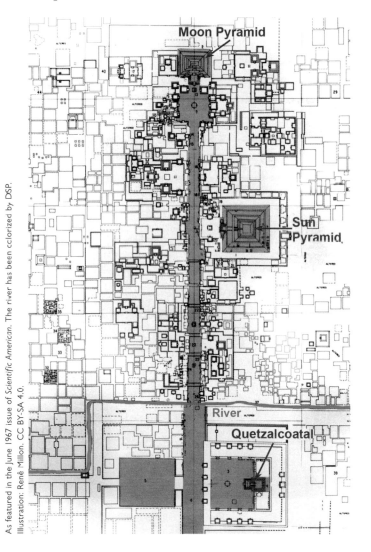

Figure 1.2. Teotihuacán complex, Teotihuacán, Mexico, as featured in the June 1967 issue of *Scientific American*. North is to the top of the image, along with the Moon pyramid, the Sun pyramid is to the east of the Avenue, and below it to the south is the sunken Quetzalcoatl complex.

In Beijing's Forbidden City, the artificial moat has been fed into the site down the western side and then deviated to cross the site at the same distance from the entryway as at Teotihuacán, but here it separates the emperor's receiving complex from Tiananmen Square and the forecourts.[6]

The phi spiral is also the foundation on which the three large pyramids at Giza were constructed. This connectivity speaks of coherence of thought regarding placement, form, and, as it turns out, function.[7]

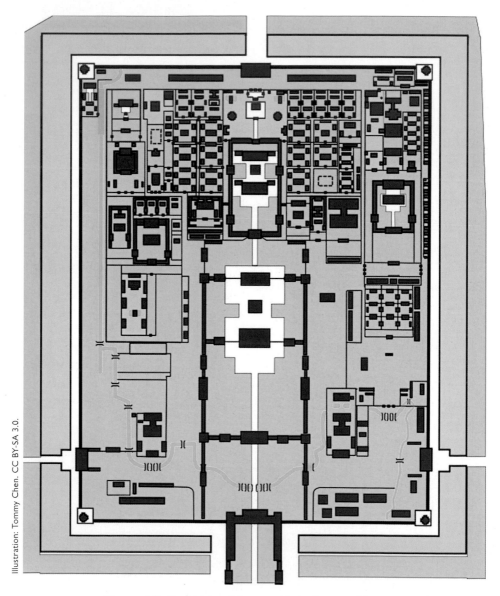

Figure 1.3. Forbidden City complex, Beijing, China.

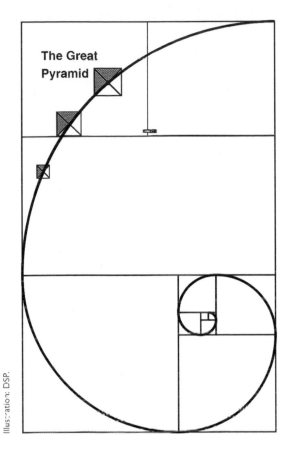

**The Great
Pyramid**

Figure 1.4. Giza plateau, Egypt, with superimposed phi spiral.

It is certainly interesting that the very same fundamental principles set out in these two ancient complexes are again present when we consider the planet's location, only this time in proportions of thirds.

GOLDILOCKS

Our solar system is located two-thirds of the way from galactic center, in the galaxy's Orion arm. Within the rocky planets of the inner solar system, the Earth is two-thirds of the way from the Sun. However, if we take the asteroid belt as the boundary between the rocky planet arena and the four outer planets, the so-called gas and ice giants, then the third planet from the Sun, the Earth, is one-third of the way from the Sun. The Earth is considered by scientists to be the most likely orbital location in this solar system where life, as a start-up, could develop from single-cell bacteria and evolve on through to human beings. This outcome is possible because the Earth orbits in what astrophysicists call the "Goldilocks Zone"—an ideal location defined by the orbits of Venus and the Mars.

Venus, although slightly larger than Earth, is considered too hot for a start-up and will never be a place for life as we know it. Mars, a planet half the size of Earth, is too cold, but again, this seems not to have been the case in the past, and speculation is rife among scientists as to the how, when, and why the environment on Mars changed. Albeit with some competition from Mars as scientists learn more and more about conditions on the red planet, Earth is just right, neither too near nor too far from its Sun. Earth also has an abnormally large Moon (nearly half the size of Mars). This keeps our axis relatively stable; it varies through three degrees over forty-one thousand years. (According to our astronomers, the range is from 21.48° to 24.4° and we are currently at 23.93°.) Mars, on the other hand, has two interesting smaller moons, but these are not large enough to affect its axial tilt, which is therefore more variable than ours. Although we all know that we are orbiting the Sun and rotating on our imagined axis counterclockwise, we tend to describe the Sun's motion from the perspective of an observer standing on Earth. As such we tend to imagine its rising in the east and setting in the west as a clockwise motion.

Earth's orbiting Moon also generates a tidal pull on both land and sea, as does the Sun. In this regard, the Moon, being much closer, has a stronger effect than the Sun. The Moon's orbital actions also generate a subtle clockwise wobble on the counterclockwise rotation of the Earth, and this wobble affects our orbital motion around the Sun. And thus our perception of where we see the Sun rising at the spring equinox changes over time. While astronomers to this day designate the spring equinox, 0° Aries on or around March 21, as the starting point of the solar retrograde precessional cycle, over very long periods of time, from here on Earth it looks as though the Sun gradually slips *clockwise,* taking 25,920 years to complete one Great Year. Ancient astronomers said of the Sun during its passage against the background stars that it "dwelt in the house": spending 2,160 years in each house of 30° arc by moving at a rate of one degree every seventy-two years. Now, as most people know, the Sun is gradually leaving the house of Pisces and will be entering the house of Aquarius.

Both ancient and modern astronomers understand that the ecliptic constellations actually vary in size, but for mapping and calendar purposes, the ancients also chose to divide the ecliptic into twelve "houses" of equal size summing to 360°. This band of twelve constellations is known as the zodiac. The tilt of our axis also means that an extra constellation is often seen sharing the house of Scorpio. As such, Ophiuchus, the serpent bearer, is referred to as the thirteenth constellation.

Interestingly, the ancient font at Avebury parish church, in Wiltshire, England, depicts this serpent bearer.

On a much shorter seasonal cycle of spring, summer, autumn, and winter, our rotating planet allows us to see the Sun move through these twelve "houses" on a monthly counterclockwise basis. The ancient Chinese also operated a zodiacal calendar of 365.24 days that recognized the influence of the Moon on Earth's seasonal year.

Whether one is for or against astrology as a methodology, it is quite possible that the combination of our position relative to the Sun, along with the Earth's axial rotation and the effects of the Moon on the planet, produce virtually the same effect on our perceptions throughout time and across cultures. It's worth recalling that while many cultures use strings of beads as a handheld tool during prayer, meditation, or simply to calm the mind, the Tibetans used a bead chain with cosmic significance. Originally made of seeds, the Buddhist chain came in three sizes, having 27, 54, or 108 beads. These multiples of nine are relevant, as to this day Tibetans consider nine to represent "all"; for example, nine thoughts would represent all thought. The nine becomes one. As the Tibetans were formerly spiritual counselors to the ancient Chinese, it's unsurprising that the ancient Han people used nine in the same context, along with the notion of infinity.[8] The fact that these three different mala bead lengths can be generated from initially squaring the number 3 leads to another insight. The auspicious number 3 is applied to many matters important to the Tibetans, among which are the cosmic triple of Earth, Moon, and stars. From all this number manipulation, the ancient Tibetans created the longest 108-bead chain. It was derived from the sum of the diameters of the Moon, Earth, and Sun and multiplying the result by 108. This figure produced the distance of Earth, a planet supporting life, from its own star—the Sun. With the addition of another single and larger bead (also present on the Tibetan chain), the 109 beads then encoded the number of times the diameter of Earth divides into its Sun's diameter. How did they know that? This stunning cultural example begs the question as to exactly how such subliminal messages might be received. And provides a clue in the function of these prayer beads as a tool for the mind. Is this information concerning a Sun, a planet harboring a sentient species and its moon, part of the specific requirements for the emergence of self-aware life within a solar system? And have Tibetan meditation practices enabled them to tap into the collective unconscious, of their own people or that of the solar system or another galaxy? And if such technical data relative to our place in the solar system have been translated into an artifact for use as a meditative tool, then it is reasonable to conclude that we might, at some stage in our cultural development, whether in Mexico, Egypt, China, or elsewhere, have included some of the same sort of technical data and translated barely consciously aware sensations or perceptions into the building of our pyramids, stone circles, and cities and, indeed, other important artifacts.

PYRAMIDS

Modern archaeologists have assessed and classified the pyramid complexes in terms of the amount of material that was displaced and the numbers of manpower hours required to build these monuments. These specialists consider pyramid building to be a demonstration of the authority of the designated "instigator" of these builds. A decline in pyramid building and in quantity or quality is attributed to social change within the culture. But this completely materialistic approach does not take into account other attributes of these pyramidal constructions.

Giza is commonly considered to represent half the hemisphere of this planet. It would be more accurate to state that its build incorporates data relative to half of a three-dimensional sphere. However, remembering that the diameter of Mars is just over 50 percent of our own diameter and that the city of Cairo is named after the planet Mars, we might be missing a trick here. Surveyors and architects have also linked Teotihuacán to luni-solar planetary matters, and it is the so-called Moon Pyramid that dominates the avenue of that city. So we might wonder if smooth-sided pyramids perform one function while step pyramids have another function. We also know that the Sumerian astronomers used a seven-step pyramidal (ziggurat) system to signify latitudes from the equator to the North Pole.[9] As each step therefore represented 12.85° (12 degrees, 51.42 minutes), the priest-astronomer at its summit was literally on top of the world. However, that number of degrees is the *average* of degrees the Moon tracks around our planet every day. Which recalls the twelve-day average length of time between a woman's ovulation and her period.[10] If we no longer build pyramids in quite such a considered and monumental form today, is it because, subconsciously, we know we do not need to because they are already in situ? Or is it because we have forgotten what they do when fully in form? Taking the square base of all these pyramidal forms, does this particular form affect the planetary environment at its location, whether under its base or in the air surrounding it? If so, how much has any subsequent damage to our truly ancient monuments affected us?[11] Even if the copies constructed more recently are but a partial reflection of an original total concept, it is clear that many of us still have a fascination for pyramids.

Photo: DSP.

Figure 1.5. The Luxor Pyramid, Las Vegas, plays on the Egyptian theme to the hilt. Built in 1993 as a casino and thirty-story hotel complex, its 111-meter height makes it similar to the Egyptian Red Pyramid at Dahshur, just south of Giza. The Vegas pyramid is flanked by an obelisk and a reproduction of the Giza Sphinx.

So much so that when pyramidal forms were first observed by the *Mariner 9* probe at 16.6° north in the Elysium region on Mars, even Sagan thought them "worth a closer look." Which is unsurprising, since the three principal pyramids on the seemingly featureless Elysium plain were remarkably akin to the layout of the Giza big three on their high plateau, while the Elysium latitude was 1.11° different from the 15.49° azimuth of Teotihuacán's Avenue of the Dead. On Mars, when a complex of pyramidal forms was found at Cydonia, some 40–41° north and just over 9° west of the Martian Airy 0° prime meridian, the cat, so to speak, was among the pigeons.

How does it happen that the northern boundary latitude of all Earth's pyramid complexes is the same as the location of a city full of pyramid forms on Mars?

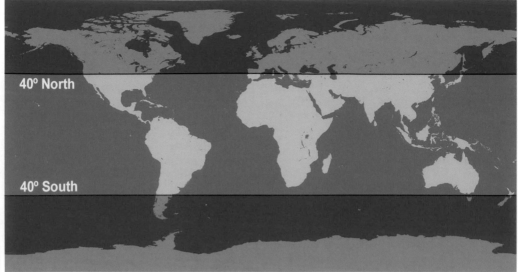

Illustration: DSP.

Figure 1.6. Ancient pyramids on Earth are found to lie between 40° north and south latitude, while 40° north on Mars finds the Cydonian pyramid complex adjacent to the famous Face on Mars.

How does it happen that the azimuth of a "lunar city" in Mexico is comparable to the first complex of Martian pyramidal forms deemed worthy of note by scientists?[12]

Bearing in mind that the designer-architects were apparently totally disconnected from each other, one might ask how such a geographical constraint came about. Although any interplanetary connection remains tenuous until we actually get astro-boots on the Martian ground, the authorities have made enough noise about the Martian discoveries to make it legitimate to inquire as to how cultures, thousands *or even millions* of miles apart, came to build complexes that have significant cartographical and highly accurate mathematical relationships to each other.

Are the observed ideas and traits that exist within all societies, and even within each human being, carried over time, across vast distances, and passed from generation to generation? Is this transference uniquely down to our genetic inheritance? Computer engineers are actively researching the possibility of storing data in synthesized DNA, so it is quite possible that the so-called junk DNA, which does not code for proteins but is used by forensic scientists when identifying an individual, has the potential to store information across multiple generations over eons. Which would mean that we can access this data via our minds and that all mysti-

Photo: Acme Photos, 1945. Prints and Photographs Division, Library of Congress, Washington, DC.

Figure 1.7. Xi'an Pyramid, China, 1945, located at 34.33° N, 109.28° E. This pyramid was unknown to the West when this photo was taken. It is the 210 BCE mausoleum of the first emperor of China, Qin Shi Huang, and is comparable in height to Teotihuacán's Pyramid of the Sun. The Terracotta Warriors were found nearby.

cal practices, whether meditation or other means used to tap into the subconscious or hyperconscious, would help in that endeavor. From a purely physical basis, the natural environment of places we intuitively choose to visit or to live potentially enhances the possibilities for subliminal data collection. And it follows that those places esteemed across time to hold special power, especially when enhanced with amazing architecture, also contain the "where-with-all" that we need.

Then other questions arise: Do we create different forms for different functions? Are square-based pyramids for example, associated uniquely with the effects of planetary rotation? Are stone circles (Stonehenge in England is a prime example) uniquely concerned with the effects of our yearly orbital revolution around the Sun and the Moon's orbital revolution around us?

At first glance, this might seem a useful way of categorizing monuments, but it only goes so far. In the 1970s, engineer and surveyor Hugh Harleston Jr. meticulously surveyed the Teotihuacán pyramid complex and surrounding area. He established that the city had a common measurement system, which he called the Hunab.[13] He also found that the distances between the principal structures along the main Avenue of the Dead, expressed in Hunabs, were virtually indistinguishable from the mean orbital distances of the planets of the solar system as we know them to be today and expressed in astronomical units from the Sun.

This apparent encoding of the solar system included Uranus, which is barely distinguishable with the naked eye. Interestingly, the Uranus structure is now called the Moon Pyramid. Harleston thought that if he had indeed found a model of the solar system and that if it included a planet generally beyond the remit of naked-eye observations, finding anything at the putative locations of Neptune and Pluto (at that time still considered as a planet in its own right) would confirm his theory. To his delight and astonishment, walking in the natural terrain beyond the ruined city complex but on the extended sightline of the Avenue of the Dead, he found significant markers inscribed on stones placed at exactly the distance predicted for the orbits of Neptune and Pluto. He felt the urge to continue his exploration, and at nearly four times farther out than the Uranus-Sun distance, he found yet another marker.

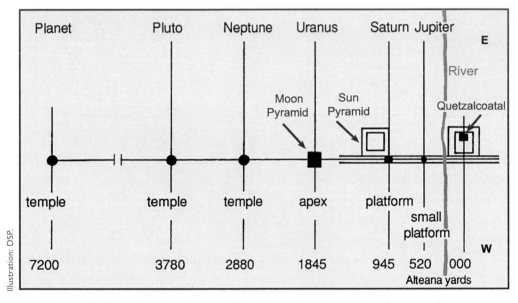

Illustration: DSP.

Figure 1.8. Teotihuacán, Mexico. This illustration shows the distances between the outer solar system and how they connect to the built city of Teotihuacán. These distances are expressed in Hunabs, the 3.47571113 feet unit of measure discovered by Hugh Harleston Jr. North is to the left of the image, and the distance between Pluto and another unknown planet has here been shortened. Crossing between Jupiter and the Quetzalcoatl complex, the river represents the asteroid belt. The Quetzalcoatl complex at the southern end of the built city represents the inner solar system. Its sunken courtyard contains the Quetzalcoatl (bird-snake) pyramid, and smaller builds represent Mercury and Venus. Mars is on the northern and southern ramparts; Earth is on the eastern rampart and the western entryway. Also note that the measurement of one Hunab is also found across the Atlantic: it is the width of the lintels linking the huge verticals of the sarsen ring at Stonehenge.

Photo: DSP.

Figure 1.9. Fellow researcher David Percy on the Moon Pyramid at the northern end of the main avenue, Teotihuacán, Mexico.

Decades later, Harleston would redefine the naming of his pyramids while keeping his Hunab measurement system. More recently, a Japanese researcher would assert that Harleston should have adopted the megalithic yard converted into meters! The differences are interesting and will come into focus later in this book; here this measurer's squabble diverts from the fact that many millennia previously, our ancestors somehow knew how to build a city accurately based on a solar system, including a barely visible planet. This "invisible" planet happens to be the only planet of this solar system that lies on its side, with its axis at about the same 5.1415° from the horizontal that our Moon has with the ecliptic. So it's even more interesting that this Teotihuacán pyramid is the focus of the main avenue of this city and that later ancestors called this pyramid/planet Uranus— "the Moon." Some have given it the nickname of "the Distant Reflection."[14]

Crossing the Atlantic and considering the Giza complex near Cairo, Egypt, there are still some enthusiasts for the notion that "the big three" on the Giza plateau reflect the three belt stars of the Orion constellation. Given the mismatch between ground features and star features, most astronomers do not find this theory (first proposed by Robert Bauval and Adrian Gilbert in *The Orion Mystery,* 1994) particularly persuasive. Even the authors themselves are less enthusiastic today, but they might have picked up on something subliminally. Our solar system is located in the Orion spiral arm. There is a multiplicity of threes on the Giza plateau, and the layout of this complex would better fit a visual reference to Earth's position as the third planet out from the Sun. This hypothesis gives another interpretation for the presence of a lion sculpture on the Giza plateau, the lion being considered a solar symbol. Furthermore, the two groups of three small pyramids are two-thirds of the total pyramid build on the plateau, and Giza's latitude is two-thirds of the way from the North Pole to the equator. In relation to each other, these two small groups of three pyramids also encode an X/Y coordinate system. It is generally considered that the ancient Egyptians did not have access to the knowledge required to encode data such as latitudes and longitudes into this location (see chapters 18–20).

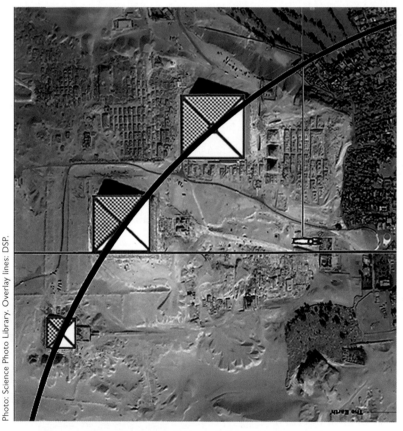

Photo: Science Photo Library. Overlay lines: DSP.

Figure 1.10. The Great Pyramid (top center) overlaid with the curve of the phi spiral crossing the big three.

CIRCLES OF TIME

It is generally agreed that whatever else Stonehenge might represent, it does reference lunar/solar positions and cycles as seen from Earth. And appearing to have been constructed in three phases, it seems to have some links to the "three motif" at Giza.

The architectural features of these three ancient sites suggest that their original designers had a profound knowledge of their environment and of the heavens: Teotihuacán demonstrates the fact that in order to encode the distances between planets, ancient man would have required full knowledge of all the planetary bodies in the solar system. Combined with the fact that even the Aztecs themselves had no clue as to who had built this city, to our modern minds this is clearly an impossibility. Today, archaeologists don't know why *sheets of mica* were used in two specific locations in the complex (one of which was the Sun Pyramid). All of which poses such profound questions about our past that while Harleston's academic qualifications are not doubted, to all intents and purposes his findings are generally denigrated by his peers, their conclusions widely avoided.

The Great Pyramid at Giza is a monument that many agree couldn't even be constructed today, nor do we really understand why *red granite* was such a vital part of the construction that it was transported all the way from Aswan in southern Egypt and installed in a chamber that, according the Egyptologists' tomb theory, was intended to be sealed forever. Further, if the "Orion hypothesis" carries any weight at all, it would also infer knowledge of our actual location *in the galaxy*.

The development of Stonehenge over time incorporates a bluestone component: stones brought from a location in the Welsh mountains some 140 miles distant. And again, experts have failed to convincingly reproduce the methods by which this transportation was undertaken at the time of construction. Nor do we understand why this was necessary. The same applies for the huge sarsen stones at Stonehenge. In 2019 a lost 42 inch (108 cm) core sample extracted from one of the trilithons being repaired in 1958 was returned to the UK. By the summer of 2020, delighted archaeologists felt able to announce that with two exceptions, the Stonehenge sarsens originated from West Woods near Avebury. Despite the fact that conventional wisdom always knew that north Wiltshire was the source of these stones, and notwithstanding the new analytical technologies available since 1958, it was admitted that no one had tried very hard to establish the exact source before. This lack of curiosity might have had something to do with the next problem such a confirmation poses: it is now necessary to explain how these huge stones were transported some eighteen miles over very undulating terrain. Stonehenge is principally laid out

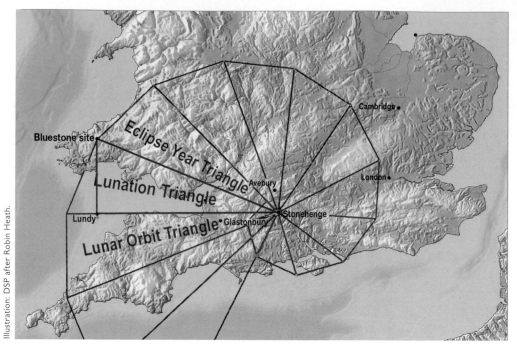

Illustration: DSP after Robin Heath.

Figure 1.11. Southern England under a lunar phi spiral
emanating from Stonehenge.

as a circular monument, purportedly to reflect the perceived motion of the Sun
and the Moon across our skies. Author Robin Heath also finds correspondence
between the phi spiral geometry and the monument at various scales. Here the
lunation triangle connects Stonehenge to the bluestones and the southern English
and Welsh landscape.

It is clear to construction engineers that building a structure like the Great
Pyramid is not possible today, even with current technology, and that from
experiments carried out by archaeologists and engineers to date using the methods extant in ancient Egypt, neither could the extant Egyptian civilization.[15] The
same observations are valid for Stonehenge. And surely the solar system layout of
Teotihuacán suggests that either its designers or its builders knew of technologies
and techniques that we have long forgotten—or have yet to access. We seemingly
have the choice between our own lost memories and the fact that another civilization from elsewhere, somewhat more technically advanced than ourselves, was
involved in the establishment of these complexes. In either case, Harleston has
shown us the way at Teotihuacán, and in recovering the knowledge held therein,
we recuperate our memories of our past and gain access to our future.

The very presence of these three major sites, as well as the circumstances of their construction, is virtually crying out to us to ask questions and poses a challenge to investigate and resolve them rather than glossing over the visible evidence by simply accepting any old myth or closing the mind and decrying any notion of outside help from elsewhere. Closer analysis indicates that these sites are multiplexed and multilayered, with meanings both exoteric and esoteric. This is the case despite the thousands of years since their inauguration, and it turns out to be highly relevant to us in the twenty-first century—as the considerable numbers of visitors to these places testify.

Other than basic guidebook information, the first principle of all ancient sites noted by those with a wider interest in their meaning is their specific layout, because the geometry defines the site relative to the local environment and the form adopted. A pentagonal shape does not "do" the same things as a square or a circle; each form harnesses the local energetic environment in a particular way.

Finding a single site using specific geometric forms to describe an aspect of astronomy, as in the three examples above, infers that the layout on the ground relates to the function. Which might be confirmed by the well-known fact that the phi spiral underlying the Giza pyramid layout is manifest in nature throughout the structure of all living things and that currently the third planet from the Sun is the only one in this solar system supporting life as we know it.

Surely any worldwide matrix of structures would have to be about more than just joining locations where single geometric forms are sited to create a visual net around the globe. Hypothetically, it ought to contain useful information in a form that conveys *something* when intelligences interact with it. Since all of us are continually transferring energy from one state to another, as is all of nature, then it is quite possible that any pattern or patterns will do the same thing. If we are receiving subliminal notions, once we become fully cognizant of what they might be and how they interact with us, we can make use of such information. Many ancient sites are seemingly related to the observed motions of our solar and galactic systems, so perhaps we should investigate the possibility that these layouts are indeed blueprints that describe new ways of harnessing the natural energies around us, but without blighting the landscape or acting in detriment to the planet or its inhabitants. Or have these sites arisen simply because our ancient ancestors used their senses differently or more efficiently than we ourselves do today?

Although scientists do not yet know enough about our DNA and RNA to be able to say whether we have inherited an awareness of form and function from our ancestors, it was confirmed in 2015 that the human optical system contains the same markers as those used by birds for flight navigation via the Earth's magnetic field. Whether navigation means the same thing for humans as it does for birds is less certain. Bird species are in the main not genetically mixed. Modern human beings, on the other hand, are already a mix of two species—Neanderthal and Cro-Magnon—in varying proportions according to their genetic inheritance.[16] And they have also merged their cultures over time. So it might be that our modern human abilities in this regard vary according to several factors: our genetic inheritance, our location at any one time, and the hemisphere in which we live, since the inflow of the Earth's magnetic field occurs through the north magnetic pole with the outflow through the south magnetic pole.

To think about this in another way, those living within the latitudes of 40° north and south of the equator might be more responsive than others to pyramidal forms and the sense of place and time given by the diurnal pattern of rotation. Farther north, those living around the western European coastline and in the British Isles, at the interface of land and sea, where stone circles abound, might better relate to the very specific tidal forces of the Moon and the Sun. And then there are those Nordic boat-shaped stone formations that consist of two parabolas of equal radii intersecting at bow and stern. These are mimicked somewhat in form by the British Isles' long barrows, variously aligned to the cardinal and quarter points of the compass.

Looking for the underlying key pattern or even fully understanding each individual site is a mind-expanding exercise in itself. However, this is far from New Age whimsy; the defense industries of many nations are just as keen to find the "keys to the kingdom," since the potential of pro-actively exploiting the effect any such patterns might have on their environment has implications for war game technologies. More peaceful people, having the niggling suspicion that ancient man was rather more in tune with Earth than most of us today, suspect that simply finding a coherent overall pattern would lead to a better understanding of our place in the scheme of things and offer more harmonious ways of working together.

And that might not be all.

NASA is set on going to Mars, but many of the aspects of this long journey are extremely challenging for the human body, to say the least. As challenging perhaps as that of accepting that ancient monuments might hold a valuable database for human space travel beyond LEO. The shock of confirmation that there are other intelligences who already know how to do this might alleviate their dismay over the fact that conventional methods of rocketry are apparently not considered best practice. Certainly, the possibility that we all have the biological potential for tuning in to the Earth's magnetic field will have huge ramifications for human space travel. It begs the question: Can we also adapt or tune ourselves to other planetary magnetic fields of different strengths?

EMERGENCE OF THE A-WORD

Although astronomers had already started scanning the Earth for clues about ancient ETI as soon as the ability to photograph the skies above was perfected, the public impetus for seeking *the* key pattern (although latent in human beings since the beginning of time) really flowered in the 1960s. The influential American Edgar Cayce, the "Sleeping Prophet," had predicted that "Atlantis would rise again" at this period. Whether he truly meant an island emerging from the Atlantic or something else is not clear from his published words, but American Egyptologists took him very seriously indeed and set off for Giza, no doubt aware of the fact that two decades earlier, ostensibly as a vital point of its World War II campaign, the US Air Force (USAF)* had created an azimuthal equidistant projection map using, as the 0/360° centerpoint, the largest mass of stone accumulated in a pyramidal form on Earth—Giza's Great Pyramid.

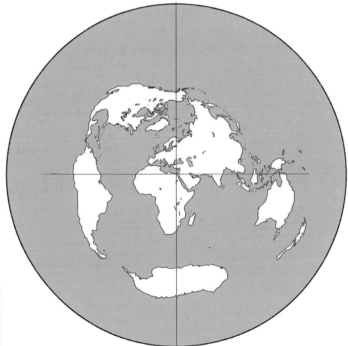

Figure 1.12. Azimuthal equidistant map centered over Cairo, modern version of US Air Force original.

Illustration: DSP.

---

*Until 1947, the US Air Force was part of the US Army known as the US Army Air Force (USAAF). In 1947, it became its own separate branch of the military, the US Air Force (or USAF). Rather than using two potentially confusing acronyms, USAF will be used for the Air Force no matter what the time period.

Meanwhile, US rocket engineers were becoming ever more adept at launching satellites and probes into space, and in the 1960s, NASA set up the Apollo space project and ended up calling a space shuttle *Atlantis*.

However, when it sought to identify space flight with the notion of lost continents and ancient civilizations, it's a moot point as to whether NASA's PR department had taken into account only the data acquired from secret military programs designed to explore and use the abilities of the human mind, or whether the pronouncements of Cayce and other American psychics had been the leading influence. Perhaps the latter was a cover for the former.

PLATES IN MOTION

A few decades later, this same monument at Giza, the Great Pyramid, would be determined by mathematician and cartographer Carl Munck (a man originally trained in the USAF) to be the 0/360° prime meridian of a worldwide geo-mathematical grid. How intimately Munck's work was connected to the previous exploration of Earth and elsewhere for lost civilizations such as Atlantis will be explored in later chapters, as will the stimulus that initiated the finding of this geo-mathematical grid. The very fact that a global grid might exist at all raises a rather big question.

Earth has very active plate tectonics shaping its geology and features, the ramifications of which are not yet fully understood. However, as a result of its tectonic plate activity, Earth's crust is constantly in motion, devolving and evolving. It follows that from the outset, any matrix consisting of notable artificially engineered locations and monuments many thousands of years old, but all relating mathematically to the same 0/360° prime meridian today, as was discerned in the twentieth century by Munck, would need to have been the subject of an overall plan from the outset. A plan that took into account the highly complex rate of Earth's tectonic plate motion together with the mechanics of rotation and orbital revolution. This is the only way in which it would have enough mathematical coherence such that it would alert us to its somewhat inconvenient existence.

GAMESMANSHIP

The process of developing any sort of space technology designed to explore the solar system has, at the very least, opened our minds to the necessity of decades-long long-term planning and the essential role played by satellites and probes in the observation and exploration of planetary surfaces. From the evidence of our own eyes, here on Earth, the fact remains that, notwithstanding any ETI hypothesis, our ancestors have provided us with opportunities to grow in understanding.

Yet, despite the best efforts of some authors and researchers, it hasn't been possible, so far, to identify anyone among our terrestrial ancestors who had access to both a global positioning system (GPS) and an appropriately sophisticated computer program to enable accurate prediction of complex plate tectonic movements over the millennia. Indeed, we humans have only been in the satellite business since the late 1950s and only very recently developed GPS capability, while computer modeling of plate tectonic movement is still in its infancy. So it's easily understandable that for the most logical among us, the notion of any such a plan has been dismissed as wishful thinking.

That being so, as we will be discussing, it would seem probable that any ancient global matrix could only have been created with the involvement of ETI, perhaps so much like ourselves they could be called our ancestral *relatives*. At this juncture, for those who cannot accept such a premise, why not consider all extraterrestrial interactions as manifestations of interaction between your own everyday mindset and aspects of your own mind so profound that you have never considered that they might affect your active everyday mind.

For those prepared to consider the totally extraterrestrial aspect, it looks very much as if the awkward questions posed by the ancient monuments and the equally awkward questions posed by the discovery of the anomalous landforms on Cydonia (which, as we will see, include multifaceted pyramidal forms and a face-like structure) are two parts of the same puzzle—or dialogue. Which might explain why these Martian landforms are publicly ignored by historians, archaeologists, and NASA. Although, in principle, NASA should fully understand the concept of data hidden within seemingly inconspicuous constructs, since that is a formula the agency adopts constantly.

There's No Business like Show Business

The Struggle to Get into Space and the Challenge of Radiation

THE DAYS OF APOLLO

Fifty years ago, the Apollo lunar missions were broadcast around the world and considered to be the pinnacle of mankind's exploratory capabilities. Since then, as already stated, human astronauts have only ventured to the ISS orbit, barely one-thousandth the distance to the Moon.

Considering the achievements of the Apollo era, one may well ask why even the relatively basic task of reaching an even lower altitude, the edge of space, some 50 miles up, is proving to be so difficult. Formerly achieved by X-15 pilots and currently being attempted by companies like Virgin Galactic, scaling up from the X-15 to a passenger-carrying craft has proven arduous. This is also the case for those involved in the redesign of what is, to all intents and purposes, a scaled-up version of the Apollo hardware intended for future crewed flights to the Moon and/or Mars.

It is becoming clear there are numerous difficulties at the root of human space travel that go way deeper than matters of funding or design problems. NASA's own publications confirm that a craft that can achieve what has allegedly been done decades previously has yet to be perfected. And there is worse: despite a lunar module apparently having left the lunar surface on six different occasions with no issues, ascent from the Moon is now described as "escaping from the deep gravity well of the Moon" and considered to be extremely risky and a major problem.[1]

As it turns out, what we knew or didn't know about the space environment back in the early days of space exploration and the manner in which those discoveries were dealt with has led to unintended consequences beyond the merely practical. These remain unresolved today. It is unlikely that human beings will be traveling safely beyond LEO until these leftovers from our early efforts have been resolved, so let's get going and start with where we are today.

STUCK IN LEO?

NASA states that it takes some two days to reach the ISS, but considering that the Apollo missions took around four days to reach the Moon, not being rocket scientists, many people naturally assume that the ISS is stationed farther from the Earth than is actually the case. They are rather surprised when they find out that the space station is only orbiting at an altitude which varies from 240–260 miles (386–418 kilometers) above the surface of the Earth.[2] In fact, the ISS orbits as near as it can get to Earth without falling back through the atmosphere. This low-Earth orbital trajectory's acronym of LEO inspires multiple images.

However bizarre it might seem to the civilian mind, NASA and its colleagues in the US Department of Defense (DOD) bestow multiple meanings and insider jokes on the terms they use. More often than not, these acronyms and project names turn out to have other meanings than those officially ascribed to them. Inevitably, the term LEO brings to mind all things leonine, including the great African Sphinx surveying the distant horizon while guarding the Giza Pyramids. NASA is also very keen on the myths of ancient Egypt and most especially that of the god Osiris and the Orion constellation; this star group is a recurring theme within the US space program, and their German rocket engineer Wernher von Braun even had a boat called *Orion*.

For the film buff, LEO might even bring to mind the roaring lion of MGM Studios, whose name was also Leo, and the studio's Latin motto is *Ars Gratia Artis* ("art for art's sake"). It would be nice to think that this is not the adopted motto of the early space program, but as it turns out, Hollywood and the DOD are surprisingly intertwined—far more than is evident to the casual observer or average filmgoer.

The "marketing" of the US space program through the use of photographs, films, and TV was a particularly effective communication tool—or Cold War weapon (depending on one's point of view). During the 1960s, *National Geographic* regularly published features on the status of the space program. Astronauts and their families were presented to the public through predominately photographic magazines such as *Life,* while those in power at *Look* were actively supporting the PR activities of the DOD—and employing talented young still photographers such as the soon-to-be film director Stanley Kubrick.

However, over and above the "normal" amount of media publicity that the Mercury, Gemini, and Apollo programs engendered in such photo magazines as well as on cinema newsreels—and excluding the rosy-spectacled "hindsight films" concerning the space industry (*The Right Stuff,* 1983, and *First Man,* 2018, are two examples)—there was a third arm to the film world specifically created for the US government by the Hollywood studios. The Walt Disney Company had been operating Graphic Studios, dedicated to making documentary films for the DOD that were notably responsible for conveying the nascent US space program's aspirations. Announcing the glorious path mankind would carve in going to the Moon and then to Mars, these productions were fronted by none other than Wernher von Braun, who explained how the United States would conquer space with his rocket designs. Although NASA historians have put it slightly differently: "Von Braun served as technical advisor on three space-related television films that Disney produced in the 1950s. Together, von Braun (the engineer) and Disney (the artist) used the new medium of television to illustrate how high man might fly on the strength of technology and the spirit of human imagination."[3] Together they would also create Tomorrowland at Disneyland in California.*

There is nothing new in the dissemination of propaganda, which has been going on since the first council of elders sat around a fire and told their tribe a story that was, or wasn't, depending on the purposes of the tribal elders, a fable. Note that one of the meanings of the word *fable* is "a falsehood."

In later times, if the principles remained the same, the development of photographic, animation, and film technologies was a gift to any postwar government in need of a fable or two to boost their nation's morale. And as a deterrent, when allied with a political or scientific agenda, these new technologies were dynamite—or in the case of the Manhattan Project, somewhat more powerful. The images of nuclear explosions and their effects on the surrounding environment were coupled with narration or captions emphasizing the dangers of such weapons, the whole package intended to cast fear in the minds of all who looked at these images. Whether friend or foe was of indifference to those who generated and then disseminated this propaganda. Cold War. Cold heart.

---

*Where possible, references to NASA sources are given, but NASA websites are subject to adjustment: documents tend to disappear or move to another section of the site, and content is also susceptible to change. It is advisable to enter a key phrase, such as "History of Gemini" and follow up all leads.

Next up for the space filmmakers after the von Braun/Disney TV films was the exploration of an environment where most of us will not travel, and it took place during the 1957 International Geophysical Year (referred to as the IGY, that particular "year" was actually of eighteen months). Three months in, on October 4, the Soviet Union launched its infamous Sputnik satellite, which orbited the Earth *within* LEO. Even though this event was entirely expected by the authorities (both the United States and the Soviet Union intended to launch a satellite during this period), in public the US government agencies made much of the allegedly surprising and fearful event. Three months later, in the United States, on January 31, 1958, James Van Allen, together with colleagues William H. Pickering and Wernher von Braun, finally managed to launch a rocket equipped with instruments to test the radiation fields *beyond* LEO.

Photo: NASA.

Figure 2.1. The three men responsible for the success of *Explorer 1*, America's first satellite, launched January 31, 1958. From left, William H. Pickering, James Van Allen, and Wernher von Braun.

This measurement project masks the fact that the results were something of a showstopper. Although data were received, *Explorer*'s instruments ceased to function due to the overwhelming strength of the space radiation beyond LEO. A fact that took some time to ascertain. The planned domination of space by the US astronauts was thrown into doubt. The two distinct radiation belts discovered by *Explorer* that protect the planet from the worst ionizing radiation that the solar system throws at us were subsequently named after Van Allen His article concerning the discovery of these belts was published in *Scientific American* magazine fourteen months later. In that article, he informed the public that crewed spacecraft to the Moon and beyond would need adequate protection in order to travel safely through and beyond the inner proton belt and the outer electron belt.[4]

Although since 1958 the ability to launch probes and satellites has at least improved, in 2019, specialists are still exploring and trying to understand the characteristics of the protective shielding around our planet, which has been operating according to the laws of nature for millennia. By its very nature, even if we did not have the means to discover this fact in 1958, the interaction with both the Sun and the Moon, let alone the effect of other planets on our system, means that this shielding has always been that of fluctuating energetic barriers. Indeed, it is only very recently that better technology has enabled scientists to realize that the extreme variability of these belts means that they can become three fluctuating belts and not the two-ring model as previously supposed. What we do know is that this region of space is hazardous for software, including life-forms such as human beings. It is not particularly good for hardware either; even within LEO, the computers on the ISS are susceptible to crashing when passing through the South Atlantic Anomaly, a region where protection from this radiation is at its thinnest.

SHOWSTOPPER? BUT THE SHOW MUST GO ON

Scientists have called the problem of cosmic and solar system radiation "the showstopper" for the exploration of space by human beings, at least with our present spacecraft technology.

Professor Clive Dyer, M.A., Ph.D., has worked in space and radiation research for more than forty years, authoring more than two hundred publications in the field, and in June 1997, he told David Percy:

> Radiation is the biggest showstopper affecting mankind's exploration of the Universe.[5]

Indeed, Sir Bernard Lovell of the University of Manchester's radio tele-scope facility at Jodrell Bank Observatory recorded the different viewpoints on this matter of radiation, and in a letter to Percy, he noted that in the 1950s and 1960s, the Soviets were very clear that they were not going to risk their cosmo-nauts beyond LEO until they had the technology to do so safely.[6] Whether the issue was sufficiently understood in the America of the 1960s and those details withheld from the general public is not known, but to put it bluntly, the fact that no space program on Earth has as yet developed spacecraft that can ensure viable human space travel anywhere at all beyond the safety zone of LEO rather throws into question the entire Apollo record.

Yet most people would assert that these missions actually happened as billed because back in the 1960s, public statements from the space scientists implied that these radiation belts presented "little or no problem" as their spacecraft would adequately protect the Apollo astronauts, and it had also been planned that they would travel very quickly through the belts, avoiding the worst bits. Unfortunately, that claim doesn't compute with what we know today. Not only is the unpredict-ability far greater than was assumed or advertised in the 1960s, but also the "worst bits" are liable to vary considerably in depth and intensity. These new discoveries surely influenced this statement by NASA chief scientist Ellen Stofan in 2014:

> NASA's focus now is on sending humans beyond low-Earth orbit to Mars. . . . We are trying to develop the technologies to get there, it is actually a huge technological challenge. There are a couple of really big issues. For one thing— Radiation. Once you get outside the Earth's magnetic field we are going to be exposing the astronauts to not just radiation coming from the Sun, but also to cosmic radiation. That's a higher dose than we think humans right now should really get.[7]

This fudge around the definition of the Earth's magnetic field seemingly exempts the Van Allen Belts from the problem. However, attempting to exoner-ate the Apollo missions from any such problems by emphasizing the long haul to Mars, ignoring the desired *Artemis* journey to the Moon scheduled for 2024 at the latest, and then massaging exactly where radiation actually starts to be a problem for astronauts is revealing. NASA's current reluctance to venture through the belts explains why no one has ventured beyond LEO in the last fifty years. However, this still begs the question as to exactly how in the 1960s Apollo astro-nauts managed the trick with practically no shielding at all—way less than that deemed necessary for the future lunar or Martian spacecraft.

Photo: NASA.

Figure 2.2. Experiment at the Lewis Research Center, Cleveland, Ohio (now the John H. Glenn Research Center at Lewis Field) making simulated Van Allen Belts. These were generated by plasma thruster in tank #5 of their Electric Propulsion Laboratory.

While on the subject of Apollo and radiation protection: to make matters worse, during the Apollo era, it was apparently perfectly safe to amble about on the lunar surface and even drill into it. Now it is publicly stated that the lunar surface is an extremely dangerous place for human beings, as astronauts will receive the full blast of space radiation untempered by a magnetic field and an atmosphere. Furthermore, while the Apollo astronauts were apparently contaminating the lunar module with the lunar surface dust they had not managed to remove from their suits, we are now informed that the lunar regolith retains energetic particles that are emanated when the regolith is disturbed.[8] As one observer put it, "The Moon is radioactive!" All of which rather puts into serious question all the joyous lunar-surface "extravehicular activity" (EVA) of the Apollo glory days.

These twenty-first century revelations make sense of the fact that NASA's future plans for lunar missions pay only lip service to establishing bases on the Moon for human beings. Primarily, operations on the lunar surface are to be conducted by probes remotely controlled from lunar orbit or Earth. And if this doesn't come about as scheduled, well, we can read all about what might have been. Andy Weir's 2017 follow-up novel to *The Martian* portrays the Moon as a mining and tourist destination. He called it *Artemis,* two years before NASA came up with the same name for its 2024 lunar foray. Whether or not this choice of name was a joint decision between Weir and NASA, clearly the book is meant to be either a vanguard for the mission or (given its somewhat dystopian outlook) its requiem.

Back in the 1960s, the Apollo missions were presented as being a race to the Moon between two opposing ideologies, and all's fair in love and war. But what was the war? With hindsight, it is clear that throughout the Cold War, scientists in the Soviet Union, the United Kingdom, and the United States were sharing information relative to space exploration. NASA historians have this to say about Soviet-US relations back in 1955:

> *Man in Space* apparently impressed one high-level Soviet space official. This is indicated by a copy of a September 24, 1955, letter from L. Sedov to F. C. Durant, President of the International Astronautical Federation. "If Disney Studios supplies us with one copy of this film on whatever terms it may put, it will help considerably the cause of promoting our contact." Erik Bergaust, von Braun's biographer, called Sedov the "front man for Russian space delegations during the Sputnik era."[9]

So it comes down to asking, if the space race wasn't such a competitive game of technological one-upmanship, as was suggested at the time, and if the technological challenges of getting to the Moon and back were too much even for the then Soviet Union, already far ahead of the United States in its space exploration capabilities, what was the primary overriding motivation for getting out into space or being seen to get men out to the Moon?

Other than the 1960s economic arguments put forward for any such collaboration (more specifically, the trading of oil/gas and grain), there had to be a primary motive for these two political ideologies to forget their differences and quietly communicate when it came to venturing into space, all the while maintaining a front of implacable hatred for the benefit of their countrymen and the rest of the world. Enter Hollywood once more.

We are all used to seeing movies that convey future aspirations with all their hopes and fears, and many of the current crop of science fiction films are either about dystopian regimes on Earth or the future colonization of Mars, with the effects of colonization off-planet highlighted in movies like *Avatar*. Audiences generally consider these movies to be the sole output of their creative writers and directors. What is less well-known is that at the advent of the crewed space program in the United States, documentaries and media promotions were not considered sufficient on their own. The big screen was also co-opted into the propaganda machine. A prime example—and again, this might come as a surprise to some—was the landmark film *2001: A Space Odyssey*. It was openly acknowledged within US government departments that this seminal film was intended to shape the public's vision of the space environment. This is the reason the production had such a great deal of assistance from NASA and help from all the organizations involved in space exploration: getting things right was an absolute priority.

Originally scheduled for release in 1967 in order to run concurrently with the Apollo segment of the US space program, this film was commissioned with another less publicized but quite specific primary aim: to introduce, educate, and prepare audiences for the presence of ETI in the solar system and beyond. It was intended that the film begin with a black and white documentary-style section featuring statements made by prominent thinkers in the domains of science, religion, philosophy, biology, and astronomy concerning the presence of ETI. The images of an extraterrestrial artifact (originally conceived as a tetrahedral pyramid but ending up as the dark rectangular monolith) were to be accompanied by a considerable amount of narration that reiterated and elaborated on this basic premise.

In a nutshell, the message to be conveyed was: we are not, and never have been, alone.

We'll pause here to note that in 1965, while *2001* was still in preproduction in the United States, Walt Disney, his brother Roy, and other Disney executives visited NASA's Marshall Space Flight Center. In an interview with the *Huntsville* (Alabama) *Times* Walt Disney said, "If I can help through my TV shows . . . to wake people up to the fact that we've got to keep exploring, I'll do it."[10]

In reality, the tour at the Marshall Center and other NASA locations did not inspire Disney to use his 1950s and 1960s television series as a model for a new film about space exploration. And of course, there was no need, because Stanley Kubrick's *2001* was already picking up the relay baton.

All this data is in a 1966 government documentary fronted by one of the head honchos from *Look* magazine and filmed partly at NASA and partly on the UK *2001* set. That same documentary also revealed Arthur C. Clarke to be speaking very much as the government science spokesman. Rather than being the put-upon "creative book/script writer" portrayed in various biographies and records of the making of the movie, in this documentary, he comes across as some sort of middleman between NASA and the director, Stanley Kubrick.

In this documentary Clarke refers to the movie's budget of $10 million. This was an excessive sum even in those days, and by 1966, Kubrick was already in principal photography. Clearly, that sum was budgeted for the *2001* production

Figure 2.3. One of many poster designs for *2001: A Space Odyssey*, 1968.

from the outset and was not due to an overrun, as was stated in later accounts of the production process. In fact, according to Frederick Ordway III, a close friend of both von Braun and Clarke since the 1950s and NASA's scientific advisor and technical consultant to the film, the budget was initially set at $10.5 million, with some $6.5 million allocated to artifacts and special effects photography.

Of course, the astute might say that the current availability of that 1966 documentary infers that it too is an educational tool masquerading as an in-house government documentary. After all, elsewhere in his writings, Clarke has claimed that very early on in the *2001* project, he had persuaded Kubrick to modify his ETI views and to discount UFOs as anything particularly useful to the film. Although Clarke's statements follow the public line on such matters as put out by the DOD, surely, the original thesis of the film, *the recognition of ETI*, belies the official DOD men's attitude concerning UFO sightings and interactions. In particular, the various narratives concerning the 1947 Roswell incident emanating from the DOD over the decades since 1947 are revealed to be somewhat dubious in their veracity.

Perhaps 1947 was a case of too much ETI contact, too soon.

However, even these exotic assertions by Clarke are not enough to warrant the difference between the movie we see today and its original concept as an educational tool introducing the reality of ETI and promoting human space exploration. We learn from various sources that Kubrick, armed with enthusiasm for a new project and with his usual meticulous research methods, had initially announced that he was going to make his next film about ETI. Good, as per the original commission. And by 1966 the experts' interviews had been filmed, with

Clarke introducing the segment and Ordway winding it up. Good, also as per the remit. However, when Kubrick's final cut was delivered in 1968, all these interviews, together with all the extraneous explanations and commentary concerning ETI expected by the studio commissioners had been removed. Bad, completely off script, literally. Indeed, Ordway was absolutely furious about the finished product and said so. Volubly.[11]

Reading between the lines of the various accounts relating to the making of this great movie, especially when taken together with the information revealed by the documentary, indicates that somewhere along the way, US government policy and Kubrick's viewpoint were no longer aligned. He left us with the contemplative masterpiece we have today, but not with the educational hitchhike through the solar system expected by the commissioning authorities. It was checkmate to Kubrick.[12]

Apart from the "technical expertise" from NASA and its affiliates that was meted out to Kubrick's production, and was much touted in the media as such, all of this fine detail was unknown to the public in the 1960s; the real extent of the official oversight by the US government in this creative work of science fiction was completely hidden.

Which leads us to the origination of that enlightening 1966 documentary (now available on *YouTube*) that revealed these connections. The facility used by the USAF and the DOD from 1947 to 1969 was to be found hidden in the canyons of Los Angeles, on Wonderland Avenue, Laurel Canyon, to be specific. And if it weren't for the trucks trundling up its narrow residential streets, the location of all this government activity could have been the private property it has since become. However, at the time of its use by the DOD, this enclave was known as Lookout Mountain Laboratory (LML). Film historians consider that LML researched projection techniques, advanced lenses, cameras, and film stock that would later make it into the Hollywood studios.[13]

Between 1946 and 1969, the LML studio produced more than 6,500 films for the Atomic Energy Commission *and other government agencies*. By 1953, it had become a hundred-thousand-square-foot facility on a 2.5-acre site surrounded by an electrified security fence. The LML had one large soundstage, a film laboratory, two screening rooms, four editing rooms, an animation and still photo department, a sound mixing studio, and numerous climate-controlled film vaults. In addition to military personnel, LML retained more than 250 producers, directors, and cameramen recruited from MGM, Warner Brothers, and RKO Pictures, all cleared to access top secret and restricted data and sworn to secrecy regarding activities at the studio.[14]

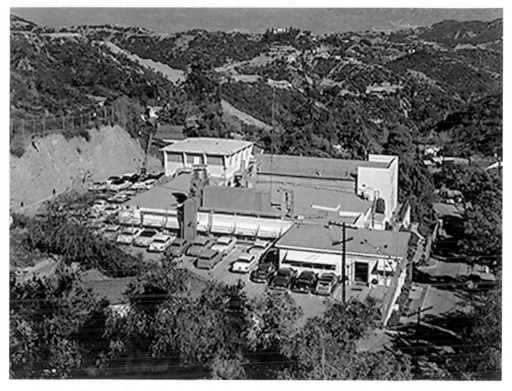

Figure 2.4. Lookout Mountain Laboratory in the early days when there were few houses built on the hills overlooking this DOD facility.

Photo: DSP.

Figure 2.5. Lookout Mountain Laboratory today.

THE WATCHERS

Taking all these facts together, along with the mass of data surrounding the actual meeting and initial arrangements with Kubrick and Clarke in 1964, and in the creative spirit in which this space enterprise is wrapped, one might surmise that when Clarke cited his short story *The Sentinel* as being the prime originator of the *2001* storyline, he had in mind both the commissioning orders and this LML facility—and quite possibly his tongue firmly in his cheek. More especially so when it is understood that it was from Disney's Graphic Studios whence came the top effects specialists, designers, and surveyors onto the set of *2001:* Con Pederson and Douglas Trumbull.

One might say that the last fifty years have been a quiet education of the public via the media relative to space and the possibility of ETI. Nothing wrong with that, except that it has been done extremely dishonestly. It is now evident that photography and moviemaking in all its forms, including film animation, have been harnessed to further the human exploration of space, a project in trouble from the outset.

These are technologies that affect the emotions of the observer through their use of color, music, and dramatic content. When does fantasy become fact, and—in the interests of government policy—is it ever justifiable to decry facts as fantasy and create scenarios purporting to be documentary fact out of whole cloth?

Here would be the place to consider another very early example of NASA's preference for trans-time operations, discreetly referencing former projects within the structure of its current projects along with a hefty dose of mythmaking. After Apollo, there was the Apollo-Soyuz Test Project. Formulated publicly in 1972, this program ostensibly united Russian and American astronauts in LEO. The mating of the two craft from rival space agencies and rival political philosophies was intended to occur between July 15 and 24, 1975. Once the two craft were locked together in orbit, a filmed handshake between crews would commemorate the event. Whether the event had connotations relative to computer terminology for a "handshake" is speculation. The fact that the camera to film this moment had to be designed for use backward, over the shoulder of the operating astronaut—thereby inferring collaboration on past events—is also speculation. What is known is that the physical handshake between these two spacefaring nations was initially programmed to take place on the very same day and month previously assigned to the Apollo 11 launch, July 16, but over which region of Earth should that momentous event take place? With regard to lunar exploration, England's Jodrell Bank Observatory at the University of Manchester was the intermediary between the then Soviet Union and the United States. Yet, in 1975, it was Bognor Regis, in the south of England, that was designated the landmark of this celestial handshake. Bognor is derived from the Anglo-Saxon name of Bucgan ora, meaning "Bucge's landing place," the rocks of the shoreline marking the boundary of her territory. Regis, meaning "of the king," was added during the reign of England's George V. With regard to the Apollo missions, the Egyptian Farouk El-Baz was involved with the selection of Apollo landing sites as well as training the astronauts in geology, and he was known as "the King."

Yet, as the story goes, in 1975, delays occurred, and in the end the handshake occurred over Metz in France. Would that place have any significance for the space agencies? Well, yes. In earlier times considered a holy city, Metz was also a Merovingian capital city. The Merovingian kings feature in the mystical and historical lore of the French, the Templars, the Masons, and other esoteric sources. In modern times, Metz's location relative to the borders of Luxembourg, Germany, and France is significant within the European Union. The uniting of the (now) Russian and the American space agencies above the triplicity of borders on the land around Metz mirrors the association of these two space agencies with the European Space Agency (ESA). All three space agencies are active in Mars exploration today.

One cannot help but think that the programmed but *unfulfilled* Bognor Regis handshake location poetically matched these agencies' earlier lunar adventures, while the later Metz location, over which the handshake *ostensibly* occurred, was intended to prepare the ground for the future of probes destined for Mars. If the timing of the Apollo-Soyuz event supposedly commemorated the July *month* and *days* (Moon and Sun) assigned to Apollo 11, the years it took from planning to the handshake itself also correspond to the *years* that saw the acquisition of the photographic images from the Cydonia region of Mars—instigated by the 1971–1972 *Mariner 9* and consolidated by the 1975–1976 Viking missions. All this subtext enforces the view that these space programs are, at one level, messages of intention, even if the goals are unobtainable realistically and practically.

This double-handed game of doing one thing in public while running a parallel private operation has turned out be par for the course when it comes to the realities of the space agencies and the roles of the media.

Back in 1976, when NASA astrobiologist Gerry Soffen had deliberately shown the first photographic image of what looked like a face structure on the Martian Cydonia complex to the press (and therefore the world) and promptly

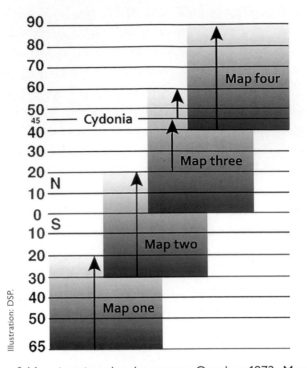

Figure 2.6. *Mariner 9* Mars imaging plot, January to October 1972. Mars latitudes from 65° south are on the left-hand column. The map blocks depict the mapping mission periods: Map one: January 2–22. Map two: January 22–February 10. Map three: February 10–March 8, which covered the anomalous region at 40° north. That total of sixty-five days completed the primary mission. Then there is a gap in imaging until June 5 when the region in Map four: between 40° north (the anomalous complex) and the north pole was re-mapped. Imaging ended on October 16. Map four imaging took place during the so-called extended mission, and it took just over four months to image fifty degrees of latitude, when the primary mission had taken just over two months to image 115° of latitude. But then, it did remap Cydonia.

explained it away by saying, "it was just a trick, just the way light fell on it," he had created a storm of interest in the media—as expected. The subsequent very public refusal to re-photograph this mound or landform at the next opportunity also created a furor in the media. Had the photo been of nothing but photons, there would have been no issue in taking another image. Furthermore as readers of our book *Dark Moon* will recall, his airy dismissal was completely contradicted by the fact that NASA had specifically re-imaged this very Cydonia complex with its *Mariner 9* orbiter, four years earlier! This particular region of Cydonia was evidently of major interest to the agency.

It is possible that some of this imaging has not been made available to the general public, and there is evidence to suggest that some of the published images of Cydonia have been deliberately manipulated prior to publication, resulting in reduced definition. Hiding in plain sight also seems to be par for the course: Google Earth's Mars section holds current imagery of Cydonia. This region has been adulterated with visual "additions" to key locations, including black-and-white photos superimposed over the natural landscape and so much "noise" across what is called the City area that it is virtually unreadable.

These "Face on Mars games" look as if they are carefully managed to produce several outcomes advantageous to the US government's space exploration program, its political stance, and NASA's own internal agenda. It is generally supposed that innate within all humanity is the desire to find answers to the key *what, when, why, who,* and *how* questions concerning our origins. Then again, beyond those in the know, if you really want to ruin a good career, get interested

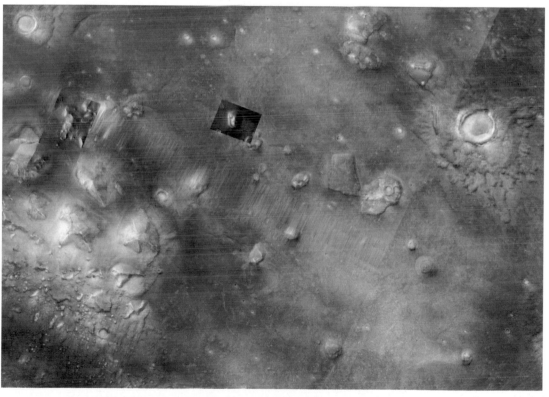

Image: Google Mars 2014, ESA/DLR/FU Berlin (G. Neukum). CC BY-SA 3.0 IGO.

Figure 2.7. The Cydonia region as portrayed on Google's Mars Map.
Note the "noise" across the central area, and how additional black and white NASA images have been added to the data supplied by the European Space Agency and Gerhard Neukum of the German Aerospace Center.

Figure 2.8. Astronomer Frank Drake of the SETI Institute.

in artifacts on Mars. Astronomer Frank Drake of the SETI Institute has certainly taken notice of that, because he completely dismisses the notion that the Cydonia complex imagery holds any signs of ETI. Which in itself is more than interesting coming from someone overtly looking for contact with ETI. Paradoxically, his opinion flies in the face of NASA's behavior once Cydonia had been initially photographed.

It is clear to scientists working within the British Ministry of Defence, and the US DOD as well as within NASA and its affiliates (whether admitted publicly or otherwise) that the exploration of Mars and the 1976 release of the first Cydonian images triggered a major cultural event here on Earth. Especially within the scientific community. However, as it stands, the majority opinion still rules and no intruder (or imagery from Martian probes) will be permitted to disturb the status quo, either when it comes to discussing ETI or when it comes to challenging the sacred cows of our scientific knowledge. Even though there is an ever more pressing need to find some clarity within the confusion of the last two hundred years of consensus science, and even though lateral thinking might be of use as we take stock of the technological challenges we are facing. According to Brian Cox, professor of particle physics at the University of Manchester and physicist at the European Organization for Nuclear Research (CERN):

> My opinion is that you are not allowed to have an opinion unless you know something about the subject you are talking about. Well, you can have an opinion but you do not have the right for it to be listened to.[15]

A blinkered attitude such as this may explain to a considerable degree why the ancient archaeological sites and their structures do not give up their secrets easily to those who spend their time looking at them from within the restrictions of their own professional perceptions. It is therefore hardly surprising that there is fierce rejection of the evidence set out before their eyes from various authorities, including Drake and his SETI friends. Yet these are the very same authorities who manifest a desire to find ETI and travel into deep space in order to expand the horizons of humankind.

Which brings us back to that alleged desire to educate the public as to the presence of ETI. Is that Face on Mars another small step in the use of imagery, just a trick of the light designed to lead the public down a desired path, or has the Cydonian landscape on Mars finally confronted the space agencies with their own worst fears—and their very own sculpted monolith, right next door?

Could it be that over the last several decades, the anomalous physical events occurring in the very public domain have come to haunt these very authorities who wish to educate us about ETI? Do the missing cylinders of earth around Lake Geneva in Switzerland, the impossible geoglyphs carved into hardpan salt lakes appearing across the planet, and specific glyphs appearing in the flattened grain fields of southern England, to name just those anomalies visible on the planet's surface, contain any details that might reveal a coherence between them? Have these events finally convinced the authorities that ETI is alive and well and in the neighborhood? Or are all of these events the repercussions of yet more misdirection by those who, while wanting to stay in charge of this planet's military and industrial agenda, can't quite manage to get themselves where they want to be—out beyond LEO? Except, of course, for the satellites arrayed around Earth out as far as the geosynchronous satellite ring around the equator some twenty-two thousand miles distant from Earth. Here's another surprise, it is now officially acknowledged that the Apollo program was a cover for the top secret satellite program CORONA.[16] The use of capital letters to describe a program indicates its high level of secrecy. With what we have discovered so far, it's more than likely that the urgency to build satellites under top secret conditions was less about the Cold War and more about keeping a lookout for ETI.

Viking Warriors

*NASA Went to Mars to Search for Life,
Mankind's Lifelong Desire to Discover What Awaits
in Deep Space, the Brookings Report, and Facing
the Consequences of Finding Evidence of ETI.*

PIONEERING

As a popularizer of science and one of the most famous US astronomers of the last century, Cornell University astrophysicist and Pulitzer Prize–winning author Carl Sagan (1934–1996) has a formidable CV. Indeed, by 2013, two biographies had already been published. As an advisor to NASA from the 1950s onward, it's Sagan's passionate interest in the search for extraterrestrial life that has relevance here. Sagan considered it a necessity for human beings to take up the search for extraterrestrial life, both from Earth-based observatories and during any future space missions. By 1966, he had published *Intelligent Life in the Universe,* co-authored with the Russian I. S. Shklovskii, which became a classic of the genre.

Sagan briefed the astronauts assigned to the Apollo missions, and notably, during a preproduction meeting with Kubrick (arranged by Clarke), he suggested to Kubrick that *2001: A Space Odyssey* should infer rather than specifically depict ETI. Here, it is of interest that eleven years earlier, in 1953, Clarke had published what is still considered by many as his best work, *Childhood's End*. Of this not-so-enthusiastic appraisal of incoming ETI, Kubrick biographer Michael Benson wrote, "It closed with the human race being shepherded through an accelerated evolutionary transformation by a seemingly benevolent alien race, the 'Overlords.' In it, humanity is depicted as obsolete—destined for replacement by a telepathi-

Photo: NASA.

Figure 3.1. Astrophysicist and author
Carl Sagan.

rally linked successor species composed, oddly, of children." At that time, Clarke thought that "aliens," as they were generally known, would physically resemble human beings. However, by 1964, Clarke had changed his mind about alien physicality and joined with Sagan's viewpoint: aliens would look extremely different from human beings. Kubrick, who preferred to speak of extraterrestrials rather than aliens, essentially considered it likely that there would be a resemblance to ourselves. Hence, the meeting of the astronomer, the writer, and the filmmaker to discuss the matter. Kubrick afterward made it clear to Clarke that when it came to Sagan, one meeting was quite sufficient, and from then he on ignored Sagan's advice on matters ETI. And Benson relates that Sagan's later account of this meeting was inaccurate in several major details.[1]

Returning to the world of science fact, Sagan was also responsible for the creation of the gold plaques that were attached to the Pioneer spacecraft in 1972 and 1973 and the content of the disks attached to the two 1977 Voyager probes. These four deep-space probes bore messages intended to signal our existence and whereabouts to any putative extraterrestrial civilization that might encounter them. Sagan then was considered to be well positioned when it came to defining the scientific criteria for finding signs (however old) of intelligent life on another planet.

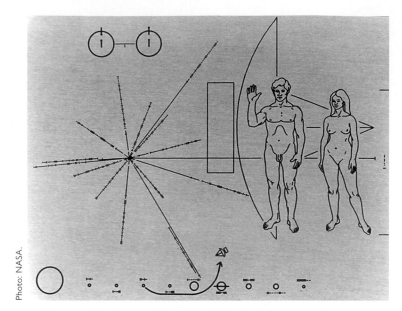

Figure 3.2. *Pioneer 10* gold-anodized aluminium plaque.

Figure 3.3. Carl Sagan, circa 1973, with a copy of the plaque placed on the Pioneer probes signaling our existence and whereabouts to any extraterrestrial civilization that might encounter them.

THE ARCHAEOLOGY OF CONSCIOUSNESS

Sagan considered that intelligent life on Earth had first revealed itself through the geometric regularity of its constructions. So when he stated in his popular 1980 book *Cosmos* that for signs of intelligent life to be compelling elsewhere, Euclidian geometry and a major reworking of the surface of the target planet should be in evidence, anyone interested in the subject should have sat up and taken notice, for Euclidian geometry has indeed been found evident, even abundant, in the layout of Cydonia, as are signs of reworking of the surface of the planet. Yet, strangely enough, NASA decided to ignore its advisor—at least publicly.

Figure 3.4. Cydonia, intimations of intelligence. This image is published on Google Mars, the same platform as the 2014 image seen in the preceding chapter (figure 2.7), and it is also generated from the European Space Agency's Mars Express orbiter. However, the diversity of landforms can be clearly seen as it is crystal clear, there is no "noise" and nothing obstructs the Face.

SAGAN'S VIKING SAGA

The Viking project was managed out of NASA's Langley Research Center in Hampton, Virginia. Each *Viking* was made up of two parts coupled together for the flight to Mars. The octagonal orbiters and hexagonal landers were built by the Jet Propulsion Laboratory (JPL) in Pasadena, California, which also managed the science missions for these two flights, *Viking 1* arriving at Mars seven weeks before *Viking 2*. Here it should be emphasized that when a Viking orbiter-lander combination was inserted into Martian orbit, it was unalterably committed to setting the lander down at a certain *latitude* on Mars.[2]

The periapsis (closest) point of the orbit was also fixed. Only the degree of longitude could be altered. Prior to the *Viking 1* departure from Earth, it was the 1974 *Mariner 9* data that had contributed to the choice of the Chryse landing site, later refined with radar-based readings of Chryse taken from Earth some weeks before *Viking* reached Mars. *Viking 1* had entered into Martian orbit on June 20, 1976, and Sagan stated that the target was 21° N. British space scientist and authoritative author on matters of space travel David Baker records the *Viking 1* periapsis point as being over the proposed set-down site for its lander at latitude 20° N, 34° W. Not bad. Present-day sources give marginally different dates and location coordinates, but wherever possible this book is using the data published by Baker. His 1995 book *Space Flight and Rocketry: A Chronology* is a reliable source and has not been subjected to the later manipulations that can sometimes occur in the digital record.

The details of the Martian orbit for *Viking 1* as described by Sagan in *Cosmos* are confusing. According to Sagan, the *Viking* orbiter's cameras were only marginally better than that of the earlier *Mariner 9* mission, which might be true of the images the public eventually got to see and might also be true if one considers the amount of dust in the atmosphere that *Mariner* had to contend with. However, Sagan doesn't mention either issue, and his statement is in fact technically misleading, since the *Viking* cameras did have much better resolution than *Mariner*'s. Albeit, the United States Geological Survey (USGS) considered *Mariner 9*'s television images to be *inferior* to those received from the 1966–1967 Lunar Orbiter high-resolution cameras.

Now here is the muddle. Sagan stated that once the *Viking* orbiter had arrived at Mars, the *Viking 1* imagery indicated that the proposed landing site *looked* unacceptably risky and that it was decided the lander should set down elsewhere. However, even with its good cameras, *Viking* could not actually tell if a location was too rough or too soft for landing safely. The orbiter carried no radar, so *Viking* had no other means of "reading" the surface. The question to be asked is this: If *Viking* could only indicate that things were "a little unsafe," but with no determination as to the extent, just how was it decided that the landing site was a problem and that another site was better? The answer, according to Sagan, was by using Earth-based radar, which was effective within 25° north or south of the Martian equator.

David S. F. Portree of the USGS's Astrogeology Science Center in Flagstaff, Arizona, who is also a space historian, noted that the Army Map Service of the US DOD had created the base map for NASA's preliminary candidate locations for the Viking landers.[3] The Army Map Service was morphed into the US Army Topographic Center on September 1, 1968, so Portree's reference to the Army Map Service infers that the Viking site map had been generated *prior* to 1968.

Portree then notes that the Army Map Service map was not itself dated and declares that it must have been readied for the Viking landing site working group meeting of December 1970. In fact, the base map in figure 3.5 dates from 1962. And it was prepared originally by the USAF's Aeronautical Chart and Information Center (ACIC).

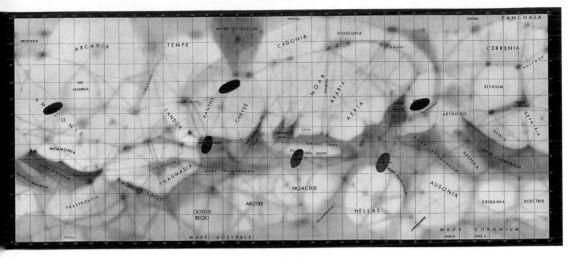

Figure 3.5. Preliminary candidate Viking landing sites 1969–1970. The regions Earth-based radar revealed to be of high elevation in the northern hemisphere are covered by the red ovals. The Lowell Observatory supplied most of the data applied to this 1962 map. It was the last time Earth-based observations would be used. From then it would be orbiters around the Red Planet.

The National Geospatial–Intelligence Agency created the ACIC in August 1952 by consolidating all the services and facilities of the Washington, DC, Aeronautical Chart and Information Service and the Aeronautical Chart Plant in St. Louis, Missouri. From 1952 to 1972, the ACIC was critical in supporting US targeting accuracy efforts in Vietnam. It also provided charts and graphics to assist the planning and execution of the Apollo 11 mission, which included determining lunar orbits and finding landing sites. In 1972, the ACIC was folded into the Defense Mapping Agency and renamed the DMA Aerospace Center. It consolidated multiple mapping agencies and assumed responsibility for producing and distributing maps, charts, and geodetic products and services.[4]

Notwithstanding its dubious genesis, this early map is of interest because the annotations referred to by Portree show that contrary to Sagan's assertion, as early as 1968, it was considered that the Earth-based radar readout was effective to latitude 30° N.

Given that radar was the only means of establishing the relative safety of the landing site, the roughness of the selected *Viking 1* landing location could have been established *prior to* leaving Earth. Was there another reason for this little exercise in site swapping? Or was this change of plan actually an intended maneuver all along?

MEANING IN THE NUMBERS

As we saw in the Apollo-Soyuz rendezvous choices, NASA has always shown a preference for using names and dates for its missions, whether human piloted or fully automated, that have poetic and symbolic meaning for those in NASA, if not for the rest of us. Lunar projects were given names related to terrestrial exploration activities, and interplanetary projects were given nautical-sounding names that conveyed the impression of travel over great distances to remote lands.

This naming practice sometimes resulted in a contradiction in terms. In 1965, NASA completely muddled these two criteria by reviving Project Pioneer with the objective of complementing *interplanetary* data acquired by probes from Project Mariner. As for projects and programs, I found this information within a section from NASA's history division concerning Mars. Buried within chapter 2 section 2 of *On Mars: Exploration of the Red Planet 1958–1978* with a somewhat contradictory subtitle of the *Evolution of unmanned space exploration to 1960,* there is this footnote: "the distinction between programs and projects was first made clear by G. F. Schilling, Office of Space Science, late in 1959. Programs signified a related and continued series of undertaking greared [sic] toward understanding a broad scientific or technical topic; programs (e.g., examining the solar system) did not necessarily have foreseeable ends. Projects were the building blocks for programs and as such had limited objectives, limited duration (e.g., Project Mariner, Project Viking)." Interestingly, NASA historians again contradict these terms when concluding that footnote with these words: "While the space science personnel at NASA tended to maintain this distinction over the years, the concept was not as clearly observed in manned spaceflight, where the Apollo project grew so large it became a program."

Back in 1976, this was not something that the general public was particularly aware of unless it was heavily flagged, as was the case for the naming of the Apollo lunar module–command and service module combos.*

Yet there is another far less overt thread running through NASA's name

---

*Here it is pertinent to note that the term "manned," used in the context of space travel, is now reserved for historical documents such as this. The term is a leftover from the early days when it referred to the ability of a human being to interface with the spacecraft technology, as opposed to an unmanned probe. Although, it is also true that until the shuttle training program in 1978 astronaut training was only given to men. And this, despite the fact that in the late 1950s, at the time of the Mercury program, female spaceflight candidates were given the same rigorous tests as the men and thirteen equaled their male counterparts. The decision to then cancel the program and exclude women from the early space prgrams was gender driven and political. Nowadays the space agencies prefer to use gender neutral terms such as human or crewed when discussing piloted spacecraft programs.

dropping concerning the Egyptian pantheon and the constellations. And it has to do with tying locations to various aspects of ancient myths concerning gods and constellations. In some ways, this preference is of little surprise, since the Egyptian myths and gods were much beloved of the Masonic Founding Fathers of the United States. Also, the solar system is to be found in the Orion arm of the galaxy, and Orion is associated in myth with the Egyptian god Osiris. It is the extent and persistence with which NASA pursues this Egyptian theme that is astonishing. It is well known that the constellation of Orion featured on the Apollo logo, but unknown to the public, this constellation was factored in to be visible above the Apollo lunar landing sites.[5] As NASA informs us that none of the Apollo astronauts bothered to take a photograph of the stars—indeed, Armstrong famously said that he didn't see any stars—one might wonder why this attention to the presence of the Orion constellation around the "dead" Moon was so important. That is, until one learns that the ancient Egyptians associated the god of the afterworld, Osiris, with the dead Egyptian pharaohs. And that the constellation of Orion was identified as being the region in the celestial Duat to which the dead pharaohs went. As an aside, author Robert Bauval's notion that this celestial Duat is strongly associated with the pyramids at Giza has more to it than the much-disputed literal mapping of the Orion constellation onto the layout of these pyramids (see chapter 4).[6] Naturally enough, assertions that there was more than purely orbital mechanics to the timing and planning of space missions have been ignored or met with scorn and derision by all those sensible people who dislike myth-making and occult ritual mixed in with their science. However, this theme has not been restricted to the exploration of the Moon.

Before it was financially curbed, NASA's twenty-first-century human space program was called Constellation, the component craft within the program were two new rockets named *Ares I* and *Ares V,* a crew capsule called *Orion,* and a crew lunar lander named *Altair.* These choices were explained thus: Ares is the god associated with Mars, and the designators I and V were a homage to von Braun's Saturn rockets. Orion represented a bright, recognizable constellation, and Altair, in the constellation of Aquila, the eagle, was both the twelfth brightest star in the Northern

Figure 3.6. Constellation, Apollo, and Altair insignia.

Hemisphere and a homage to *Apollo 11*'s *Eagle* lunar lander. When budget cuts and technical challenges limited the scope of this program, *Ares I* and *Ares V* were incorporated into one single heavy lifter, the *Orion* capsule and service module went into development, the lunar lander *Altair* was cancelled, and the Constellation Program became Orion. The earlier working title of *2001: A Space Odyssey* was *Journey to the Stars* and the Earth to LEO spacecraft was called Orion III. The orbit to lunar surface shuttle was called Aries 1B. So with all that in mind, scrolling back to the Martian *Viking 1* mission reveals interesting and historical facts that only become relevant when seen in the context of the overall mission.

HISTORY'S MYSTERIES

NASA historians state that prior to its launch, the *Viking 1* lander was intended to be released onto the Martian surface on July 4, the two-hundredth anniversary of Independence Day in the United States. Then sometime between June 20 and July 4, as a result of the designated landing site being declared unsafe, it was decided to migrate the orbiter longitudinally to arrive at 51°W by July 16—the anniversary of the launch date of Apollo 11. (As we are in Egyptian mythological territory with these games, this longitude virtually recalls the 51.84° slope angle of Giza's Great Pyramid.) This is much the same procedure that was applied to the change of location for the Apollo-Soyuz handshake. Sagan says that it was also agreed that a crash landing on July 4 would be an inappropriate anniversary gift. A point that surely would have been considered pre-launch? Anyway, this migration was begun on July 9 and then—surprise! It was stopped at 47° W on July 13. By then, there was enough overview to conclude that the ground looked increasingly rough farther west.[7] Remembering that we had already been advised that the orbiter could not register the roughness of the ground sufficiently well to inform any landing decision, it is surprising to read in *Cosmos* that the decision for this new landing site was made on the basis of *Viking 1* orbiter photography together with the Earth-based radar scans. *Viking 1* spent the next seven days in orbital imaging before releasing its lander, though not without problems for Sagan's readers. Having been told that the lander had an *unalterable landing site latitude* fixed at some 20° N, apparently that was not an issue when this lander actually touched down nearly 2° farther north! On land, this would equate to a distance of 138.46 miles, or 222.8 kilometers. And by the way, those seven days of waiting made it another memorable date for this landing. It finally happened on July 20, 1976, and the seventh anniversary of the Apollo 11 lunar landing.

How timely.

TIME LORDS

As we have seen, the only possible ultimate decision-making technology always rested with JPL, so all of this protracted maneuvering was totally unnecessary, unless something else was going on. Just how much of those four weeks from June 20 to July 20 was due to capturing images without having to publicly state that they had in fact been taken is open to inquiry. Or were these four weeks of dilly-dallying not only a ritual commemoration of Apollo 11 timetables but also intended to be a covert ETI communication—should anyone out there be watching? Did the longitude of 47° W have anything to do with any putative ETI interaction at Roswell in 1947? Whether or not such speculation is a step too far for some readers, it is noteworthy that the first Viking lander managed to reiterate both the calendar ascribed to the Apollo 11 Earth-to-Moon trajectory as well as the astronomy: as it had been in the lunar "sky," Orion was prominent in the Martian "sky" during this Viking mission. This desire to incorporate Orion into its space exploits leads me to wonder if Egyptian myth is sufficient reason for these astronomical tie-ins. In a nod to another starry ancient culture, NASA might well be adopting star rituals such as the "tying of the Sun to its hitching post" celebrated notably at Machu Picchu. This led me to look at all the other numbers featuring in this orbital saga to see if any other such memorials were encoded within the actions of the Viking probe.

Looking up the start date of the *Viking 1* migration, July 9, and the stop date of July 13, the difference between these two numbers is 4. The originally proposed July 4 Mars landing date is also the date at which the Sun is farthest from the Northern Hemisphere's axis, which might be a contributory factor in its choice of date for the US Independence Day. And July 9 is the date that the Declaration of Independence was read out to the members of the Continental Army. This army group was created at the outbreak of the American Revolutionary War to coordinate the military efforts of the *thirteen* colonies in their revolt against Great Britain. Thus, the original landing date of July 4 proceeds to July 9 and finally arrives at, and sums to, that special number 13. It's pretty clear that these dates were relevant and important to someone, whether the US Viking project leaders or those further up the chain of command. The association of these dates with particular events and places important to its own history rather suggests that the United States has quietly set aside any pretense of planetary exploration "for all mankind."

TRISKAIDEKAPHOBIA

It is also a timely reminder that if the number 13 is viewed with morbid fear by many Westerners and if such fears are considered to be superstitious mumbo jumbo by many, for the Founding Fathers to present-day Freemasons and all those in the know, this number has more to do with origins than with simple notions of good and bad luck. As authors such as Stan Gooch, John Michell, and Christine Rhone have noted, the concept of "12 around the 13" has deep roots. The Osiris myth, the number of zodiacal constellations, the division of land into twelve regions around a central thirteenth core, the twelve knights of Arthur, the twelve apostles of Jesus, the twelve or thirteen moons we experience per annum, all are concerned with group origins wherein the thirteenth part is the core—the essential node around which the other twelve orbit.[8] A fair number of US buildings do not have a floor named the thirteenth floor: if that is true for many buildings, then it may not be because it is bad luck but rather because the number is too special. In which case finding a thirteenth floor would be an auspicious powerbase. When the former convent of Berlaymont, in the Rue de la Loi, Brussels, Belgium, was converted into the European Commission headquarters, the president's offices were located on the thirteenth floor. Coincidence? Perhaps. Then I ask: Is it merely a question of aesthetic convenience that no matter how many nations join the European Union, the stars on the union's flag remain a circle of twelve around an empty center? A leading question? Perhaps. Then note that one of the original flags of the United States had a circle of thirteen gold stars against a dark blue background. These things are culturally important. And the European connection is important for space projects: the establishment of the ESA was strongly encouraged by NASA. Although a cynical study of its genesis indicates that space politics, not altruism, was the driving mechanism behind NASA's enthusiasm.

Even so, Europe is also important to the United States due to its philosophical links with France. If some records state that instead of July 13, July 14 was the stop date for *Viking 1*'s migration westward, this is also a convenient way of incorporating the birth of the French Republic into this Martian ritual, as July 14 is Bastille Day, their equivalent of the US Fourth of July. For, in addition to the political and historical ties between Washington and Paris, both nations perpetuate to this day the Masonic traditions of subtly encoding their philosophy and aspirations through civic architecture and esoteric landscaping.[9]

Returning to *Viking 1*'s migration, only five minutes after it finally hit the Martian deck, the lander started taking its first images. These were relayed via its orbiter back to Earth and took just over nineteen minutes to travel the

212 million miles. Eight days later, the *Viking 1* lander started searching for signs of life via soil analysis probes. The ensuing discussions on the finer points of soil reactions certainly helped shift the general public's focus onto bacteria and away from any further thoughts of complex life-forms having existed on Mars—that's for sure.

THE ANCIENT MARINER

As did the shift of location for *Viking 2*. For up in orbit, *Viking 1* was also in the business of imaging the planet so that the JPL team could finalize their landing site for the *Viking 2* combo, due to arrive into Martian orbit on August 7 and destined for 44° N, Cydonia. Ostensibly, this was because signs of water might be traceable at this location, and as Sagan also noted, the Viking biology tests were strongly biased toward seeking organisms comfortable in liquid water. Sagan infers in his writing that decisions were being made only weeks from the arrival of *Viking 2* into Martian orbit, but in reality the implied last-minute urgency was not by force of circumstance, it was the norm. With an on-site orbiter already in place, the JPL team was able to refine its decision for *Viking 2* just days before its arrival. It was always known that this latitude was outside the range of the Earth-based radar systems, and it is not a feature of the Viking B1, B2, and B3 landing sites, as can be seen on the 1969–1970 map established from *Mariner 6* and *Mariner 7* flyby data (see figure 3.7).

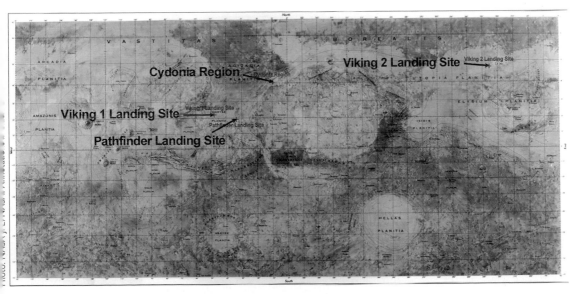

Figure 3.7. Viking and Pathfinder landing sites.

Therefore, the decision to land *Viking 2* at 44° N was seemingly based on other information, yet the only information of any detail must be from the *Mariner 9* probe.

Mariner 9's imaging timetable (seen in figure 2.6) reveals that *more time was spent over the area of Cydonia, which contains all these paradoxical landforms, than anywhere else on the red planet. Mariner 9*'s map three session covered 40° of latitude, from 20° to 60° N, this session covering the future *Viking 1* landing site and Cydonia. Nearly three months later, during *Mariner*'s so-called extended mission starting June 5, 1972, Cydonia was remapped when *Mariner* covered the latitudes from 40° N to 80° N, taking *four months* to do so.[10]

So it would be fair to say that the Viking missions were not only a direct result of interesting imagery returned from *Mariner 9* but also that the search for biological traces of life by the Viking landers was essential in establishing whether the outrageous notion that intelligent life might once have been extant on the red planet was a valid hypothesis. Incidentally, those biology test results were later stated to be inconclusive but strongly veering to "no" according to NASA, yet strongly inclining to "yes" according to the designer of the experiments. With the benefit of hindsight, this confusion would now appear to have more to do with the greater political decisions associated with the Cydonia area than with any actual biological analysis taken at the Viking lander sites. For as it happened, *Viking 2* finally landed at 44° N, but over in Utopia. Certainly to the Earth-bound scientists the ultimate choice of landing site for *Viking 2* was illogical in the extreme, because despite stating that the Cydonian site was found to be too rough (and on the basis of cameras that were hitherto deemed not good enough to evaluate the terrain with any great certainty, remember), in the end *Viking 2* landed in Utopia *at an even rougher and tougher location,* its only apparent advantage that, at 225° W longitude, it was well away from Cydonia's *structured landforms.* Although, with the reticence displayed by NASA when it comes to informing the public of the reasons behind its decisions and actions, what you see is not necessarily what you get. Whether *Viking 2* really made it to Utopia or was actually landed at its selected Cydonian destination is another moot point.

Scoffers may cry "conspiracy theorist" at this last observation, but the disparity in the record and in the history of these events makes this observation more than reasonable. Nor does the amount of information withheld from the public, either through scientific superiority (they don't understand the technology and we cannot be bothered to explain it) or through a sense of national security (we cannot reveal how good our cameras, radar, lasers, and so on really are) dispel such doubts. A NASA insider informed me that there is a really serious problem with data organization and that it is indeed shocking how badly the historical record of early space exploration has been archived. This source believed it to be a case of too many cooks contributing to the same broth from different time zones and different levels of the organization—the classic screw-up theory. This might well be an issue, but organizations work from the top down. If the historical space exploration data has always been badly kept, then it is because it's not a priority for NASA administrators and their managers. Nor is this lack of archiving limited to the human exploration of space—NASA's window dressing. The real business end of NASA, the automated exploitation of space, is just as badly archived. Which is a hugely convenient reputation to have. The excuse that there is an endemic disorganization of data across NASA and its affiliates facilitates the keeping of secrets, and if "Big Org = Lost Data" doesn't cut it, all information concerning space exploration is liable to the massive redaction of documents or the withholding of information on the grounds of some extended notion of that old and convenient chestnut "the national security of the United States."

In the United Kingdom in March 2013, a TV docudrama was broadcast portraying the *Challenger* space shuttle accident that rather bears out this two-faced approach. Made from the point of view of the theoretical physicist Richard Feynman, one of the Rogers Commission investigators into the accident, the drama was a well-made production and virtually 99.4 percent accurate. It was shown that during this investigation, General Donald J. Kutyna, an official in the USAF who was also on the Rogers Commission, had been openly critical of NASA's decision to allow the shuttle to keep flying despite knowledge of the flaw in the O-rings that would eventually cause the *Challenger* disaster. He took Feynman to one side in a secure Pentagon briefing room and explained why NASA had been under pressure to launch that fateful day, come what may. General Kutyna stated that NASA had to launch *Challenger* despite the icy weather, because not long after the space agency's inception, NASA officials had gone cap in hand to the DOD soliciting contracts on the promise of being able to launch at any time that was requested. NASA had then secured all the DOD launch contracts, together with additional budgets. Thus, when it came to *Challenger,* they were under an obligation.

Historical records demonstrate that while essentially this narrative is true, the timing is wrong. NASA never did have to go cap in hand to the DOD and say "We can launch any time, give us your contracts" because it is stated in the 1958 founding document that NASA would undertake both civilian work *and* US DOD work, for which additional funding was allocated to the already extant civilian budget—from the get-go

This is part of the record, and it has been that way ever since.

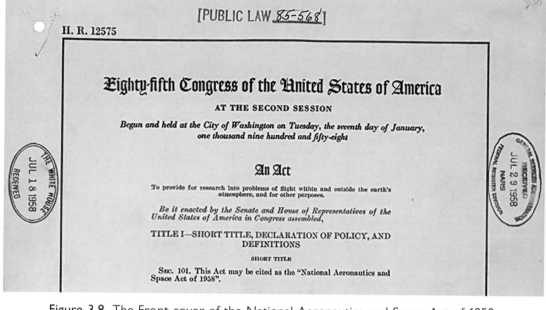

Photo: NASA.

Figure 3.8. The Front cover of the National Aeronautics and Space Act of 1958. Signed by President Dwight D. Eisenhower, intended to provide research into problems of flight within and without the atmosphere and for other purposes. Note that wording.

NASA is an entity with two heads on one body. The agency's structure has been compared to the failed separation of monozygotic twins called "state" and "civilian."[11] This convenient double identity enables NASA to be whatever the situation demands—to the public a civilian agency with a civilian budget and to the government a defense contractor, again with its own budget. Thus, when we read of NASA being worried about continuing space exploration because its budget is up for review, the first thing to remember is that the "budget up for review" won't be the defense budget but the icing-on-the-cake civilian budget. So the reason to launch *Challenger* as given to Feynman is a little wobbly, to say the least, until one factors in the defense situation: if there was, for example, a satellite onboard the *Challenger* that needed to be in space within a specific time frame dictated by military requirements, then there would have been an overweening urgency to launch. This also confirms that the exploitation of space for the purpose of satellite communications, both military and civilian, and the ability to develop and use laser technology and manage off-planet space stations are all potentially more important to NASA than crewed trips to the Moon and Mars.

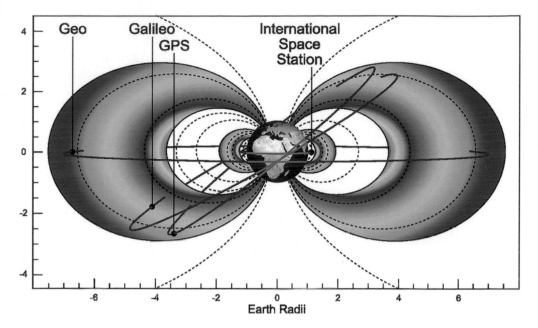

Figure 3.9. Four principal satellite systems and their approximate distance from the surface of Earth: the International Space Station (ISS) orbits at some 250 miles altitude; the US Global Positioning System (GPS) is orbiting some 12,550 miles altitude; the European Union's Galileo Satellite constellation orbits at some 14,121 miles altitude, and the geosynchronous ring is 22,237 miles altitude.

KEEPING A WEATHER EYE

Unless it was, and perhaps still is, a fear of ETI driving this split agenda. In which case, the initial establishment of a satellite ring around Earth might have been more pressing and executed for other reasons than those presented to the public.

It is known that the Kremlin and the White House shared data on UFO sightings and anomalous events such as Tunguska in 1908 and Roswell in 1947, and whatever the public spin, these two events were of particular relevance to these two governments. So it is not unreasonable to speculate that in the interests of planetary defense, the two superpowers at that time sought to collaborate—without causing a public outcry or revealing why. Political differences between the two nations could be highlighted to procure funds and support for the Cold War and the alleged space race to the Moon while they were covertly fostering mutual aims in space, including the eventual establishment of the geosynchronous satellite ring some twenty-two thousand miles from Earth.[12]

Returning to the Viking probes, it would seem that Sagan made much of the fact that the radar was a driving force in the choice of sites, and historian Oliver Morton states that it was Sagan's preferred method of site analysis,[13] but it looks as if *Mariner 9*'s photography was actually the prime motivator in the choice of both Viking landing sites. But again, there were other, mythical connections to be made for these Viking adventurers.

When recounting the arrival of *Viking 1,* Sagan, weaving serious astronomers and fiction writers together, had taken care to inform his *Cosmos* readers that the Viking cameras revealed "no signs of canal builders, no Barsoom aircars or short swords, no princesses or fighting men." Now we find that the Utopian landing site actually brought fantasy into reality, because *Viking 2* landed just northwest of a location identified by Edgar Rice Burroughs as "a city called Horz located on the prime meridian of Barsoom." We will be back there in later chapters. Commonly known as ERB by his fans, Burroughs's science-fiction adventures, set on Mars, had influenced many a space scientist's boyhood dreams. Sagan said that as a child he used to spend countless hours standing with his arms outstretched, wishing himself onto the surface of Mars, reasoning that if it worked for John Carter, hero of *A Princess of Mars,* it ought to have worked for him. Decades later, perhaps those boyhood memories come surging back when the Viking landers were programmed and when Carl wrote his comments on the Elysium pyramids.

In Burroughs's books, Carter and the people on Earth call the distant red planet Mars, while the inhabitants of Mars thought of their planet as Barsoom. As did Michael Malin, the renowned space scientist and camera expert.*

Back in the Viking days, Malin was NASA's principal investigator for the Planetary Geology and Geophysics Program from 1975 to 1993. In 1990, he set up his own company, Malin Space Science Systems in San Diego, but during Viking, he was at JPL and his archived Viking images were originally filed on the Malin Space Science Systems website. Today that segment of the website is "unobtainable," just like the precious mineral of the Na'vi's planet, Pandora, in the movie *Avatar.* It would seem to this author that when space probes began returning images from Mars, creating either the illusion or an illustration of humanoid presence on Mars sometime in its past, instead of following a policy referred to by a NASA consultant as "keeping an open mind, but don't let your brains fall out," NASA's space scientists preferred to camouflage, ignore, and manage out of existence the Cydonia region of Mars. And if that is the case, then the reasons for such behavior surely lie with orders issued back at the dawn of the space age.

---

*A. C. McClurg initially serialized ERB's *A Princess of Mars* in *All-Story Magazine* in 1912. It was not published as a book until 1917. Arthur C. Clarke also credited the Barsoom novels as a source of inspiration.

BROOK NO ARGUMENT

In November 1959, the newly born US space agency had commissioned a survey from the Brookings Institution concerning *Proposed Studies on the Implications of Peaceful Space Activities for Human Affairs*. It was prepared in 1960 for the Committee on Long Range Studies of NASA and written by Donald N. Michael with the collaboration of seven others.

Despite the fact that the Brookings Institution has created many reports on many subjects for various organizations, this particular report for NASA has come to be known simply as the *Brookings Report*. Its collaborators met for two-day conferences once a month with an extra two-day special meeting at the halfway point of their deliberations. Two hundred people were interviewed for the report. The time allocated for this study was short: thirteen months of independent work from its contributors with twenty-eight days of conferencing seems small potatoes compared with the outreach of space exploration. Such brevity might well indicate that it was initially commissioned as lip service to the funding authorities, for if the *Brookings Report* was actually sent to NASA on November 30, 1960, it did not reach the politicians in Washington until mid-April 1961, during the 87th United States Congress.*

The findings of the *Brookings Report* were estimated to be of value for anything from ten to twenty years after its publication, but with the hindsight of

Figure **3.10**. The *Brookings Report*, part of front cover.

Image: NASA.

---

*Proposed Studies on the Implications of Peaceful Space Activities for Human Affairs, known as the Brookings Report, was commissioned from the Brookings Institution in Washington, DC, by NASA in November 1959 and was delivered on November 30, 1960. The report, by authors Donald N. Michael, Jack Baranson, Raymond A. Bauer, Richard L. Meier, Aaron B. Nadel, Herbert A. Shepard, Herbert E. Striner, and Christopher Wright, was dated December 1960.

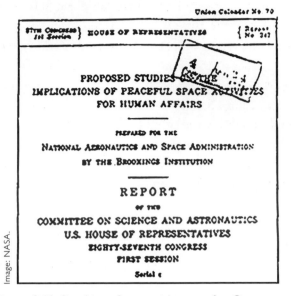

Image: NASA.

Figure 3.11. *Brookings Report* title page for Congress, March 24, 1961. The twenty-eight days of conferencing fitted with mean orbital period of the Moon, while the monthly time period allocated for this study, that sacrosanct number 13, was relevant to the political life of the United States, the lunar cycle, the thirteen viable "body parts" of the defunct Osiris, and so much more.

the present day, it is clear that the guidelines set out in the report have affected developments far beyond that sell-by date. And also far beyond the remit of the space agency. What emerges is that on the whole, the technical aspects of becoming a spacefaring civilization might be coped with, whether by probe or through eventual crewed missions. However, not everyone would be able to cope with the emotional and mental ramifications of human beings leaving Earth's atmosphere.

It also stated in the footnotes to section nine of the 279-page report sent to Congress that more than any other by product of space exploration, signs of extraterrestrial life would have the most impact on two principle sets of communities other than the general public: the religious and the scientific. Religious leaders and their congregations (the more fundamental, the worse it would be) would be "electrified"—their word. The scientists and engineers would be the most "devastated"—their word again. Brookings might just as well have written: "Such discoveries will lead to complete disaster, the right hemisphere of our brains becoming overexcited and uncontrollable, the left hemisphere depressed and thereby incapable of the appropriate analysis and response." Then *Brookings,* with more to say about the scientists, extrapolated on this gloomy prognostic:

> The scientists and engineers of this world are concerned with the complete mastery of nature rather than with the understanding and expression of man. Advanced understanding of nature might vitiate all our theories at the very least. If not also require a culture and perhaps a brain inaccessible to Earth scientists. Nature belongs to all creatures, but man's aspirations, motives, history, attitudes etc., are presumably the proper study of man.[14]

This inference that man is unique to Earth, that everything else "out there" is a creature, virtually requires that "proper scientists" stick to the intellectual and cultural status quo. When seen in this context, the control of scientific knowledge via funding mechanisms makes more sense—at least to the fearful.

Quite apart from the massive assumption that all humans are unable and/or unwilling to diverge from this Western notion of the scientific separation of man from nature, the *Brookings Report*'s assertion that those in charge of the science bit are constitutionally incapable of handling the discovery and/or contact with ETI takes us to the nub of the problem. Clearly, all of humanity will crash and burn. And with that, any understandings of "how things work" along with us. The fear of being crushed by both the ramifications of such cultural contact and the ensuing technological and scientific knowledge gaps that might be revealed has seemingly become the foundation of the policies established for the dissemination of any information concerning ETI contact to this day.

AUTHORIZED GOSSIP

The above section of the report is especially illuminating when taking into account the "authorized gossip," the discussion and dissemination of anomalous events, that has surrounded the space program since the publication of this report. Referring again to the *Brookings Report* we find this:

> Evidences of its [ETI] existence might also be found in artifacts left on the Moon or other planets. The consequences for attitudes and values are unpredictable, but would vary profoundly in different cultures and between groups within complex societies; *a crucial factor would be the nature of the communication between us and the other beings.* Whether or not Earth would be inspired

to an all-out space effort by such a discovery is moot: societies sure of their own place in the universe have disintegrated when confronted by a superior society, and others have survived even though changed. Clearly, the better we can come to understand the factors involved *in responding to such crises* the better prepared we may be.[15] (emphasis added)

The wording here infers that time and effort should be made in seeing how people would respond to the notions of those who are most certainly "out there." For if their presence were only a hypothesis, this report should have read "communication between us and other beings."

Allegedly, there have been UFO sightings during many space missions and supposed conversations about them between NASA and its astronauts. These reports are neither denied nor overtly acknowledged. So the general public, thinking that NASA would strenuously challenge any false statement, is left to conjecture that, indeed, such rumors must be true. Thus is set in motion a sociological experiment from which public opinion can be evaluated. The Cydonian "show and not tell," the allegations of leaked images concerning enormous structures on the lunar surface, indeed, the literally outlandish rumors about bases filled with extraterrestrial or human personnel on the Moon and/or Mars, all fit into the same category. In the upcoming chapters, we will see how the SETI adventures at Green Bank, Virginia, movies such as *Close Encounters of the Third Kind,* and those rumors spread by NASA's PR departments have all been employed to help these authorities understand how we, the public, would respond to such crises.

WAR OF WORDS

It is probable that the collective of NASA scientists, while enthusiastic perhaps for the *notion* of ETI, can better accommodate such matters within the right-brain framework of science fiction. Left-brain logic would control any excessive imaginings when evaluating actual data.

As far as Mars is concerned, the current understanding of how the solar system evolved does not allow for the development of intelligent life on Mars within a time frame that would leave significant traces today. Hence the images of the Cydonia region from the Mariner and Viking projects leave NASA with no other option but to deny that they are what they seem to be—signs of intelligently positioned artifacts. The *Brookings* section cited above goes some way to explaining this attitude, and it also informs the robust put-down toward those scientists prepared to buck the cultural scientific paradigm, such as the highly qualified and innovative image analysts Vincent DiPietro, Gregory Molenaar, and Erol Torun. Despite being given helpful nudges in the right direction by these specialists, the preferred response was to publicly state that these professionals were unable to adequately interpret the Viking data.

What nonsense!

SOUNDS OF SILENCE

Yet to respond with anything other than dismissal could potentially be so disruptive it would result in cognitive dissonance—the inability (or sheer unwillingness) to reconcile what one has seen with what one considers as the "known."

Cognitive dissonance would go some way, if not the whole nine yards, toward explaining many of the various authorities' responses to anomalous phenomena and signs of intelligence having been at work on a rocky sphere other than our own. But from the point of view of those whose brains have not entirely set into two rigid hemispheres incapable of change and are therefore free from cognitive dissonance, those authorities would be wrong.

If the *Brookings Report* was correct in its conclusions, the decision makers, threatened in their own worldview by the discoveries on Mars, half-desired and half-dreaded, would have been more likely than not to undergo cognitive dissonance, and as such, in matters of ETI exploration, the ends would justify the means. Obfuscation, deliberate confusion, dissemination of misleading or partial data, manipulation of the public mindset either via appointed emissaries or through both factual and fictitious media information—all of these would be acceptable response strategies.

Taking all of the above into account, it is probable that NASA knows exactly what DiPietro, Molenaar, and Torun have found at Cydonia. Suffice it to say here that from its own comments, NASA has shown that it cannot support any interpretation of Cydonia other than its own. This blinkered approach appears to be limiting the agency's ability to adequately interpret or translate what its experts are seeing. The refusal to change an opinion even in the face of evidence leads to this sort of remark: "Even if you show me the elephant in the room, because I don't believe that elephants can exist under the present paradigm, I cannot accept what I have seen."

Let's go and find that elephant.

4

Nemes or Nemesis

Early Imaging of Mars, the Ongoing Search for Evidence of Intelligence beyond Earth, and the Significance of the Martian Cydonia Complex

KEEPING AN OPEN MIND

Even before receiving the Viking images from Cydonia, data from the *Mariner 9* probe had alerted NASA to anomalous landforms on Mars. So much so that in 1980, Carl Sagan wrote of "the beckoning pyramids of Mars" in his book *Cosmos*. However, a footnote to this specific comment refined his poetic generalization: he was only referring to three big pyramidal landforms sitting on an apparently smooth plain at Elysium Planitia at 16.6° N and 198.4° E of the Airy 0° Martian prime meridian, and Sagan became quite chatty in a footnote regarding these pyramids:

> The largest are 3 kilometers across at the base, and 1 kilometer high—much larger than the pyramids at Sumer, Egypt or Mexico. They seem eroded and ancient, and are, perhaps, only small mountains, sandblasted for ages. But they warrant, I think, a careful look.[1]

The physical exploration of Mars brings out comparisons to its analog in science fiction, namely Burroughs's Barsoom, in many a level-headed American scientist. For Barsoom aficionados such as the two Carls, Munck and Sagan, it is certainly convenient that the location of these very first pyramidal shapes brought to the public's attention—and wittingly associated with pyramids of our planet—bore comparison with the location of Burroughs's Barsoom city of Dusar, noted for its honey production. So one might legitimately wonder if the fantasy worlds of

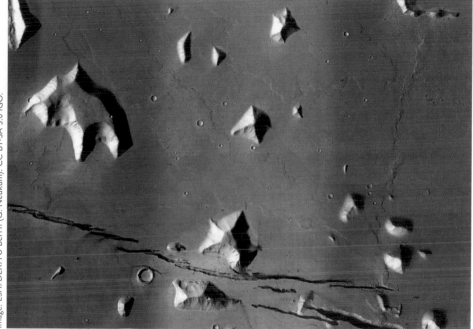

Image: ESA/DLR/FU Berlin (G. Neukum). CC BY-SA 3.0 IGO.

Figure 4.1. Elysium Pyramids (center area) with the largest of these to the south. These three Elysium pyramids are the inverse of the Giza Big Three.

Burroughs affected the initial planning of the *Mariner 9* photographic coverage and the public relations exercise that went with it.

And how much did these remarkable Elysium pyramidal landforms influence the decisions of NASA when it came to designing the Viking missions and deciding the official position concerning their photographic coverage of the Cydonia complex? Rhetorical questions aside, despite publishing *Cosmos* four years after the Viking project, when it came to Cydonia and the Viking imagery, Sagan was less enthusiastic. Cydonia's putative pyramids were not on the call sheet; in fact they didn't beckon him at all. There is no specific mention in his book of any of the Cydonian landforms.

For the space agency orbiters to have picked up the Elysium artifacts mentioned earlier and for their shapes to be remarked on as warranting a careful look by an authority on Mars such as Sagan, invites the public to consider these outstanding items as artificial constructs, like the pyramids on Earth. As humans are unable to engineer the surface of Mars to any degree, without being specific Sagan is also asking us to consider the fact that an earlier intelligent civilization has been on Mars. Deliberately evoking the cultural pyramidal past, specifically Sumer, Egypt,

Photo: NASA.

Figure 4.2. Cydonia complex principal landforms, Viking orbiter image 35A72, 1976.
Numbering code: orbit 35, Viking A, frame no. 72.
Note that due to "noise" across these pyramidal landforms, it is
impossible to reproduce this area from the imagery on Google Mars.
The dark circle top center is not on the planet, it is a technical fault.

and Mexico with regard to the Elysium Pyramids, he also draws the reader's attention to unexplained linear markings on the Tharsis Plateau (for which, read: wide dead-straight lines meeting at angles). The *Mariner 9* image of these lines is too small and without context, but Sagan placed this image just above the Elysium pyramid image, which strongly infers that these lines are not considered entirely natural. Indeed, they should remind us all of the mysterious Nazca lines in Peru. But then, having brought up the subject of ETI, Sagan demonstrates a talent for sitting on the fence by making this statement toward the end of his chapter on Mars:

> Human beings have a demonstrated talent for self-delusion when their emotions are stirred, and there are few notions more stirring than the idea of a neighboring planet inhabited by intelligent beings.[2]

Notwithstanding ETI, just restricting the discussion to landforms that look like pyramids and are laid out in interesting configurations at two different locations on Mars should have been enough to warrant the full attention of all the scientists, yet unfortunately, this was not the case. Another paradox, but perfectly in line with cognitive dissonance.

HERE'S ONE WE PREPARED EARLIER

Perhaps it is both the location and the mathematics of the Elysium placement that have upset the applecart. The Elysium big three sit on the other side of Mars, virtually opposite the massive stand-alone five-sided Cydonian pyramid established as the prime meridian of ancient Mars by cartographer Carl Munck.

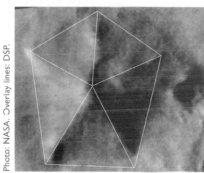

Photo: NASA. Overlay lines: DSP.

Figure 4.3. Stand-alone five-sided pyramid located on the Cydonia complex, Viking image 70A13, 1976.

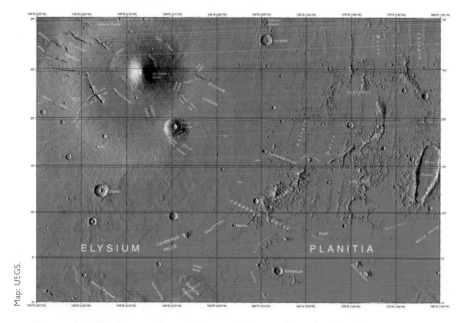

Map: USGS.

Figure 4.4. Elysium region of Mars. The transfer of orbital data to maps is carried out within the USGS by its Astrogeology Science Center.

Considered against locations on Earth, the Elysium complex echoes data from the pyramid complex at Teotihuacán in Mexico, referenced by Sagan in his footnote. The principal Avenue of the Dead connecting the Mexican big three pyramids is aligned with 15.49° east of true north. According to cartographer Carl Munck, the number 15.49 also features in the proportions of the sunken Quetzalcoatl complex at the southern end of the avenue and in the math relevant to the Great Pyramid. The longitude coordinate of the Elysium pyramids on Mars is the same: 15.49° E.

Even without the cartographic number crunching, taking the Martian time line (as understood by geologists), the Elysium remains or pyramidal structures are considered to be many hundreds of thousands of years old at the very least. By comparison, whoever built Teotihuacán is unknown; the Aztecs came upon the big three pyramids already in situ. As for their age, modern archaeologists only commit themselves to dating them in the region of several thousand years old. Yet when comparing Earth and Mars, the older pyramidal forms on Mars have not succumbed to the elements, as is the case with many here on Earth.

As Sagan pointed out, from observational studies of Mars, we know that there is significant local dust storm activity, with planet-wide storms estimated to occur every eight years or so.*

Figure 4.5. Teotihuacán Pyramid of the Sun in 1905, with the Moon Pyramid in the background at left.

---

*Decades of Mars observations show a pattern of regional dust storms arising in northern regions in the spring and summer. Global dust storms have been recorded in 1971, 1977, 1982, 1994, 2001, 2007, and 2018.

Any constructions by ETI intended to be spotted from space would need to have been extremely well-built in order for there to be *sufficient remains* with enough of their geometric forms preserved in some way to make them relevant to scientists, architects, mathematicians, and philosophers when examining images taken by space probes.

For the artificiality of these Martian landforms to be noted, they would also have to be large enough to be significant when imaged from orbit and their configuration would have to be sufficiently anomalous *relative to the surrounding terrain* to arouse curiosity when later analyzed. Both Cydonia and Elysium fit the bill in this regard. Indeed, the mere presence of this apparent geometry on the smooth plains at Elysium makes it highly likely that these three pyramidal forms have been specifically designed to endure and attract, ready for the day when the nearest neighbor pops by with a camera. If the arrangements of the pyramid forms at Elysium and at Cydonia were found to have structure, geometry, and reasons for existing that chime with our understanding of how things work and how our own cultures have proceeded architecturally, then it would surely follow that we share common values with the former inhabitants of the red planet. It also suggests that either the Martian planetary conditions were once very different or that the civilization responsible for these structures had the technological means to transform their environment in such a way as to protect life.

BACK TO THE FUTURE

While ostensibly ignoring these signs of ETI, Sagan actually comes close to acknowledging this once and future connection of like minds when he suggests that we will be engineering the climate of the red planet to make it habitable and that the Martians will be us. Working out how to do that on Mars could lead to innovative thinking in terms of managing our own climate and pollution issues here on Earth; unsurprisingly, manipulating and preparing the Martian environment in order for human beings to live there has been a hot topic among scientists and writers for decades.

The general public had to wait until 1994 for Arthur C. Clarke's nonfiction book *The Snows of Olympus: A Garden on Mars*.[3] He stated that he had used three components for explaining the transformation of Mars through the creation of an atmosphere. Two of these were JPL and the Big Island of Hawaii. The third ingredient, as he put it, until recently existed only in science fiction. He was referring to the development of personal computing and virtual reality software. Writing his introduction in 1993, he stated that "the concept exploded in my head one morning and by nightfall the synopsis was complete." Remarkably, he had his explosively inspirational moment the very year that *Two-Thirds: A History of Our Galaxy* was first published, in which the authors Myers and Percy had already described the means by which the atmosphere of Mars had been acquired some half a million years ago. This is not to say there was any overt plagiarism going on, for neither party knew of the other's intentions. However, there are three notable facts.

An acclaimed science-fiction writer publishes a factual book concerning the future of human beings on Mars. Two unknown authors publish a book reading like science fiction but full of facts, many concerning past events on the red planet. Both books appeared within a year of each other—the past preceding the future. The genesis of both books was due to inspiration and concerned the ratio two-thirds. All of which might have something to do with the transmission of consciousness, both collective and individual, across space and time.

SPECULATION OR FACTS?

From a scientist's point of view, any conversation about the merits and intentions of extraterrestrial civilizations is speculation, and most try to avoid the subject as there is (in their view) insufficient proof. When they do indulge in a little speculation, some spend their time seesawing between the idea that it is a good thing to contact ETI and, then again, perhaps not. Although they say that as our presence in the solar system is detectable, it is a bit late to try and put the genie back in the bottle. The astrophysicist Stephen Hawking sat on the fearful end of that seesaw, and back in his day, Sagan occupied the other end. Although Sagan could also move from the seesaw to again sit on his fence by making statements such as that previously noted on the Elysium pyramids.

This flip-flop attitude is also the approach taken by most scientists toward the subject of UFOs. There is a stigma attached to being interested in such matters, and this affects the reporting of these events. Those who report UFO sightings are generally considered to be unreliable witnesses, and those who are within the military services, civilian pilots, and policemen, all of whom are considered reliable witnesses, tend toward avoiding reporting them or, if obliged, these reports are more carefully kept under wraps. It was generally agreed that while it is probably a good idea to look for ETI there is not sufficient evidence for the existence of physical nuts and bolts UFOs and that even if a few cases are unexplained by any known phenomena, whatever it is that ordinary people are seeing is demonstrably not an issue of national defense.

The above was the case until April 25, 2019, at which point there would be a change of direction: UFOs were now considered to be UAPs (Unidentified Aerial Phenomena), and military personnel from all the forces were from then on positively encouraged to file their findings. This announcement was backed up with footage of sightings made prior to 2019 by *military personnel*. These reliable witnesses, in charge of the technology recording these events, also provided an excited commentary.

Eight months later in December 2019, this so-called new initiative was followed by two *New York Times* articles revealing the existence of a TOP SECRET program conducting research and analysis into unidentified aerial vehicles (yes, UFOs). Led by Bigelow Aerospace, the clumsily named Advanced Aerospace Threat Identification Program was funded from 2007 to 2012 to the tune of $22 million in "black money"; those in charge ensured it was never mentioned in budgetary discussions on Capitol Hill. However, the subsequent release of documents obtained from the Defense Intelligence Agency under the Freedom of Information Act reveals that this program was concerned with many other aspects of advanced flight technology research, both in the air and in space. Furthermore, the program was originally named the Advanced Aerospace Weapon System Application, which might be more relevant but less attractive to the public than the threat of alien invasion.*

The change of title for this program neatly diverted the public from the hard-core research and onto the sensational footage featured in the media. And how did this footage even reach the media? It was thanks to the director of the Advanced Aerospace Weapon System Application/Advanced Aerospace Threat Identification Program, the very busy military intelligence officer, Luis Elizondo. Asserting that the subject was not being taken seriously enough by the Pentagon, he insisted that the UFO footage was made public and then promptly resigned from the program and swanned off to join the To The Stars Academy! This company was fronted or founded (depending on your point of view) by rock musician

---

*After a Freedom of Information Act request from physicist Steven Aftergood, director of the Project on Government Secrecy for the Federation of American Scientists, the Defense Intelligence Agency listed thirty-eight "products" (the agency's term for a research program) available for public scrutiny. Multiple media sites covered this story, among them *Politico,* the *New York Times,* and the *Atlantic.* Christopher Mellon, who from 1985 to 2004 was the deputy assistant secretary of defense and staff director of the Senate Intelligence Committee, wrote an op-ed published in the *Washington Post* on March 9, 2018. Mellon is also a member of the To The Stars Academy, as is the military intelligence officer Luis Elizondo,and the physicist Harold E.(Hal) Puthoff, who was actively involved with some of those thirty-eight "product developments" for the agency.

Tom DeLonge in 2015. Although its 2018 financial statement stated losses of over $37 million, the organization continues and is running a documentary TV series on the History Channel titled *Unidentified* while also informing the public of the military interactions with UAPs and flagging the potential threat to national security. This mix of rock and rollers, entertainment personalities, and those in the sci-tech and sci-fi worlds is not unknown in the intelligence world. The To The Stars Academy group also includes high-level people from the military-industrial-intelligence complex, such as Mellon, who has retained his contacts with these authorities, is still an academy member, and in March 2019 he had this to say:

> We know that UFOs exist. This is no longer an issue. The NAVY itself has publicly acknowledged the fact that they exist, and NAVY pilots—active duty pilots—have gone on the record in the *New York Times* acknowledging the fact that they exist. So the issue now is: why are they here, where are they coming from and what is the technology behind these devices that we are observing?

Yet this insider within the defense and intelligence communities also said:

> Nobody wants to be "the alien guy" in the national security bureaucracy. Nobody wants to be ridiculed or sidelined for drawing attention to the issue. This is true up and down the chain of command, and it is a serious and recurring impediment to progress.[4]

On the back of such comments, this supposed new tack simply confirmed what most of the public already knew: any anomalous events occurring around Earth had always been monitored by the authorities, and these same authorities had no compunction about lying to their public. Yet caveat emptor, with no independent images from civilian sources, it is equally valid to speculate that all of these UFO confessions are less to do with an ETI interface and more to do with political and budgetary matters concerning the covert development of yet another Star Wars–type space technology program hiding behind the threat of ETI interactions.

Therefore, while the public is now being asked to think of UAPs, we shall continue using the term UFO, as it was used for most of the period we are writing about: 1947–2019. So going from here to back there, when peer pressure and approbation held the military charade together and most academics joshed about such matters among themselves, how did it ever come about that in 1984, a group of scientists could even consider forming an organization dedicated to looking for ETI?

Figure 4.6. (From left to right) the original SETI logo,
the second logo, and the third (current) logo.

WHERE IS EVERYBODY?

They called themselves the SETI (Search for Extraterrestrial Intelligence) Institute. The original SETI group, Carl Sagan, Frank Drake, Seth Shostak, and their colleagues, were influenced by a conversation that had taken place at the Los Alamos National Laboratory (LANL) much earlier, in the summer of 1950.

Enrico Fermi, walking to lunch with Edward Teller, Emil Konopinski, and Herbert York, had been lightheartedly discussing a cartoon that had appeared that May in the *New Yorker* magazine that linked a recent wave of UFO sightings with the ongoing disappearance of municipal trashcans from New York City streets.

All were agreed that, obviously, UFOs were not real. By the time they had reached this point in their discussion, they had arrived at their lunch destination and the conversation turned to other matters. Later during the meal, Fermi suddenly asked, "Where is everybody?" and even though his question didn't fit the ongoing conversation, everyone at the table laughed; they just knew that Fermi meant ETI. York remembers that Fermi then did his calculations and drew the conclusions as to why (despite that current UFO wave) we were not being visited by alien craft. The skeleton of this discussion, now called the Fermi Paradox, has supported the fragile body of scientific attitudes toward the subject of ETI and UFOs ever since. It can be summed up in five points:

1. According to astronomers' calculations, our 4.54 billion-year-old Sun is actually very young and there are billions of suns much older than ours in this galaxy alone

2. There is therefore a strong chance that some of these solar systems will contain an Earth-like planet.

3. Presumably, these civilizations might develop interstellar travel technology to the level where it becomes possible to colonize the galaxy in—at the very slowest rate of travel—some five to fifty million years (as conceived by humans in the 1950s and in the twenty-first century).

4. Following this line of thinking, Fermi and his friends then concluded that Earth should either have been colonized or at least visited already.
5. Despite these assumptions, these scientists thought that no convincing evidence existed of either aliens or their artifacts. In their view, the lack of observational evidence to date also failed to support this conclusion.

What is never mentioned when discussing the Fermi Paradox is the question Fermi asked Teller during that preprandial walk.*

Fermi's question concerned the speed of light. As Teller recalls in a letter to space historian Eric Jones:

> I remember having walked over with Fermi and others to the Fuller Lodge for lunch. While we walked over, there was a conversation which I believe to have been quite brief and superficial on a subject only vaguely connected with space travel. I have a vague recollection, which may not be accurate, that we talked about flying saucers and the obvious statement that the flying saucers are not real. I also remember that Fermi explicitly raised the question, and I think he directed it at me, "Edward, what do you think? How probable is it that within the next ten years we shall have clear evidence of a material object moving faster than light?" I remember that my answer was "10^{-6}." Fermi said "This is much too low. The probability is more like ten percent" (the well known figure for a Fermi miracle.)[5]

From which it might be concluded that the report on the birth of the Fermi Paradox is somewhat incomplete.

FOOTPRINTS IN THE SANDS OF NEW MEXICO

What is really interesting is that while critically thinking scientists generally acknowledge that any ETIs we might encounter will most likely be far more advanced technologically than ourselves, it also seems that such entities will nevertheless be using propulsion technology similar to our own and will also be limited by the observed speed of light—because, of course, that's what Einstein claimed.

---

*The SETI Institute's evolution since the original concept in 1984 is documented on their website. The SETI web page referring to the Fermi Paradox does not tell the story as it happened, instead locating the entire conversation at the lunch table. It is nevertheless worth a look at this web page as it has links to other aspects of the ETI question—again, as defined by SETI.

Lots of sticky questions are raised by this single issue, not least how to square the status quo of Newton and Einstein with quantum theory and the emerging speculation as to the precise nature of so-called dark matter and dark energy. It is not evident that science is necessarily on the right track in trying to squeeze these different theories into the same box. However, for those who are grappling with them, these big issues become overwhelming when taken together with the question of how astronauts can travel safely and swiftly in the vacuum of space. The fact remains that in 2020, we still don't know how to *protect* a bio-organism sufficiently well from all forms of ionizing radiation, provide a gravity field inside a craft to be able to voyage into space well beyond LEO, design and build such an advanced craft, and launch it into space. Nor can we yet deal in a timely fashion with the vast distances between here and Mars. With the result that even if ETI and UFOs are perceived as being interconnected, despite Fermi's speculations on the matter, the waves of cognitive dissonance can be heard crashing onto the shoreline near the LANL.

In 2008, during filming of the BBC six-part TV series *Stephen Fry in America,* in episode five, "True West," Fry visited the Los Alamos National Laboratories. Terry Wallis, the LANL director, informed Fry that the LANL existed for one reason only: "to do national security science." He also asserted that research into dark matter and dark energy was the next "big thing."

As the first "big thing" had been the Manhattan project to create a weapon using nuclear fission, one must surmise that dark matter and dark energy are equally important for the United States. On the one hand, it is patently ridiculous that a matter of hypothetical physics can be annexed by one nation, while on the other hand, considering such a notion acceptable is hardly surprising, given the track record of that same nation as the primary secret developer of an energy force that we still do not know how to manage successfully.

If the LANL is now inclined to the idea that future technologies are as potentially dangerous as their previous nuclear efforts, one can understand their paranoia. Especially if such energies were deemed useful as a replacement for chemical rockets. Although, on past form, it is more than likely that any dark matter and dark energy research is being undertaken with weapons development in mind, rather than understanding that the current blood-and-thunder methods of brute-force rocketry is not fit for long-term purpose and long-distance travel. Which is where ETI-UFO sightings and energy development merge. While it has been asserted that none of the ETI-UFO sightings listed by the US government are threats to its national security, if the Manhattan Project is anything to go by, another nation harnessing any technologies that are space compatible would also be a threat. The LANL paranoia and selfishness over research into dark matter and dark energy indicate that any new technology will not be shared willingly with other spacefaring nations.

However, at the time of this writing, both dark matter and dark energy are still pie in the sky, and to date, sadly, statements made by most scientists concerned with space travel make it clear that their imaginations do not stretch to anything other than conventional brute-force rocketry of some description. Even though conventional rocket technology is still neither safe enough nor sufficiently powerful to withstand any accidents that might occur, either at launch or from launch through to traveling a safe distance away from Earth. For decades, NASA has used nuclear devices on space probes and satellites, and no doubt the other space agencies do likewise. Littering the solar system with dangerous junk with scant information available from all these agencies as to the amount of nuclear technology that may be carried on these spacecraft. Back in the 1960s the original NASA program researching nuclear-powered rockets was abandoned, mainly for reasons of safety. Guess what? It had the project name Orion! The use of the name Orion to this day for spacecraft programs may have more meaning for NASA than is understood by the general public. NASA's space historian David Baker, author of *Spaceflight and Rocketry: A Chronology,* uncharacteristically references the genesis of Project Orion out of sequence at the end of his chronology for 1955. He notes that during 1955, Stanislaw Ulam of LANL (then called the Los Alamos Scientific Laboratory) proposed the use of thermonuclear fusion to drive spacecraft across the solar system. Baker, from his perspective, described thermonuclear fusion as the most bizarre form of rocket propulsion envisaged. But in 1958, DARPA funded General Dynamics to the tune of $1 million, and the project was called Orion. The project was extended (funded) twice more, in 1959 and again in 1963, but that same year Kennedy signed the Test Ban Treaty and that apparently halted the Orion project. Apparently. Because at the time of this writing, NASA experts are still prepared to consider incorporating nuclear devices with future rocket drives, and research into fusion, as it is termed in the media, continues.[6]

The whole issue of how best to travel in space is linked to what has actually been achieved to date, and unsurprisingly, because the Fermi discussion was based on the premise that the speed of light is a constant, issues regarding the capabilities of current space travel also color the debate on UFOs. Indeed, everything astrophysicists evaluate is based on the premise that the speed of light is a universal constant.

If the speed of light actually turned out to be *variable* according to location, then mankind's assumptions about our space environment would need total reappraisal, and that would include the way ETIs actually travel and therefore how we might travel through space farther afield than LEO—our very, very local "backyard."

To make an analogy with generations of swimmers, NASA and its scientists are like the modest Victorians. Transported to the shoreline in their beach huts on wheels, they emerge clothed in bathing costumes designed by the House of Einstein. Weighed down by water, they merely dip their toes into the speed of light and space travel discussion, announcing that there might be variables but that this is not adequately proven; these modest swimmers then do a few strokes, constructing experiments that are designed to be sufficiently inconclusive so as to maintain the status quo. Then it's a case of everyone back into the beach hut to change and home in time for tea.

The rather more daring swim a little farther out and dabble with nuclear propulsion systems, although their bikinis and swimwear still have Einstein designer labels. From splashing around on the shoreline and in the shallows to actually going deep sea diving would require another type of garment altogether—an individually tailored wetsuit.

Or even nothing at all, because entirely different and faster speeds for interstellar and intergalactic space would mean that everything we think we have understood about measuring light speeds and therefore vast distances in space, along with the speeds attainable, may be wholly inaccurate. Virtually everything

Figure 4.7. Typical Victorian bathing hut.

that we currently measure in deep space relative to local solar system light speed would probably change, as would our basic understanding as to how the universe works. And as would the current estimates for traveling on crewed missions within acceptable time frames for the human body.

This is a big ask, so it's not at all surprising to see that while we are standing at the edge of this vast ocean of understanding, leading scientists hold back from changing their preferred designer label. After all, not everything is wrong, we have simply reached the limits of where we can swim using our current kit.

In this regard, it is also of interest that in April 1972, NASA's New Projects director George Pezdirtz had organized a conference on speculative technology. Trialing an ESP (extrasensory perception) machine designed by laser physicist Russell Targ, NASA administrator James Fletcher and Wernher von Braun, scored well. After the conference, in an interesting parallel with that earlier LANL lunchtime walk, although this time the discussion was on human psychic faculties and related phenomena (PSI), not UFOs, Targ went for a walk along the Atlantic shoreline. He was acompanied by Fletcher, Pezdirtz, von Braun, the astronaut Edgar Mitchell, and author Arthur C. Clarke. Targ had already found that the very top people in the organizations he had encountered already accepted that PSI was a reality. Of this particular gathering, Clarke was the only skeptic present, which was noteworthy, given that his book *Childhood's End* was very much concerned with PSI.

Subsequently, Mitchell and Pezdirtz were influential in Fletcher's decision to ask Targ and Hal Puthoff, both from the Stanford Research Institute (SRI), to design a program to help NASA's astronauts develop an intuitive psychic contact with their spacecraft. Titled *Development of Techniques to Enhance Man/Machine Communication,* the final report from SRI was delivered to NASA in 1975. At the time, nothing about such matters reached the press or the public, of course.[7]

KEEPING "FERMI-LY" IN CONTROL

The fact that the present scientific paradigm is virtually cast in stone also informs the attitude of most scientists toward the inquiring public over sky-related issues such as UFOs. It is nonetheless true that there are UFO cases that have not been resolved to the satisfaction of scientists. Whether these sightings consist of nuts-and-bolts spacecraft or are reflections of energetic systems that manifest to us as such, being unable to understand the process or reproduce the technology and unwilling to openly admit that anything or anyone might be beyond Earth informs the scientists' attitude toward any truly unaccountable sightings. These are classified as "not for real," and those who would consider them so are castigated for their uneducated and wild imaginations. In order to retain some form of authority, the scientific community generally infers that most sightings are due to poorly understood natural phenomena and have nothing to do with any intelligent insertion by A. N. Other into our locality. Yet no valid explanation is offered as to what these genuinely anomalous events might actually be.

Instead, the media cries "alien invasion," the thinking public laughs, having been well programmed for decades, the question of any real anomaly is brushed under the proverbial carpet, and everyone goes back to sleep.

Or nearly everyone.

Circa the mid-1990s, anomalous light events abounded in the county of Wiltshire, England, home to the ancient stone circles of Stonehenge and Avebury and the largest artificial mound in Europe—Silbury Hill. This region is also the primary location for the relatively recent phenomenon of crop circles or crop glyphs. Upon the arrival of these anomalous events into the Wiltshire fields, the tourist population was boosted by the arrival of various governments' thinktank geeks, officials, and scientists, all discreetly monitoring these events on the ground, in the skies, or even from space. Given the SRI contract in the 1970s, this is less surprising than it might at first seem. On one occasion during an experiment in group meditation spread out across various ancient sites around the Avebury area, I was sitting in the northeast quadrant of the Avebury stone circle, along with two other women, one of whom was a principal organizer of this PSI experiment. We three women all saw a multicolored light show animated against a large tree trunk. At the postevent meeting, we reported in meticulous detail this event, which had taken place only 127 feet (38 meters) from where we were seated. Yet when the official version of the report emerged months later in the United States, this light event had become a solid nuts-and-bolts spacecraft. I asked the organizer, who had herself witnessed the whole event, why the report had been fundamentally altered, and the reply was, "Because it was wanted that way." Further information as to whom it was who decreed such a deception was not forthcoming, but it was clear to me that intelligently manipulated light is more difficult to deal with (either for the funders or for those directly involved) than interstellar hardware. More paradox, until one remembers Fermi and his friends. Years later I discovered that the green, gold, red, and pink colors seen near to the ground at Avebury that night were found in the Sumerian illustrations of meteors. The ancient Sumerian rulers and their astronomer-priests used all such anomalous sightings as useful tools for the manipulation of public perceptions, to the benefit of themselves.

Which then reminds one that NASA and the alphabet agencies treated the collision of the 1994 Schumacher comet with Jupiter in exactly the same way.[8] The manipulation of the facts concerning this light event also reflects this practice.

BOXING IN THE AWKWARD SQUAD

For all the reasons stated above, it seems to be necessary to dismiss anyone who dares to consider that UFOs might be *any* sort of manifestation of ETI, and since ETI and extraterrestrial transport systems are two sides of the same coin, the scorn poured on UFO groups by SETI and their colleagues then smears any discussion of potential ETI artifacts found on Mars, or anywhere else. Those who do speak out publicly, such as SETI astronomers Sagan and Drake, while being exceptions to the rule, were still only inclined to discuss contact in forms that appeal to their own worldview: radio telescope searches for ETI = good; ETI showing up in any other form = bad. It could be said that in matters of extraterrestrial transport systems, the scientific community generally considers UFO spotters to be at the lower end of the social scale, while those who participate in the astronomical search for signs of ETI (SETI), such as Sagan and Drake, are at the high end of that very same scale.

This intellectual climate of scorn, built as a defense system around serious matters that cannot be openly addressed, has bred a gang of foot soldiers who loudly harass and (since the advent of electronic media) cyberbully those who dare ask questions of the scientific fraternity regarding these delicate matters. Sagan considered himself to be a skeptic, yet even he was critical of such actions. He wrote:

> And yet, the chief deficiency I see in the skeptical movement is in its polarization: Us vs. Them—the sense that we have a monopoly on the truth; that those other people who believe in all these stupid doctrines are morons; that if you're sensible, you'll listen to us; and if not, you're beyond redemption. This is unconstructive. It does not get the message across.[9]

Indeed, it does not, but in the light of such categorization and consequent behavior, it's hardly surprising that in the interests of avoiding these inquisitors, the subject of anomalous landforms on Mars or anywhere else is not top of the list and that most scientists have avoided the ETI issue altogether. That is, until 1976.

CYDONIA

If the Elysium pyramids slid quietly under the public radar and no one else apart from Sagan seems to have mentioned them, when the *Viking 1* orbiter images were brought to the public's attention in July 1976, the die was cast and the controversy over the anomalous landforms found in that area began in earnest. It's hard to remember that in those days, there was no internet, no social media, and no color images of the Martian landscape available to the public. Everyone outside of NASA was evaluating monochrome pictures of the anomalous complex of Cydonia.

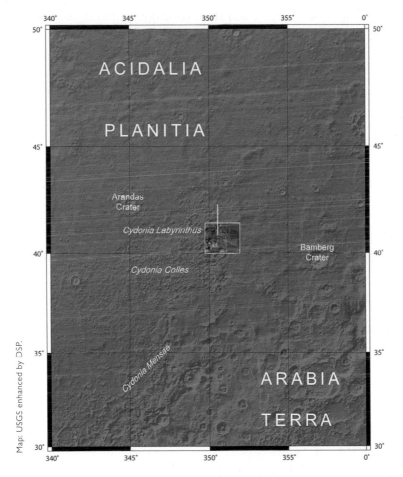

Figure 4.8. The location of the anomalous complex and its distance from the Mars prime meridian. Note that this map, from 2015, divides Cydonia into three different geological regions: Labyrinthus, Colles, Mensae. In the 1970s the anomalous complex was referred to as being in Cydonia Mensae, but on this map its surroundings are allocated to Cydonia Labyrinthus.

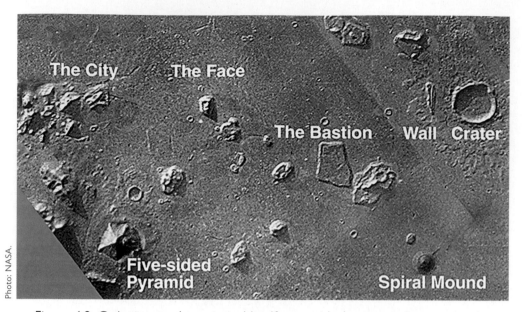

Photo: NASA.

Figure 4.9. Cydonia complex principal landforms with the names that came to be associated with them. The City is that grouping of distinctive pyramidal shapes, with a varying number of sides. Within this grouping, four smaller shapes define a square. Mars Viking orbiter image constructed from 35A72, 73, and 74, 1976.

Sitting at the interface of a northern sea basin and higher land to the south, this Cydonia complex would turn out to be of particular interest thanks to its several distinctive and very odd landforms.

Taking a closer look at these landforms, across the plain to the bottom right of this image lies a lone circular mound. Enhanced by the digital image processing and satellite remote sensing expert Mark Carlotto, this close up image clearly shows it to be cone-shaped with traces of a spiral path around it.

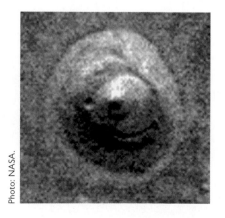

Photo: NASA.

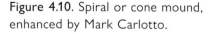

Figure 4.10. Spiral or cone mound, enhanced by Mark Carlotto.

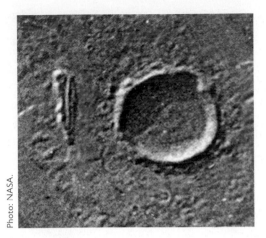

Photo: NASA.

Figure 4.11. Wall feature and crater, Viking orbiter, 1976.

Due north of the spiral mound, sits the enormous crater. On its western flank a long rectilinear outcrop is set very straight for what should be a natural formation On the rim of the crater's northeastern quadrant, there is an apparent triangular or tetrahedral outcrop located above a gap.

Southwest of this crater, two other landforms are remarkable for their contrasting conditions: an extremely *flat* landform juxtaposes an extremely *rough* landform in such a way that it positively invites observers to use their imaginations. The rough eastern portion looks as if it is made of the material removed from its neighbor in order to create what appears to be an artificially leveled surface for the western mesa.

The whole complex as presented displays so many weird and wonderful contrasting shapes within such a relatively small area that one would have thought that since NASA went looking for signs of life, these extreme contrasts would have surprised and delighted the scientists involved. However, it is not quite so

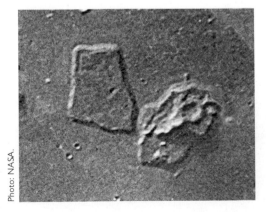

Photo: NASA.

Figure 4.12. The rough with the smooth. The flat mesa was named the Bastion by authors Myers and Percy in *Two-Thirds*. Viking orbiter, 1976.

simple as that, because as discussed earlier, the choice of orbit and landing sites for the Viking probes rather indicates that the agency knew something special awaited them out at Cydonia even before the craft had left Earth.

At the time of Viking's first scan, there was a particular landform serendipitously illuminated in such a way that it would become the most visually arresting image from Cydonia. Located northwest of the flat and rough paradoxical landforms, it really pushed our collective buttons, for staring up at *Viking* as the craft passed overhead was something that looked very like a face.

Even a human face.

The story goes that Toby Owen, a member of the Viking imaging team, found this image while searching for a safe landing place for the *second* Viking lander. The image, labeled 35A72, had been returned to NASA on July 25, 1976. The first two numbers designated the orbit, then a letter designated the spacecraft, and the last two numbers designated the frame number. At a press conference in JPL's Von Karman auditorium on July 26, NASA's Viking project scientist Gerry Soffen casually introduced frame 35A72, then dismissed it in the same breath.[10]

"That face," the audience was told, "was only a trick—just the way the light fell on it" (it had been taken at 10° Sun elevation). The NASA spokesman then stated that a photo of the area taken "a few hours later" showed no trace of this face.

Naturally, NASA's exploration of Mars benefited from the publicity when the press chose the spectacular over the scientific explanation and dubbed this particular landform the Face on Mars. All subsequent and sometimes halfhearted attempts to dismiss this image as nothing but a circumstance of fortuitous timing have failed to dispel its potency, and the name has stuck in the public's mind ever since.

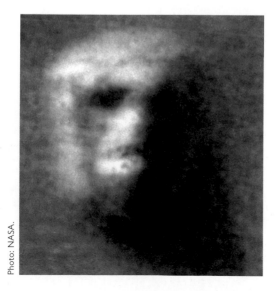

Photo: NASA.

Figure 4.13. The original Face image 35A72 taken in 1976 by the Viking orbiter.

Figure 4.14. Chilbolton, Hampshire, England, August 14, 2001. In the background is the Chilbolton radio telescope, in the foreground the "face" crop glyph. Five days later the "Arecibo" glyph arrived. It was taken to be an inaccurate copy of the 1974 digital binary code transmission sent into space from the Arecibo dish in Puerto Rico by SETI.

It is worth noting that at the time of Soffen's reveal, NASA had just been put into the freezer by President Nixon, its budget on hold with no increase in spending for five years. And it is also worth noting that this "trick of the light" image has been trotted out on a regular basis over the last four decades, not necessarily at budget review time, but often enough, lest we forget. Decades later, in 2001, a "face" would appear in the crop fields of southern England, handily situated adjacent to a radio telescope.

Claimed by some to be "the real thing" and by others to be flattened crops manufactured by human beings, it nonetheless showed that the idea of a humanoid-ETI link was being maintained by interested parties, although whether for or against is a moot point. It was possible for people (as well as gear such as cameras, compasses, and the like) to experience a definite charge in many of the crop glyphs of the early 1990s. Visiting this one on the day it was discovered, I found that it did not seem to affect people. Nor was it as difficult to make as the design suggested. This was not a popular view with others who visited it, so I left them to it and walked off to talk to a tractor driver I could see working a couple of fields away. He told me that the field was used by the University of Southampton as a part of its biological field study program, so he had not been curious when he saw several people in that field over the days prior to the discovery of this crop event. A group from this university did frequent the crop circle area in Wiltshire, but they could hardly be accused by association, even if they were vociferous in their skepticism. It was beyond them that anyone intelligent could imagine that ETI was involved. Discussing the matter with one of their members, it became clear to me that cognitive dissonance and fear were running his program.

VIENNA, AUSTRIA

Coming back to the events surrounding this facial recognition, NASA itself soon developed a severe case of selective amnesia, as one European researcher[11] found out the hard way.

At home in Vienna in 1976, Walter Hain had seen a TV documentary on the Viking space probes. He immediately wrote to the German studio responsible for the program asking for more information, and he also asked for a print from their broadcast. While he waited for the reply, he also wrote to various branches of NASA: the Marshall Space Flight Center in Hunstville, Alabama, the Martin Marietta Aerospace in Denver, Colorado, and the home of Viking, JPL in Pasadena, California.

At the time of Viking, making a request to the agency for a copy of any of their images required the proper reference as supplied by NASA. Hain thought that getting hold of a picture of the Face should be fairly straightforward. After all, it had been displayed to the press on a specific date and spoken of as having a likeness to a face, and it was stated there was only the one image. How hard could it be?

The print from the TV studios eventually arrived on May 27, 1977, and although a bit blurry due to it being a copy from the studio's transmission tape, it was sufficiently interesting for Hain to write to Erich von Daniken, author of the bestseller *Chariots of the Gods,* about it. Bearing in mind that von Daniken had spent a considerable amount of time writing about the evidence for aliens having visited Earth, it was amazing that it took him five months to come up with a reply and doubly amazing was the content: on October 26, 1977, he wrote, "Unfortunately, the Face on Mars is something we have no use for." Which beggars belief and makes one wonder who the "we" might refer to. Two decades later in another interview around the time of the Mars Global Surveyor Project (which began orbiting Mars in September 1997), von Daniken was asked, "What do you think we would find on Cydonia, where some people think there's a face and pyramids?" He replied that he was definitely convinced they would eventually find traces of primitive life, then, ignoring the pyramids, repeated his dismissal of the Face, saying, "The Face on Mars? I was never convinced of that. I am not sure if this is a serious thing or not."

From these replies, it appears that von Daniken is not quite so keen on extraterrestrials after all. Just like NASA.

NASA also replied to Hain in October 1977. The JPL's public information officer, Don Bane, stated that "the nicknames you give the pictures 'MarsFace' [*sic*] mean absolutely nothing to me—they are your nicknames, not mine." After a further nagging from Hain, by March 2, 1978, Bane felt able to write, "At last I know what you mean when you refer to the 'mars=Face [*sic*] picture!'" He kindly added the coordinate details for its location but then managed to get them wrong. He wrote, "The latitude of the feature is 40.89 North. Latitude [*sic*] is 9.55." This should have read: *longitude* is 9.55° West and both should have been stated as degrees. Bane then stated that the picture is "in the correct orientation when the sun is coming from the left. That is, with the chin at the bottom and the hair at top. The sun is about 10 degrees; no we have no other picture of that feature."

Six months later, on September 12, 1978, presumably after even more nagging from the persistent Hain, Bane wrote again, this time stating, "There is nothing in any photo from Viking that suggests to any scientist here at JPL that there are ruins of any kind on Mars. Even the 'Face' as you seem to insist on calling it (having previously described it as having a chin and hair, this is a bit rich) is a combination of wind erosion and sunlight angle."

And here's the cherry on the cake: Bane concluded by writing, "Since you already have a picture of the Face, I see no need to load you down with another, and we have no picture that shows that region at closer range." Bane's wording suggests that he knew that indeed there was another picture of the Face taken from an equivalent orbital distance or farther. He was surely thinking of frame 70A13, as it had been found that same year.

Not entirely buying into Soffen's "trick of the light" comment, scientists Vincent DiPietro and Gregory Molenaar decided that by verifying all the images for the Cydonia location, they should either find something or there would be effectively nothing and Soffen would be proven right. These two scientists would use the same skill sets for assessing and analyzing the Cydonian images as they did for their government day jobs. After a painstaking search through the entire Viking data set, checking the images frame by frame and number by number, in 1978, they came across the second image, which had been mentioned by Soffen at the 1976 press conference. Labeled 70A13, it was taken exactly thirty-five orbits after the first "Soffen Face on Mars." The fact that these 35 orbits are the same as the reference number of the first image: 35A72 makes life very confusing for the uninitiated. This second image's frame number 13, coincidentally (or not) is a very special number for US culture, and with a Sun elevation of 27° this was very special picture. As both images clearly show the same facial details, there was *no* trick of the light on July 25, 1976, when the first image had been taken.

However, the facts that this second image showed a landform that still looked like a face and that the different lighting angle enabled even more detail to be seen—not less—might explain why, through misfiling, it had been "lost" within the archives. So completely so, that without the assiduous mining by DiPietro and Molenaar, it would have escaped the public eye altogether.

All rather pointing to the fact that NASA and JPL knew that they "had a problem, Pasadena."

This landform was and is indeed significant. Which is hardly surprising because whatever its true origins might be, for most people that apparently mega-sized sculpture of a head immediately struck a chord. Whether intended by NASA as a publicity stunt, whether actually the result of lighting conditions at the time of photography, or whether truly the remains of an apparently humanoid sculpture with all the connotations of intelligent design that infers, the image captured by Viking echoed our own ancient past in Egypt.

The Face on Mars seemed to be wearing a headdress familiar to us and also present on a large sculpted landmass here on Earth: the half-lion/half-human Sphinx at Giza, whose head is clothed in a Nemes headdress while s/he contemplates the blue-green river Nile and the eastern horizon.

Or perhaps s/he is contemplating the idea of Nemesis, which originally described the distribution of fortune, neither good nor bad, but allocated in proportion to each, according to what was deserved. What New Age Westerners tend to interpret as karma. For whatever reason, since 1976, NASA's PR people have taken every opportunity to reiterate that there's nothing to see at Cydonia, move along folks, while serving up ever-worsening images of that landform.

Photo: DSP.

Figure 4.15. The Great Sphinx, Giza, Egypt, wearing a headdress similar to that worn by the Egyptian boy king Tutankhamun.

Photo: Roland Unger. CC BY-SA 3.0.

Figure 4.16. The Mask of Tutankhamun.

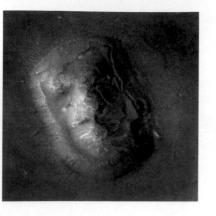

Photo: E03-00824, 2001, NASA.

Figure 4.17. The Face as seen by *Mars Global Surveyor* (MGS). Note the sharp right angle at top left of the mesa, and the straight line across the forehead.

Figure 4.18. In this *Mars Global Surveyor* stereo image that telling right angle in the top left corner of the Face has been cropped out. Craters located just under the chin (to the left of this image) were also deleted from the MGS images.

Photo: M16-00184, NASA.

These two Surveyor images were released to the public in 2001. But that is not necessarily the year that the orbiter actually took the images. Malin Space Science Systems states that *prior* to the primary mission slated for 2001, the Mars Orbiter Camera was pointed at the Cydonian landforms in a test run of its capabilities. It then states that the so-called catbox images were only taken during the extended mission, which began February 1, 2001, but that it was only on April 8, 2001, that the first opportunity arose to turn the spacecraft and point at "the popular Face on Mars." And if you can believe that, read another page from the same Malin Space Science Systems website. Adopting the term "City" for the twelve pyramidal forms, on this page it is stated that images of the massifs in Cydonia near the City square were taken on April 14, 1998, the City square landforms on April 23, 1998, and the Face on Mars landform on April 5, 1998! The gap from 1998 to 2001 was apparently taken up with image processing/amalgamating. And if you can believe that . . .

This Surveyor effort was so distorted it was dubbed "the catbox," an American term for a cat's litter tray. In a poetic aside, that's not a bad nickname, given that Cydonia researcher David Percy had previously shown that by mirroring the left and right sides of the 35A72 image, the results revealed, respectively, a monkey/hominid face and a big cat/lion face.

The monkey/hominid and the lion/feline are potent symbols for many of

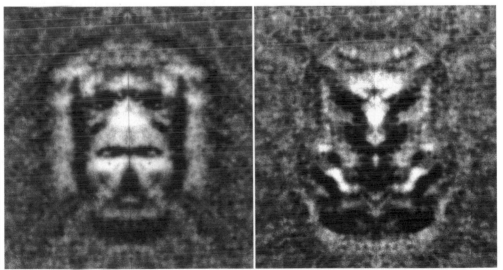

Photo: NASA. Derivation: DSP.

Figure 4.19. The mirror images originally created by David Percy in London for a major presentation in February 1992 at the headquarters of the United Nations in New York City. L: monkey/hominid, left side mirrored, R: lion/feline, right side mirrored, derived from Viking image 35A72.

our own Earth cultures, not only for the Egyptians. While readers of *Two-Thirds* might note that "cat box" is also golfing slang for a sand bunker on a golf course,[12] some will recall that TOM, the ETI transmitter in *The Only Planet of Choice,* said, "We do not belong in boxes."[13] It looks as if the *Surveyor* catbox image was a particularly blatant attempt at shutting up those (including their own imaging experts) still asking questions about "that face." And for good measure, finally bury the several cultural associations the public had already made with this Martian landform:

1. Evidence of intelligent ETI intentions set out on Mars
2. Visual similarities with the iconic Egyptian sphinx
3. The idea of ETI having actual connections to humanity.

To push the catbox analogy a step further, in publicly denying that Cydonia has anything interesting to offer the search throughout the solar system for signs that we are not alone, NASA could even be said to be burying their own bodies up to the neck in the Martian sands.

ATTEMPTS AT AN EXPLANATION

The Great Sphinx was rescued from the sands that had drifted into its own cat-box by a man who was inspired by a dream. Or so the story goes. NASA could have taken inspiration from Cydonia, but instead the agency would stick with its opinion that the 2001 *Surveyor* image definitively demonstrated that there never was and never had been a face sculpted on Mars: "We can show you what it is, what is really here folks, and it's a naturally eroded mesa." This statement is true, in as far as it describes the state of this landform *now*. It is indeed naturally eroded over many thousands of years, and geologically speaking, it is indeed a mesa. While words unspoken and unwritten can hide a great deal, this rather desperate attempt at management actually sabotaged itself thanks to the technology used to take the image. No amount of photographic hocus-pocus could disguise the fact that even in the hideously deformed catbox image, evidence of intelligent intervention with the surrounding landscape is clear. What the NASA PR machine did not say was this:

> Thanks to this high-resolution Surveyor *image, it is obvious that this landform appears to have been artificially worked. We can see remains of this in the right angle at the northwest corner on the higher level, in the curve of the southern edge*

Figure 4.20. Perspective view of the Face
derived from a *Mars Express* image, September 2006.

where it joins with the western straight-edged base, and again at the base of the
northwest corner, and we can detect the slope angle of the sides.

Later images from the ESA sided with NASA in the interpretation of the Face
as "a remnant massif," and in 2006, the ESA published the image in figure 4.20 with
attempts at explaining away all that troublesome detail on the western side.

Yet even these images do not remove the impression that we are looking at
the semblance of a face. And even the worst photos still reveal the Nemes headdress effect on this Martian Sphinx, as it has also been dubbed. For those still
insisting that the Face was of importance, another method of denial other than
geology was then adopted by the agencies. Human beings are particularly adept
at recognizing the traits of the human face among any amount of background
clutter. It was now asserted that those who thought they saw a face were merely
experiencing a simple case of pareidolia, a psychological phenomenon in which an
image (or even a sound) will trigger the mind into perceiving a familiar pattern
where none actually exists.

The sheer nonsense and the implicit irony of that statement was too much
for NASA to recognize, but no doubt it pleased the velvet-gloved, iron-fisted
cassock wearers.

THE GIFT THAT KEEPS ON GIVING

If Cydonia has been waiting patiently for millennia, the manner in which we dealt with the place when we did reach it with our probes was left up to the authorities in charge of space exploration. The glossing over of the *Mariner 9* Elysium pyramid forms and the philosophical discussions on the merits of photographic interpretation of this face image might have been considered a primary measure in dealing with the challenges Mars has produced. However, Cydonia is more than this single, highly anomalous landform.

From the mapmakers point of view, there was added piquancy in the fact that this anomalous face was looking up from a region of great interest to Earth geologists. This area of Cydonia lies at the interface between two different planetary surface conditions, considered by some geologists to be evidence of an old shoreline. The difference between the northern hemispheres smoother plains and the rougher southern hemisphere is called the Mars Dichotomy by geologists.

Looking closer at this 2D mapping of Mars, the four snow-topped cones of

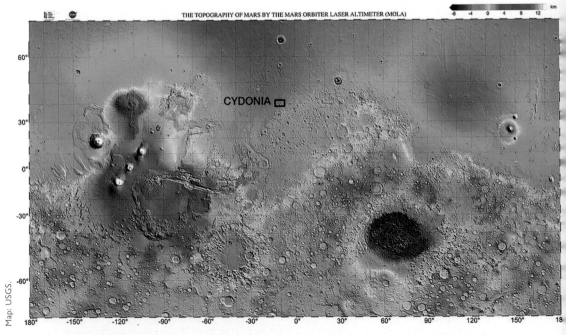

Figure 4.21. Location of Cydonia on the Mars Topography map established by laser readings from the Mars orbiter laser altimeter. The blue-green northern section indicates the land considered to have been smoother ancient seabed and shoreline while the orange, yellow and reds are the rougher higher lands.
The location of Cydonia has been marked by DSP.

the principal volcanoes on Mars suggests an equiangular triangle with the two sides fractionally longer than the base. That suggested triangle described by these volcanoes will come into focus later. Here is it is of interest that on this flat mapping of Mars, the very small snow-capped cone located halfway along the left side of this triangle, between the southernmost of the three Tharsis volcanoes and the huge Olympus Mons volcano, forms an angle 90° with its nearest and farthest larger neighbors. The shield volcano Olympus Mons, at the top, extrudes onto this northern plain. The other three volcanoes lying diagonally across the equator recall the big three pyramidal forms at Elysium, Giza, and Teotihuacán, while the northernmost of these forms an angle of 30° with that small cone volcano.

Do we get more information about Sagan's Elysium pyramidal landforms when we find out that these numbers appear in the details of built structures on Earth? The slope angle of 30° is same as that of the Silbury hill flat-topped spiral mound in Wiltshire, England. The angle of the Teotihuacán avenue and the angle of the Elysium pyramids are both 15.49° degrees east of true north. Furthermore, as the height of Olympus Mons is variously estimated from just over 13 miles to just under 16 miles, these numbers suggest that the relevant height of Olympus Mons is likely to be 15.49 miles. Finally, noting that each side of the Great Pyramid is concave at its halfway point and that the buried ceremonial boats were found marking the southern halfway point, it's even more remarkable that the base diameter of Olympus Mons is estimated at around 375 miles; this number is very close to the half-side length of the Giza Great Pyramid, which is on average 377 feet. There is more: the fact that the Martian Face landform marks this Martian boundary while the Great Sphinx at Giza marks the boundary between the Nile delta, the Nile valley (formerly with its annual rising waters), and the Giza plateau on which the greatest of all our ancient Earth pyramids stands can hardly have escaped the notice of NASA geologists. Nor could the other geological anomalies waiting for them at Cydonia be missed.

How was Cydonia ever going to be swept under the carpet? Basically it's Catch-22 and from a NASA perspective might be like this:

Our science cannot yet manage human space flight to Mars, so we cannot verify our received data with boots on the ground. Nor are we likely to manage it anytime soon, given the limitations of current technology. So we have to buy time because admitting that structures exist or are manifest as ETI artifacts on Mars opens too many cans of worms for science—and for NASA—concerning how to manage getting there and back safely with human beings on board. But that's all right, we can control all these aspects of the matter because we are in charge.

We are going to say that any image captured by a probe that resembles constructs familiar to us—especially a face, for goodness sake—is open to emotional and wishful thinking when interpreted by human beings on Earth, who are hardwired to make such attachments.

We are publicly going to ignore any potential geometry found on Mars as subjective, being open to interpretation according to personal bias.

And identification of anomalous landforms will be proclaimed as being ignorant speculation from those who do not understand geology or weather patterns on Mars. After all, no image is proof; its interpretation is just an opinion.

Via the media, in the meantime, we're going to play both sides and keep ETI speculation in the loop in order to keep your interest in funding us to actually get there.

Over time, we might dilute earlier findings by pointing out images from Mars that "look like" things we find on Earth.

And all of the above applies to sightings of UFOs around Earth and to the notion that extraterrestrials have visited or affected matters on this planet.

Of course, when we can actually send astronauts to Mars and officially verify the probe data, then it will be a case of "we knew this all the time." Which we did, way back when.

Bearing out Russell Targ's observation mentioned earlier in this chapter, it is a generally known fact within the space industry that the majority of NASA staff also give credence to the idea of ETI, but the agency itself declines to discuss the subject seriously with the wider world. The official policy toward ETI on Mars (let alone elsewhere) therefore must originate further up the chain of command, beyond NASA, and farther back in time, back to that *Brookings Report*.

Measure for Measure

Exploring the Martian Complex,
the Geometry of Cydonia, New Findings,
and the Continuing Search for Evidence of ETI

The very special nature of the Cydonia region surrounding the Face mesa could be fully appreciated once the USGS started producing geology maps of this region. The map in figure 5.1 covers the area of interest seen in the photographs in the previous chapter. The spiral mound at lower right of the rectangle, marked "Akr," is depicted in pink. This mound is of a totally different geological makeup to all the other components of the anomalous Cydonia region.

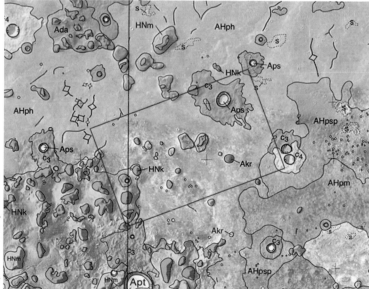

Figure 5.1. The geology of the Cydonia anomalous region as mapped by Tanaka and others, 2003. The vertical black line is true north. Note the mound marked "Akr."

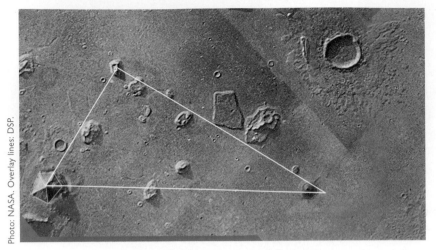

Figure 5.2. Viking image of Cydonia
with a 30°-60°-90° special right triangle overlaid.

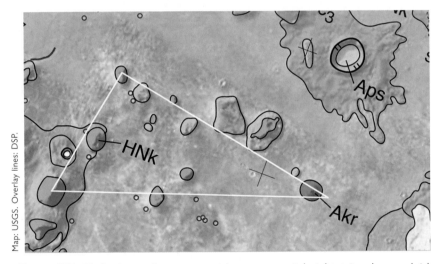

Figure 5.3. Cydonia geology map with same special right triangle overlaid.

Close study reveals the geometric relationship between three outstanding items in the area: the huge five-sided pyramid, the Face, and the spiral mound form a *special right triangle* with a line running from the Face past the two large flat/rough mesas to the spiral mound forming the adjacent side (see figure 5.2).

A special right triangle has angles of 60° (at the five-sided pyramid), 30° (at the spiral mound, and 90° (at the Face). Its sides are formed thus: the opposite has a length 1, the hypotenuse is twice that length, and the adjacent is side length 1 multiplied by 1.732 (the square root of 3).

This triangle is one half of a two-dimensional tetrahedron, a form that will become increasingly important as we continue.

Returning to the spiral mound, it is not only anomalous geologically, it is also the anchor of the special right triangle. The Face and the five-sided pyramid are generated from local mesas, but this special right triangle could not even appear unless the spiral mound was placed exactly where it is, at the 30° angle. How coincidental can it be that on our own planet, the largest prehistoric mound in Europe, Silbury Hill at Avebury, England, also has a spiral on its surface and a *slope* angle of 30°? Here is an interesting collection of facts: the photo of the spiral mound at Cydonia compared with Avebury's Silbury Hill makes it clear that these two artificial constructs share similar visual aspcets. Much of Silbury Hill is built of chalk dug from its immediate surrounds. Loose chalk, when left to fall naturally from a height, will settle at an angle of 30°. The majority of the crop glyphs that predominated from the 1970s through to the early 1990s occurred on the farmed chalk downlands of southern England. In

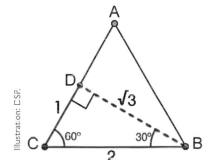

Illustration: DSP.

Figure 5.4. 30°-60°-90° special right triangle geometric construction extrapolated to an equilateral triangle, the two-dimensional view of a geometric tetrahedron.

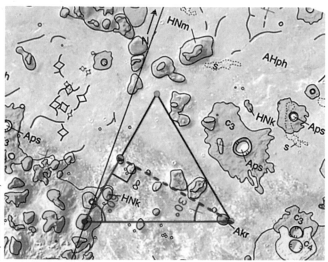

Map: USGS. Overlay lines: DSP.

Figure 5.5. Cydonia geological map with the special right triangle construction and tetrahedron.

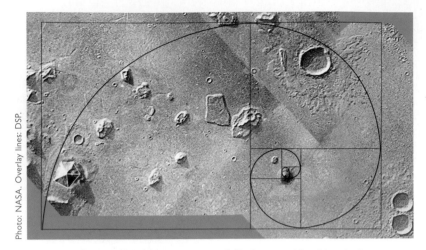

Photo: NASA. Overlay lines: DSP.

Figure 5.6. Viking image of Cydonia with phi spiral showing anchor point around the spiral mound.

those early years Silbury Hill was the primary seat for geometrically significant glpyhs and anomalous EDI events over the hill and the fields to its south. The Great Pyramid—which itself is sitting on a virtual phi spiral (see figure 1.4, page 15)—is also located at virtually 30° N latitude, and this latitude will take on even more significance later in this book.

During our research for this chapter, David Percy and I decided to place a phi rectangle over this part of Cydonia, with its spiral generated from this mound, for all of the above reasons.

The mound's geometric placement recalls the philosophical aspects of ratios with which many are familiar. We are talking here about *intelligent* design.

We can see that the phi spiral tangents the tetrahedral mound on the upper right-hand rim of the large crater before picking up on the special right triangle. It tangents the 90° and 60° components of the special right triangle, the Face, and the five-sided pyramid. This triangle's 30° angle also tangents the spiral mound.

Considering the caliber of advisors and consultants to NASA, it's hardly credible that having studied the results of the Mariner and Viking images, the scientists concerned failed to figure out the basics and significance of its very obvious geometric layout. Indeed, the level of obfuscation concerning Cydonia would indicate that the agency is perfectly well aware of the site's uniqueness.

The management of the public perception of Mars takes us back to the Viking mission control rooms at JPL and restraining orders.

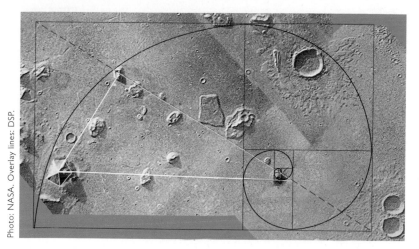

Figure 5.7. Viking image of Cydonia with phi rectangle and spiral with special right triangle. The phi rectangle diagonal (dotted) corresponds with the 90°-to-30° line of the special right triangle.

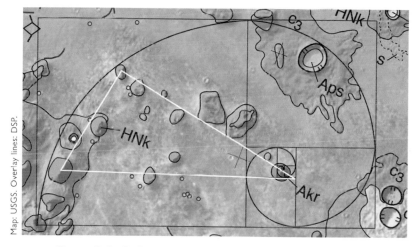

Figure 5.8. Cydonian geology overlaid with the phi spiral and the special right triangle.

RED SKY AT NIGHT, NASA'S DELIGHT, GREY SKIES IN THE MORNING . . .

When the first Viking images arrived at JPL, their color monitors initially displayed a normal greyish-blue sky. However, a technician then went up to each console, adjusting the settings until the sky was red tinted, which then threw out the other colors. One member of the JPL staff had brought his young son along for the day. This innocent child (now a physicist) observed this maneuver and then went around the room resetting all the consoles back to their default settings, which to him was an obvious thing to do, since the control room screens had been specifically calibrated to receive these images from Mars. Whereupon he was told off by the technician, who instructed him to put everything back to the red sky setting or he would be banned from the building.

It is interesting that NASA wanted to perpetuate such a fiction, because even though red-tinted skies do occur as a result of dust storms, as it turns out, when NASA probes were sitting out the Martian dust storms, they generally had their cameras covered.

The fact that the permanently red-tinted Martian skies beloved of science fiction and therefore in the public consciousness as "a Martian sky" were more important than the reality arriving from the probes out around the red planet and that the information provided to the public (and JPL staffers) is altered to suit NASA policy is very, very significant. These reddened Martian images, entirely different from the skies of Earth's northern latitudes, neatly circumnavigate any suspicion that the images were not from Mars but from somewhere much closer, such as the NASA facility on the largest uninhabited island in the Arctic—the remote Devon Island.

This is what NASA says about the location:

> Polar regions offer good Moon/Mars analogs because they provide extreme environmental conditions along with relevant geologic features and stressed microbial habitats as may exist on the Moon or Mars. Polar sites are also remote and isolated, with little or no local infrastructure and resources. Devon Island in the Canadian Arctic is set in a polar desert, with a cold climate, a

Figure 5.9. Devon Island location, Canada.

Photo: NASA.

Figure 5.10. K10 rover at Haughton Crater, Devon Island, Canada.

frozen subsurface, and a high ultraviolet flux during the 24-hour summer sunlight. Haughton Crater is relatively large and exceptionally well preserved, and contains fluvial, glacial, and periglacial geological features along with microbial niche habitats.[1]

But Devon Island has only been officially used as a Mars analog base since 1996, the year before *Surveyor* arrived at Mars, which leads one to wonder why the color of the Martian sky during the mid-1970s Viking missions would be of such a concern—unless someone, somewhere, was suffering from a guilty conscience or a surfeit of insider knowledge over Cydonia. On the basis of this Viking screen incident, it is legitimate to wonder whether this policy of altering images and control screens suits a hidden agenda. Other images of Mars have indeed been altered by NASA operatives,[2] therefore, it is legitimate to consider that this policy of altering control screens and images infers a long-standing NASA policy of the end justifying the means.

WEATHER

Should this Viking color adjustment be the result of a policy designed to refute any claims of inauthenticity, then it's beyond ironic that in order to achieve this aim, even though the Mars mission control teams generally adopted grey-blue shirts, these Viking images of Mars *were themselves adjusted* at the receiving end prior to being shown to the public. Later missions have reverted to the "normal" grey-blue Martian sky, but then so has NASA also tried to orient the exploration of Mars away from ETI, leaving such matters to the media, UFO online sites, and the like, while itself focusing on the climate and geology. It is still in the business of pursuing life on Mars but prefers the very primitive kind: bacteria.

In order to manage this change of direction, it was necessary to explain away the Cydonia region landforms in the eyes of the public. This requirement was attempted by asserting that the weather was responsible for producing such things as multiple variations on those distinct pyramidal forms in the City just to the west of the Face. Below there are a number of three-sided, four-sided, and five-sided pyramids virtually next to each other.

And then there's that huge five-sider to the southeast of the City, which came to be known as the D&M pyramid by many researchers and as the Tor by Myers and Percy. Going forward in this book, it will be referred to as the D&M Tor.

Unfortunately for the weather proponents, if the public was ill informed about the effects of weather on landmass, their own experts were not. As Erol Torun, a geomorphologist at the US Defense Mapping Agency has stressed, an object with five straight sides cannot be formed, or at any rate cannot be *maintained,* by the

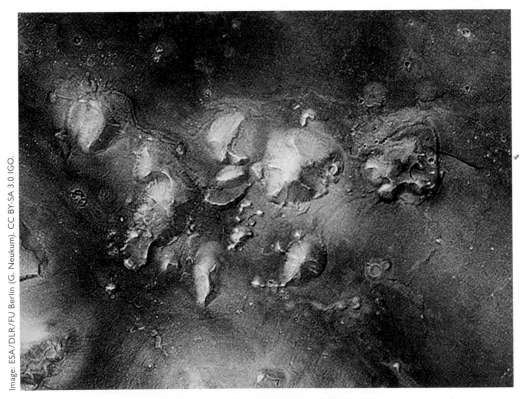

Figure 5.11. The City area of Cydonia.

Figure 5.12. D&M Tor pyramid with guide outline.

action of wind and weather. The force that is sharpening one face will at the same time be causing any existing opposite straight sides or edges to erode.

Undismayed, the weather merchants changed tack and had another go. With regard to intentional design revealed by the geometry, it was decreed that even expert analysts crawling over these Cydonian images with their straight edges and their protractors—and all who agreed with them—were creating patterns easily found between any set of objects if one is predisposed to find them (that is, draw the pattern and then find the fit) and that seeing meaning via alleged geometric relationships between the various Martian landforms at Cydonia (or anywhere else on Mars) was therefore irrelevant.

Sagan's actions post-Viking rather belied that particular put-down; having remarked volubly on the Elysium pyramids, he *avoided* those at Cydonia. Paradoxically, the Elysium pyramidal forms have largely been forgotten, and Sagan's opinion that the Elysium's three were *sufficiently anomalous* to merit further investigation is absent from any current discussion of Martian landforms. It is tempting to speculate that by mentioning one half of the Cydonian/Elysium opposing pyramid groups, Sagan was giving everyone a nudge without rocking the boat in which he sailed with his fellow scientists.

Obviously, both locations are worthy of further observation and intense discussion here on Earth. As stated in the first chapter, we can all agree on one thing: human beings are endlessly fascinated with our origins. How we all got to be how we are now and where we came from are drivers of our species and rightly so. If we had found traces of a pyramid field somewhere unexpected on our own planet, undoubtedly, we would have sent a team of archaeologists to have a good look, although it's fair to say that the relatively recent Bosnian pyramid dig has also divided opinions as to its authenticity. And if those arguments are the result of the politico-cultural perceptions of the Bosnian authorities and other interested parties, then it's likely to be the same for Mars. Although finding signs that locations on Mars fulfill Sagan's criteria for intelligent design is such a jaw dropper, it has apparently been deemed necessary to stop the conversation altogether. When analogs of the Cydonian problem site on Mars are found here on Earth, then surely the conversation has to be taken up again.

INTERPLANETARY TRAVEL

It has been conjectured by some that if there are indeed similar sets of architecture on two different planets, since *we* do not know how to get out *there,* then those responsible for the Martian artifacts must have known how to travel easily from Mars to Earth. This reasoning goes that whoever they were, they either built our pyramids and other ancient structures, given all the complexity that is involved in the construction and accurate placement of such structures, or they taught human beings how to do so. That is the most comfortable fit for most people since it chimes with a lot of our ancient mythology.

An alternative hypothesis centers around the idea that the knowledge and skills required may have been transferred to us by other means, either through our inherited DNA or via our consciousness, whether the subconscious, via dreams, or during high meditation states. These PSI conjectures are of little comfort to those organizations, both religious and profane, which have concluded that human beings are *the* dominant intelligence and the only existing species made with a body designed to walk upright on two legs, using two arms, having hands with opposable thumbs, a head equipped with an amazingly efficient brain, and a selection of senses with which to interact with the physical world. Yet, we are only in the foothills when it comes to fully understanding our own brains. So what if we also have the senses with which to interact with the non-physical world as well? What if, through developing and using our innate ESP and PSI faculties, individual practice can lead to accessing an "internet without wires" or a "universe wide web"? What if our ancestors have left all the clues and the display of their abilities for us to find and use, once we can accept that we are more than we think we are when it comes to using our minds?

Figure 5.13. Nazca lines, Peru. The trapezoid forms strongly evoke landing pathways, and the mountaintops suggest artificial flattening. For comparison, on the lower left of the photo the defining ridge of another, lower range of mountains can be seen.

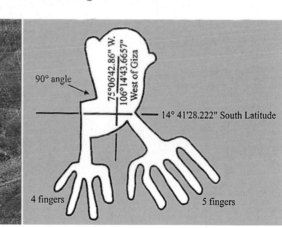

Figure 5.14. Manos, Nazca, Peru. The longitude data refers to location of Manos relative to both the Greenwich prime meridian and the ancient Giza prime meridian. Note the five- and four-finger arrangement as well as the 90° angle at the "shoulder." In the ancient world, odd numbers were considerd as male and even numbers as female. Or positive and negative, or yang and yin. The ratio 5:4 is also the musical tone of a major third (the tone E if using the diatonic scale of C major, and the tone of C if using the A minor diatonic scale). The geometry and relationships between these geoglyphs at Nazca reveal intelligent planning, and silent contemplation of these figures can bring deeper understandings.

Should it be publicly acknowledged that these landforms on Mars were intelligently placed, the questions from the scientific and academic establishments would be, "What happened to the inhabitants? Are their descendants still around?" From the governments of the world would come, "If so, what are their motives? Are they friend, foe, or wolves in sheep's clothing?" Then from those military organizations capable of developing spacecraft and satellites, "We must get hold of their technology before anyone else does because it can seemingly do what we cannot, and in any event it will be superior to those of our designated enemies here on Earth." With such consequences, playing mind games with the general public has a role in all these scenarios.

RETURN TO PAREIDOLIA

In July 2009, another attempt was made to play up, and then play down, the alien aspect of Martian exploration when yet another location was announced to be an area of "artifacts." The Ruell Vallis site supposedly includes at least nine statues of human and animal faces clearly visible from space, including very large ground drawings featuring the profiles of primates and canines. Like the long-

Image: ESA/DLR/FU Berlin (G. Neukum). CC BY-SA 3.0 IGO.

Figure 5.15. Perspective view of Ruell Vallis, Mars, looking southeast.

discussed Cydonia complex, this apparently intriguing new site also contains a large, pyramid-shaped landform.

This information was released by NASA on the occasion of the forti-eth anniversary of Apollo and picked up a month later online by the *Seattle Exopolitics Examiner* (SEE), which stated that the data was originally acquired three years earlier, in 2006. Conveniently, this Ruell Vallis story takes the pres-sure off any imperative to explain or explore Cydonia, trivializing the Cydonia region by inference. After all, if these Vallis images aren't in any way convinc-ing, then neither should we consider the Cydonian images seriously. But just in case, once again the public was told that it's a natural human trait to see familiar forms in a landscape and that over here in Ruell Vallis, there were lots of them! Including images of the very animals used for testing conditions in space by the Americans and Russians prior to sending a human being into space (primates and canines, respectively). From which we can deduce that the timing of this image release was in part governed by the need to reinforce the public's memory of earlier human space exploits, coupled with a desire to keep the pot boiling concerning our abilities to travel in space, while managing to exclude the possibility that *they* ever came here. In this regard, SEE, with its apt acro-nym, ignored much of the scientific appraisal of Mars climatic conditions over eons and instead went for the emotional jugular beloved of many—the end of the last ice age—and published this commentary:

> My speculation is that Ancient Mars was a colony of Ancient Egypt, rather than vice versa. I say this because Earth had the abundant biosphere that could have sustained ancient civilization in its attempt to reach Mars. The Great Pyramid at Giza and the Great Platform at Baalbek, which may have been an ancient launch pad, may be evidence of this effort. I find it far less compel-ling that human beings on Mars reached Earth first, given Mars' far less fer-tile environment. The human beings on Mars are probably the descendants of colonists from Earth, separated from us by the solar system catastrophe of 9,500 BC.[3] [Note the use of the present tense, inferring that there are currently human beings on Mars.]

TIMING ONCE AGAIN

At the height of interest in Cydonia's face, the hold-back period on images from acquisition to public release was said to be at best six months, at worst a year, and these timescales were considered a scandal by all those wishing to study the data. The Ruell Vallis release after three years created no scandal at all, either because no one knew the images had been taken in the first place or because these images were so unconvincing. Yet even that didn't matter if they were simply designed to support another epic in space exploration. This particular commentary triggered numerous books, articles, and ever more speculation concerning aliens on Mars (and the Moon). When it comes to "Who did what to whom, when it came to exploiting a planet for its wealth?" even more confusion is threaded into the Sitchin oeuvre by this clever SEE article; linking ancient Middle Eastern monuments with *outgoing* travel, it juxtaposes these ineffectual images with Zecharia Sitchin's powerful writings.[4] Sitchin associated Baalbek, the Sinai, and Giza with *incoming* travelers intent on using the indigenous Earth peoples as their slaves in order to extract the wealth of the planet for themselves. This SEE posting also opens the door to all the rumors and speculation concerning NASA's so-called secretly developed advanced technology that resulted in hypothetical secret space fleets and underground military bases, but the current travails of NASA in developing crewed spacecraft do not support the SEE speculations. NASA has been relying on the Russians to get astronauts up to the ISS since the demise of the Space shuttle in 2011 and by May 2020 only a single demonstration flight of a new Space X crew capsule has taken place. Which blows apart any ability to be out on Mars with super secret space fleets. However, no doubt cognizant of these issues, the SEE commentary also infers that at some time in the past we have had the ability to travel safely in space and that therefore the knowledge for doing so can exist still, albeit held in secret. Deliberately taking ancient myths out of context in order to play on deep-rooted emotions is an attempt to control the mind. In this case, the seeding of fear of "the other"—even if a relative, one now so distant as to be considered alien—looks to be an attempt to steer the thinking of the public in a particular direction. That of military defense against ETI. Mars, already linked mythically to war (without that state being understood in its most profound sense as a dimension of consciousness), and its moons, become purveyors of aggression and tap into subconscious deep-rooted fears. Hence the names of its moons, Phobos and Deimos: fear and panic.

Discussing President Trump's desire for a space force with *Politico* journalist Bryan Bender in June 2019, USAF Colonel Peter Garretson (retired) stated categorically, "There has never been a strong voice in NASA for space industrial development or space settlement." He added, "There has never been a strong camp in NASA that really wants to build sustainable infrastructure and technology that enables a broader segment of society to follow."[5] This com-

ment should put paid to all those who are stretching SEE's Ruell Vallis speculations beyond their already fatuous limit. But it also tells us that the mining and
settlement aspects of the future Artemis and Ares space projects are window
dressing for the public and the accumulation of funds, while the real intention is more likely the development of military defenses out around the Moon
and Mars. In August, two months after the publication of Garretson's interview
with *Politico*'s Bryan Bender, the DOD announced the formation of the US
Space Command.

FOOL'S GOLD

Those who promote the fear factor relative to interaction with ETI are buying into
an argument often mooted by sociologists looking at our own history. Namely, that
a technologically superior culture would annihilate the lesser culture and that it
would therefore be unwise to try to link to, or even make ourselves known to, a
culture that is obviously superior and possibly aggressive. The three examples used
to bolster this opinion are the Spanish conquest of the South American Indians, the
colonizing of North America, and the colonizing of Australia. Yet this argument
ignores the fact that the only item to worry about is *intention*.

In all these cases, the indigenous cultures suffered when the invaders, primarily motivated by greed, were indifferent to the fate of those encountered
along the way. The South American indigenous peoples were decimated on
account of the greed for gold. The dominant powers, both secular and religious,
disguised the enslaving of people as "the saving of souls" in order to generate
wealth. In their greed for the land and its wealth, the incoming colonizers of
the North American continent were implacable in their destruction of those
they found already living there. For instance, they deliberately killed off the
bison so that the Indians who lived from their bison herds would be easily controlled and reduced in population. Incidentally, the US conquest of the Wild
West is considered a plan to follow for developing space colonies.[6] In Australia,
the colonial incomers thought they knew better than the indigenous Aboriginal
race as to how to live in an extremely alien land. Having extracted some basic
knowledge of the lay of the land from them, the Aboriginal population was
marginalized and corralled. Young children were separated from their parents
in an attempt at their forcible integration into another culture. In every case,
whether in the Northern or Southern Hemisphere, disdain for the indigenous
by the newcomers exists to this day.

The basic reasons for all this atrocious behavior are now glossed over by the descendants of all those colonizers and because these issues are distasteful and uncomfortable to live with, some of them remain unresolved to this day. It is therefore held as an axiom when discussing space colonization that *all* invading cultures, irrespective of motivation, destroy the local population. And this reasoning is applied to ETI contact by fear merchants who really should know better. Not only has their thinking not gone beyond our past history but they also contradict themselves. First they claim that to have survived long enough to build such advanced spacecraft, extraterrestrials who can physically reach us here on Earth are likely to be more intelligent and peaceful than we are, then they contradict these assertions by presuming that extraterrestrials think and act as we do. The confrontational path most governments are walking in the first decades of this century belies the notion of us as a peaceful species; we are not behaving intelligently when it comes to managing our own environment, and we do not behave with tolerance toward others of our own kind who are perceived to be different from ourselves in their orientation, politics, or ethnicity. So if extraterrestrials are like us, then we should indeed be quaking in our boots, or feel guilty, since we are of the same DNA. But since both of these assumptions cannot hold at the same time, if they are like us, then it also stands that their belligerence and stupidity relative to the harmonics of the universe they inhabit renders them incapable of long distance space travel.

There may be some justification for this argument on mental and spiritual grounds—like attracts like—but not on a physical level alone. If we are looking at extraterrestrial civilizations fully capable of traversing interplanetary distances that our visionaries can only dream about, then if such a conquest had occurred during the days of our prehistory (as asserted by Sitchin), plenty of time has elapsed for Earth to have been sucked dry of any material wealth relevant to a passing extraterrestrial civilization. On the contrary, without invoking extraterrestrial civilizations, we seem determined to trash the environment ourselves—both on Earth and in near space. International space agencies are not only littering LEO orbits and satellite orbits with space junk and leaving a trail of nuclear junk in the solar system, they are also intent on mining the asteroids and other planets for their wealth. When taking all this into consideration, the fearmongering might be a reflection of a collective nightmare. And when it comes to seeking wealth from afar, knowing whether ETI is still around Mars and/or the solar system might therefore be a concern.

Regarding Earth-sourced exploration of the solar system with a view to the acquisition of mineral wealth, at the time of this writing, NASA has started exploring asteroids with robots.

GOLD GROW THE RUSHES

On September 8, 2016, the space agency launched a sample return mission expected to cost a minimum of $1.16 billion for its mission to the carbonaceous asteroid 101955 Bennu (formerly designated 1999 RQ36). The spacecraft duly arrived in orbit around this 1,614-foot-wide (492 meters) lumpy rock in December 2018. Since then, the probe has registered a surface covered in gravel, pebbles, and boulders, along with eruptions of plumes of gas and dust. So drilling is not going to be as easy as was hoped. Sample returns are expected back on Earth in 2023. The probe is called OSIRIS-REx. This is an acronym for the Origins, Spectral Interpretation, Resource Identification, Security-Regolith Explorer. If you are going to surmise as to which came first, the acronym or that ungrammatical tongue twister, based on past NASA form, it's going to be the acronym. Although the choice of the word *security* is of interest, selecting OSIRIS-REx for the project name also informs us that NASA is mainly sticking with Egyptian mythology for its inspiration. Indeed it now acknowledges that on its website. This asteroid actually belongs with the near-Earth asteroid group Apollo. In another piece of the ritual messaging, both outward and back to Earth's pragmatists, the sample is going to be taken from the asteroid in 2020 (the year that the sitting president initially wanted everyone back to the Moon), and the results of this asteroid hunt are scheduled to arrive back on Earth the year before Artemis, Apollo's twin sister, is scheduled for a 2024 Moon landing. As Greek myth tells it, Artemis was in love with Orion, and in Egyptian myth, the constellation of Orion is associated with Osiris and Osiris is associated with the Bennu bird. All of which provides more information about NASA's desire to comingle Egyptian concepts of origin, rulership, death, and reincarnation with aspects of Greek myth. And with their adventures off planet Earth.

Prior to announcing OSIRIS-Rex, NASA had been monitoring another near-Earth asteroid group for some years and had stated that Eros, as the near-Earth asteroid 433 is also known, contains more gold than has been extracted from Earth so far. Which would make that asteroid worth more than $3 trillion. However, that was the PR spin. In fact, the analysis of the asteroid revealed it to contain gold (at 1999 prices) to the value of about $1,000 billion. As it also contains platinum and other minerals, its overall worth is estimated at around $20,000 billion.

Eros, which crosses Mars's orbital path, shares its Greek linguistic origins with the red planet's moons Deimos and Phobos and its physical origins in the Amor asteroid group with those two small Martian moons and an asteroid named Pan.

This half-goat/half-human Greek mythological creature is most fittingly associated with seekers after gold, since panning for gold using sheep skins stretched over a frame and placed in mountain streams is linked with the myth of Aries and the story of Jason looking for the Golden Fleece at Colchis. So perhaps the stated ambition to get robots to Phobos is more about seeking gold than about that little moon being a stepping-stone to Mars. In that vein, if it weren't for the preference for Egyptian project names, it might be that NASA would name any future mining craft the Argo, but that would require a crew on board, not robots. Or are we back to *2001: A Space Odyssey,* and the naming of the Aries spacecraft? Of course, successfully mining Eros or even Phobos might affect the price of gold, but the prospect of so much potential wealth fires the imagination and opens the wallets of those funding the machinery required to eventually gain access to it.

The juxtaposition of these two asteroid events, wherein the assumed presence of valuable minerals might justify or even finance the expense of an exploratory trip prior to the actual recuperation of any gold or other minerals (as our technology advances, so do we need more and more rare minerals to sustain it) mirrors the arguments put forward to the secular and religious authorities *prior* to the exploration of South America. And just as those seekers after gold were at the mercy of pirates on their return home, the signing of the 2020 National Defense Authorization Act and the establishment of a US Space Force likely has more to do with getting the goods home safely without interference from any other space nation, despite earlier avowals concerning the peaceful exploration of space for all mankind.

THE MIND TRAP

Since we have no way of independently controlling or even verifying the true origin of any images from space, it is of course perfectly possible for NASA and its affiliates or other space agencies to produce images of anything an agency might wish, at any time, for the purposes of driving space policy. Over sixty years of space exploration have led millions of people worldwide to trust in the US government's authority within NASA, and therefore to believe that the statements issued by the space agency are absolutely true.

Over time, the ability of more and more people to gain access to the internet enabled them to analyze NASA images for themselves and to read documents posted by NASA and other space agencies. For better or worse, this widespread dissemination of information has also freed up the thinking of many people and led to the evaluation of the opinions posted or promoted by the space agencies from a different perspective. The agencies themselves have not factored in this possibility nor the fact that people today are less inclined to accept what is said by any authority without very good reasons. The relegation of the Face on Mars to the status of a kitty litter tray effectively removes the necessity to even mention the rest of the Cydonia complex for NASA. But to the thinking public, back engineering this remark, it is also obvious that such a ploy was an equally valid reason for initially bringing that mesa to public attention in 1976: ridicule and explain away one of the most symbolic anomalies at Cydonia, reinforce it over time with ever-worsening images, and cultural pressure should take care of the rest.

The subsequent assertion/denial games, incidents such as the Viking red-screen/blue-screen episode, or seemingly suspect information such as the Ruell Vallis event all destroy the trust between the public and these explorers. So that now, such actions imply that the space agency is withholding information—or playing mind games with the public. The upside is that with the advances in technology, we are able to become more discerning. Aware of this, NASA has used its PR departments to bolster the preference for a Martian mystery, and this subtle PR process has made it increasingly fashionable, nay intellectually essential, in any discussion concerning ETI and any of its attendant manifestations to laughingly tolerate a mention of the subject but then to dismiss it, to opt for a mystery. *To prefer not to know.* How many books on interesting but offbeat subjects promise everything on the front cover, but by the last line of the final chapter, we read that phrase: "We can never know and that's the wonderful part: it's a mystery." As far as the management of tricky subjects associated with exploration of the solar system and our relationship with ETI, officially, it would seem that matters are to be maintained as a mystery. Upholding Einstein's opinion that "the most beautiful thing we can experience is the mysterious. It is the source of all true art and science." A lack of inquiry becomes an intellectual virtue. To request more information, a sin.

As a mind control mechanism, this is a formidable tool. It has been used for millennia by influential authorities, mostly religious, but the high priests of both science and academe are not against using it when it suits them. This technique is particularly useful when they don't know the answer or when they prefer to avoid having their statements questioned, examined, probed, or rebelled against. To step outside the peer group and say something along the lines of "This I have seen, this I have experienced, this I know for myself to be true, but I want to learn more" would be tantamount to self-destruction and social ostracism from one's fellows. Followed these days by a potentially lethal dose of online trolling. And if that doesn't work, since it is no longer as effectual as in the early days, all statements are to be so contradictory that no useful conclusions can ever be drawn from them.

Curiosity Killed the Cat

*More Evidence, Further Findings,
More Questions, Growing Fears of Extraterrestrials,
and Remote Viewing: Keeping the Dream Alive*

CONTRADICTIONS

Every time the Mars probe *Curiosity* is mentioned in the press, the old adage "curiosity killed the cat" springs to my mind. Whether this *Curiosity* rover was intended to bury forever the memory of the sphinx-like Face at Cydonia, along with any evidence of other pyramidal structures at Cydonia and Elysium, is a moot point because some thirty-six years after imaging that Cydonian Face, NASA was now *reminding* everyone of an Egypt-Mars connection.

Curiosity landed at Bradbury Landing, in Gale Crater on August 6, 2012. Its landing site was named after Ray Bradbury, the science-fiction author whose words and ideas were considered an inspiration by many of those involved in designing today's space vehicles. Deploying its cameras, *Curiosity* trundled across the Martian surface for some 925 feet (282 meters) before fortuitously encountering a solitary four-sided rock that NASA promptly called a pyramid. Named Jake-M after a recently deceased JPL employee, Jake Matijevic, this "pyramid rock" was pronounced anomalous relative to other findings from previous Martian probes.[1] Although looking at the official NASA photo of this rock, it also appears anomalous in relation to the background in/on/against which it was sitting, and with a height of 10 inches (25.4 centimeters) and a width of 16 inches (40.64 centimeters) relative to the massive *mile-sized* pyramidal landforms at Cydonia and Elysium, it is certainly anomalous in size.

Co-incidents: According to NASA and the JPL Curiosity progress report, it

Figure 6.1. Pyramid rock, Mars, *Curiosity* image.

Photo: NASA.

took forty-five days for the probe to come across the first good-sized rock suitable for trying out its ten scientific instruments. The number 45 recalls those four and five sided pyramids, while 40.5 recalls the latitude of the five sided D&M Tor at Cydonia. The number 10 is the height of this igneous rock expressed in inches. Relative to previously analyzed igneous Martian rocks, this particular rock turns out to be geologically anomalous in its element ratios.[2] Although rarely found on Earth, the same type does exist on the island of Skye. This Scottish location is 16 miles (25.74 kilometers) west of mainland Glenelg, a distance that numerically recalls both the base width of this pyramidal rock in inches and its height in centimeters. *Curiosity*'s target on Mars is named Glenelg. NASA stated that the name was chosen because it was a palindrome and that *Curiosity* would visit it once on its outward trajectory and once on its return. On October 21, 2012, just over a month after Jake-M was found, in the presence of NASA astronaut Bonnie Dunbar, the village of Glenelg was officially twinned with Glenelg on Mars. Hidden communications galore and a coincidence too many.[3]

Photo: US Signals Corps, 1957.

Figure 6.2. The Great Pyramid, Giza, Egypt. This photo of the northern and western faces of the pyramid was taken in 1957. The north face is distinguished by the entryway offset from its centerline.

It would later be announced that Jake-M's pyramidal form was due to erosion by grains of sand carried on the wind. How appropriate, since it is sitting in Gale Crater, whose central peak is Aeolian Mons. If this substantiated the statements of NASA relative to Cydonia and Elysium that pyramidal forms were the norm on Mars due to its weather systems, it went against the opinion of geologists. Notwithstanding these niceties, *Curiosity*'s pyramid rock is remarkable for another reason: just looking at this rock, its striations are all over the place. One facet has vertical layering, yet the next facet around has horizontal layering and a convenient dimple that poetically resembles the north face entryway of the Great Pyramid as we see it today, without its casing stones.

In case the public was not paying sufficient attention to rocks, NASA amped up the memorabilia exhibition a few days later by publishing a picture of *Curiosity*'s caterpillar track marks while drawing attention to the comparison with an Apollo *footprint* photograph. Frankly, this was over egging the pudding, but given that it was NASA making this link, this juxtaposition had to mean

Photo: NASA.

Figure 6.3. *Curiosity* tracks.

Photo: NASA.

Figure 6.4. Apollo footprint.

something—to someone. Was there a comparison to be made between the robot tracks and the ridges on the Moonboot? Some forty-three years after the event, why did NASA find it necessary to link its *Curiosity* adventures with its Apollo mythology? Thinking logically would not be of much help here, but doing some lateral, associative thinking does reveal hidden layers of information.

Taking these two curious events together, were we supposed to be thinking about a five-toed human foot and a four-sided pyramid? Viking had landed on Mars decades previously, so the analogy to "first steps on another planet" doesn't fit the bill. But thinking about time does provide some resolution: the American cartographer Carl Munck has published extensively on his conviction that there is a connection between Giza's four-sided Great Pyramid and the huge stand-alone five-sided Cydonian pyramidal landform. His mapping studies find 0°/360° for the longitude of these pyramids, and he states that these two pyramids were the ancient prime meridians of their respective planets. Munck also noted that *both* these pyramidal meridian markers were constructed on the basis of the British twelve-inch *foot* (or the root foot, as those who study ancient metrology would prefer), which is the basis of the statute mile system.

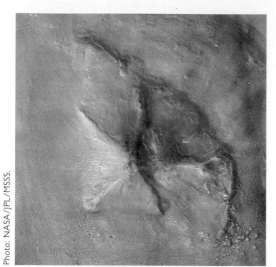

Photo: NASA/JPL/MSSS.

Figure 6.5. D&M Tor five-sided pyramid, Cydonia. Taken in 2003 by the Mars Orbital Camera 2-484, then processed by Malin Space Science Systems (MSSS).

Photo: Harry S. Truman Library & Museum. Accession Number 2017-614.

Figure 6.6. The Giza Great Pyramid seen from the north in 1943.

CURIOUS REASONING

Whatever underpins NASA's reasoning, it so happens that in the very early weeks of a two-year stint on Mars by its robotic rover *Curiosity,* the agency produced these two "curious" items. Granted the *Curiosity* rover is the template for the Mars 2020 expedition rover (*Perseverance* lifted off from Earth on July 30, 2020, and is intended to arrive at Mars February 18, 2021) but that's not enough explanation for the loaded symbolism in these two pyramid and rover track photos.

At the very onset of *Curiosity*'s mission, the authenticity of the photos returned from Mars had been questioned by the public, Certainly, the NASA PR machine was sensitive to such matters, and some four months after these two images were produced, the authenticity of the *Curiosity* imagery again came into prominence when the BBC devoted a series of three TV shows to astronomy in January 2013. *Stargazing Live* was fronted by the CERN physicist and broadcaster Brian Cox together with Dara Ó Briain, a stand-up comedian who had read mathematics and theoretical physics at the University of Dublin. Aimed at "everyman," this pairing represented a "let's make science fun" perspective, and the show's presentation in the round to a standing studio audience was blatantly modeled on the original format of the BBC's *Top Gear* motoring show.

One of these three *Stargazing Live* programs was entirely dedicated to the exploration of Mars in general and *Curiosity* in particular. It summarized the history of Mars exploration from its very beginnings, including the more fanciful notions, such as the early conviction of Italian astronomer Giovanni Schiaparelli, Percy Lowell, and others, that artificial canals existed on the surface of Mars (later disproved by better telescope observations), but somehow those apparently equally fanciful ideas concerning the Face on Mars didn't make the cut. This was a remarkable "tell." After all, NASA's technology had revealed nothing special about Cydonia to its official eyes. So why not treat that face mesa in the same manner as the artificial canals? This omission suggested that deep inside NASA's ring-fenced bureaucracy, there was enough doubt about this landform or even enough conviction that the Martian face was not considered to be a trick of the light that the best option was to say nothing about it, rather than something that would later turn out to be wrong.

Cydonia was not mentioned at all. But then again, nothing much was said about the Viking missions either. Way back in 1971, planetary scientist Bruce Murray had stated, "The focus of the entire US planetary program is on the Viking missions to Mars in 1975–6." By 1976, Murray was the director of JPL. Despite this, by 2013, this BBC show managed to deal with the entirety of the

Viking missions by using a single brief shot of the Viking control room at JPL in Pasadena, taken at the moment *Viking I* first reached Martian orbit, some eighty-five miles off the surface.

We saw JPL technicians, some sporting Viking helmets complete with horns, all wearing grey-blue shirts, leaping up from their consoles and shouting "hurray!" Jubilation was followed by man hugs and NASA's traditional and ostentatious lighting of cigars. It was interesting to see that in this 2013 BBC TV broadcast, the Martian sky was shown as the color it is generally conceded to be: just a shade darker than the shirts of the Viking hordes back at JPL. Then for the *Stargazing* presenters, it was onward, swiftly onward to the next decades of exploration.

The Curiosity project having been explained, viewers were encouraged by the jovial Ó Briain to participate in feedback regarding this current phase of Martian exploration by leaving their observations on a dedicated BBC website. At which point, the professor added his tuppence worth. Moving forward and looking straight to camera and with no jollity whatsoever, Cox said that while he welcomed any questions about the program and about space, if anyone thought that the Mars images from *Curiosity* were hoaxed, they could just "keep their opinions to themselves." Marginally losing his cool as he said this, he then stepped back from the camera. Notwithstanding his personal views about the qualifications required to even have an opinion (quoted in chapter 2), his announcement stood out as infelicitous in the extreme.

Both presenters then wrapped the show by happily agreeing that *Curiosity* would probably find signs that life had been on Mars or even that we were once Martians. Dramatic pause for the audience to take that in, then just in case anyone was getting carried away by their inferences, they clarified this last remark by saying that they were referring to the presence, survival, and possible transfer of life from one planet to another via very, very, very small bacteria. Condescending laughs all round.

Despite their best efforts, that elephant called Cydonia remained in the studio.

MORE QUESTIONS

What do the organizers of Martian exploration expect us to conclude from their curious photos and their TV shows? What is it they want from us, the public? Are we now being set up to accept the idea that intelligent life was present on Mars at some point prior to our arrival? Cydonia Mensae certainly seems to have the elements required to create the kind of curiosity that stimulates public backing and therefore funding. But getting humans to Mars is going to be an exceedingly long technological haul, as NASA has stated, because present technology would expose human crews to the vagaries of long hazardous journeys and planetary surfaces insufficiently protected from dangerous cosmic radiation, the solar wind, and solar proton events.

The research and development of robotic technologies remains ongoing within NASA. If this is a sensible interim strategy on many levels, it is far less inspiring for the fundraisers and the taxpayers, unless some cultural relevance is added to the mix. As an argument, that justifies these two *Curiosity* photos. They fit with a blatant publicity campaign to keep the public interested so that the funding for the civilian arm of the agency remains intact while NASA tries to work out how to get to Mars with a human crew. The campaign fails when the anomalies at Cydonia are factored in along with the behavior of NASA associates. The brevity with which an intrinsic part of humankind's early Mars exploration was treated by the BBC, allied with the final comments from its presenters, suggests that the US space authorities are jittery about something. The fact that those promoting the exploration of Mars feel the need to preempt any criticism to the point of rudeness *on national television* means that all is not well. Odds are that the very suggestion of any ETI involvement in the Cydonia region of Mars is rattling those aboard the good ship NASA and its flotilla of representatives—in this case, professor Brian Cox.

THE HOME RUN

The oddity and clumsiness of those *Curiosity* images are not stand-alone examples. NASA has form when it comes to cultural/politico dissonance in their messages to the public. In seeking to make its programs relevant to the public, NASA's PR department wittingly exploits symbols of perceived cultural relevance and adopts the myths of the ancient hero's journey and the gods of the past when naming its machines and describing its space exploits. But sometimes there is what appears at first glance to be a disconnect. NASA chose to call the ill-fated Apollo 13 command module *Odyssey*. The name was ostensibly a reference to the adventures of Odysseus as recorded by the Greek poet Homer. The dramatic period of the Trojan War was encapsulated in his two epic works, the *Iliad* and the *Odyssey*. These became the two founding literary works of Western civilization. In the *Iliad* Homer recounts the background to the war, Odysseus's outward-bound journey, and the future after the war. The *Odyssey* is all about the adventures that befall Odysseus during his return home. So the Apollo 13 command module is misnamed, its accident is on the record as occurring on the outward-bound

Illustration: NASA.

Figure 6.7. Apollo 13 insignia. "Home, James, and don't spare the horses!"

journey and in order to return home, technically the rocket trajectory required the craft to continue on its outward bound journey. Did the NASA culture vultures get it wrong, or was there no mistake at all—for reasons unspoken. The theme in Homer's works of going out and returning certainly finds echoes in the justification for the choice of the palindromic Glenelg for its *Curiosity* rover. As it does with the Apollo 13 storyline.

Even before human spaceflight was possible, von Braun had made a list of essential items that would have to be accomplished in order to one day reach Mars with a crewed spacecraft. Yes, Mars. The red planet has always been the ultimate prize, and travel to the Moon was considered to be a practice run, simply another item on von Braun's little list of things to do.[4] This list included the necessity of conducting a full-blown space rescue of a crippled crewed craft—*before* ever attempting a crewed mission beyond low-Earth orbit. So the Apollo 13 command module was aptly named *Odyssey* after all because the whole mission was *always* going to be *about the return home*. It's an extra nice touch that the Apollo 13 crew badge depicted the horses of Apollo pulling the Sun across the sky while the Apollo 13 commander was called James.

HOME TOWN

History has shown that long odds can lead to big payoffs. Many years ago, back in 1865, Heinrich Schliemann accepting the *Iliad* and *Odyssey* as recorded fact rediscovered Homer's Troy. Historians of his time for the most part scoffed at Schliemann's idea. They considered that Homer had set down the remnants of an entirely mythological oral tradition and that, like the stories of Atlantis, there would be no actual city to find. Schliemann thought otherwise, and being a wealthy man, with Homer in one hand and a bag of money in the other, he set

Photo: DSP.

Figure 6.8. The walls of Troy, Turkey, 1992. Heinrich Schliemann, in excavating Troy, came across nine levels of habitation. Cataloguing intelligently and listing his strata instead of simply plundering, he became known as the Father of Archaeology.

out for northwestern Anatolia to find and then dig up King Priam's lost city.

The ancient Greek name for the city we call Troy was Willion (Wilusa in the ancient Hittite language), and the first part of Homer's epic poems was titled "Ilias poiesis" (The poem of Ilion). The *Iliad*'s format resonates with NASA's cultural/political aspirations, incorporating the preparations for the Apollo project's "within the decade" limit set by President John F. Kennedy on September 12, 1962, but it also works with Odysseus's return journey, the *Odyssey*, which also took ten years. Perhaps Homer's *Iliad* and *Odyssey* also mirror conditions of the human mind from which the search for our essential selves takes place. This could be why such ancient tales stick with us across time, why the principle of the hero's journey still has meaning, and one of several reasons why NASA chose the name *Odyssey* for its Apollo 13 command module.

An in-depth discussion of the *Iliad* and the *Odyssey* is not needed here, but Homer's works are relevant in that, exoterically, these stories convey common themes relating to both the downside and the upside of human behavior in the physical world. The fostering of strong connections to the greater esoteric outer world was considered an essential requirement for the business of staying alive, thriving in the unknown, and ultimately getting back home safely and becoming complete—the whole comprising the trajectory of the hero's journey. NASA's attention to mythmaking and its adherence to the principles of the hero's journey tell us that these criteria are as applicable to present-day space travel as they were when first spoken at the dawn of time, before Homer wrote them down. So we shall come across this theme again in later chapters.[5]

HAPPY RETURNS

NASA's PR departments were not alone in turning science fact into science fiction (and vice versa).

Since the very first storyteller sat around that original hearth one dark evening, enthralling those listening with tales made from Moon-spun silver threads, through the star-filled gateways in *2001: A Space Odyssey* and beyond, into the greed-driven saga of *Avatar,* human adventures have been adapted, redacted, edited, and embellished by the storyteller to reflect the aspirations and the responses required of their audiences of the day. As a result, over time much original knowledge was lost or altered due to various cultural preferences. And this is where it gets somewhat paradoxical and rather confusing for the observer. Generally, this sort of storytelling is deemed pure fiction, and in the case of Kubrick and Clarke, excellent science fiction. Such stories are seen to be the inner processes of the authors' imaginations, set either far back into the past (*Star Wars*) or way into the future (*2001* in its day and *Avatar* now), at any rate out of time with the present. Modern humans assign more importance to the everyday, logic-driven, left-hemisphere brain mode—assigning anything that doesn't fit their own criteria of reality as a walk on the wilder side by the more creative right hemisphere. "Real life" is generally linked to the present day. And it is this time dislocation way back into the past or far forward into the future that makes these stories safe; they are viewed simply as entertainment. The notion that time does not work in such a linear fashion and that, at varying levels of awareness, everything is present goes unnoticed. That these fictitious sagas might also relate to contemporary events and contain political and cultural aspirations, motivations, and propaganda is generally not considered important by most modern commentators, just as nineteenth-century historians considered the reality of Homer's locations and storylines to be impossible, unthinkable. And if *2001: A Space Odyssey* had hidden poitical motives, consider the US government highjacking of the name *Star Wars* for its own purposes.

IN THE SERVICE OF THE MASTERS

On March 23, 1983, President Ronald Reagan announced the idea of a space shield, to be called the Strategic Defense Initiative. The defense of the United States via space was not a new idea; it had been around in one form or another for some time and with a variety of names since the 1960s. And it is back again in 2020 with the establishment of a human US Space Force, so all of this commentary becomes even more relevant.[6] Back in 1983, the day after Reagan's announcement, Senator Edward Kennedy called Reagan's proposal "a reckless *Star Wars* scheme," and the name stuck. In 1985, two political advertisements promoting the Strategic Defense Initiative used the name Star Wars, at which point, that movie's director, George Lucas—by all accounts a peaceful man and a creative genius—took legal action in order to disengage his film from the program. Lucas had used the mythic structure in order to create his *Star Wars* stories (to such good effect that Joseph Campbell, author of *Hero of a Thousand Faces,* called Lucas "a star pupil" for his understanding of Campbell's thesis), and he resented the profound misuse of the function of myth manifested by this purloined title. Nevertheless, the judge decided that the name Star Wars had become a part of the public lexicon and therefore could be used for social comment.

As for the public, the attribution of a fantasy name to this Cold War scheme, however reckless, also sowed a seed into the public's subconscious: if space shields and assorted weaponry really were being developed, then Darth Vader and the Storm Troopers, traveling in moon-like grey death stars, were more likely to turn up than Yoda and the Djedi.

STOLEN DREAMS

This misalliance of ancient storytelling (after all, the *Star Wars* movies are about past events) with modern politics underlines the potential for abusing creativity (even unwittingly) in the pursuit of political or economic advantage. It is abused when drama (the creative element) is used to disguise the truth of the documentary (the actual event). The disparity between drama and documentary, between storytelling and reality, can be used poetically and constructively to enhance the information. This drama/documentary split is mirrored by the human brain, whose job it is to decide how to interpret and then act on the streaming information acquired by the creative and linear brain functions.[7] In other words, the choice for the observer is one of discernment.

When it comes to the subject of space exploration, with much of the space program occurring way out of sight of the general public, it is really important to clearly signal the demarcation between the drama component (the advertising-funding spin) and the reality component (the actual problems of space engineering). The inability to portray much of space travel adequately, except through today's computer-generated imagery (CGI), combined with the knowledge that talking about the actual dangers and the difficulties of crewed space travel will not endear such enterprises to the public nor win the required funding, means that the temptation to overload or even misuse the drama department is ever present. NASA's organizers and PR departments are obviously well aware that it is the human component of space travel and a dose of danger (viz Apollo 13) that truly stimulates the taxpayer and ultimately ensures funding for future space programs. Even if it is somewhat understandable from a DOD PR perspective that classical names and myths of yesteryear would be adopted for such aggressive projects in order to soften the reality, living authors such as Lucas should not have their works appropriated against their will by the warmongers. The idea that a nuts-and-bolts organization such as NASA would use aspects of the hero myth in order to validate its conquest of the final frontier in the public's collective consciousness is even more alarming, until one remembers that NASA is also under contract to the DOD.

The Star Wars rhetorical questions are:

1. Who are we arming against: them or us?
2. Notwithstanding the NASA photo analysts, were the Cydonian and Elysium ruins on Mars placed with such evident signs of intelligence that NASA considers it necessary to develop such defenses prior to going to Mars with human beings?
3. Were the *Stargate* movies designed to fix in the public's mind a link between Egypt, Mars, and ETI, helped along by the two *Curiosity* images discussed earlier?
4. Was the robot *Curiosity* named as a pun on Curio City—those twelve anomalous forms dubbed "the City" by researcher Richard Hoagland? Or was it another Egyptian link, this time to a cluster of pyramids at the edge of the city of Cairo.
5. How are we to absorb such "drama-doc info-tainment"?

TIMING

Important events receding into the past are like some lumbering ghost ship sailing out of sight into the mists of time, its precious cargo the memory of the event.

Back on land, the event is maintained in the collective consciousness by timely annual reminders of past glories. Eventually, the reminders become decades apart and not annual, and the ship gradually sinks into the ocean of our collective minds, fragmented on the seabed of our subconscious. Later, wishing to explore or revive those memories of the event, seekers after knowledge recuperate parts of the scattered shipwreck. Then we often find all is not well: the materials recovered or the circumstances differ from the accounts in the historical record.

Photo: DSP.

Figure 6.9. Saturn V F-I engine on display in the Sydney Museum of Applied Arts and Sciences, Australia. Seeking to revive the same excitement for space travel in future generations that he had felt as a five-year-old in 1969, Jeff Bezos led an expedition in August 2013 to search the Atlantic seabed for jettisoned artifacts from the Apollo launches. Equipped with NASA radar coordinates of the impact sites, located at a depth of 3 miles (15,840ft/4.8km) and despite the fact that the definitive NASA data has these impact sites listed as "theoretical"—it took just three weeks. The recovered items were restored and installed in various US museums and traveling displays in time for the 50th anniversary of the Apollo missions in 2019. Jeff Bezos is of course also developing his own spacecraft under the name Blue Origin.

In the case of extremely ancient events, there is little or no record on which to base our information. Researchers generally extrapolate on the findings relative to contemporary understanding and experience of the world. Those who try to peer into the future generally extrapolate on past events as the prognostic for future events. It seems as if we are rather firmly stuck in our present, and indeed, mental health specialists recommend that we should live in the NOW. But that doesn't mean simply the present reality; it also means being fully aware of each moment, wherever that moment may be situated. If the past and the future are treated with the acuity with which we tend to consider the present, it becomes possible to navigate the choppy interface between these "time zones" with relative ease, to observe, to discern, to be involved but without bias. This understanding, along with a grasp of metaphor, allegory, and punning wordplay (all used by the ancients to convey information) might not have been on von Braun's little list, but it can be an essential tool when using stories from ages past to decode present conundrums. As it happens, it is not a tool that some of those who first "officially" studied Cydonia felt to be an important item for the public's toolbox.

DISCUSSING CYDONIA

Since that fateful 1976 press conference at the JPL, many books and papers have been published concerning Cydonia, and of course, the internet is now replete with opinions and commentaries on the topic. Mostly disparaging. However, the first person to publish a considered study of the subject was the Austrian Walter Hain. Following his considerable difficulties in obtaining images of Cydonia from the JPL's Don Bane, he wrote *Wir, vom Mars* in 1979. The book was translated into English in 1992 as *We, from Mars: Old and New Hypotheses about the Red Planet.*

In the meantime, the Americans got going in 1986 when the *The Face on Mars: Evidence for a Lost Civilization?* was first published.[8] This book was the record of a computer conference held between six specialists from the space industries. Their email sessions were collated by the anthropologist Randolfo Rafael Pozos who also conributed his own commentary on Cydonia. The other five particpants were:

John E. Brandenburg: Star Wars program scientist, Sandia Laboratories, New Mexico

Vincent DiPietro: Digital image processing, Goddard Space Flight Center

Lambert Dolphin: SRI consciousness researcher
Richard Hoagland: Journalist with NASA and JPL associations
Gregory Molenaar: Computer scientist at Lockheed, contracted out to NASA

DiPietro and Molenaar were the assiduous miners digging out that second Face image from the archives. The others weave in and out of this Martian tale, so we shall keep tripping over them.

The Pozos report was written in a breathy, *Boys' Own Adventure* style beloved of those who write about space and NASA, and despite proclaiming that other mainstream scientists, as well as official organizations including NASA, did not necessarily join with them in their conclusions about Cydonia, it read more like propaganda than the unofficial musings of a group of computer conferencing Mars enthusiasts.

The book's foreword stated that this conference had been set up following the initial 1979 study of those two Face images by DiPietro and Molenaar and that now, in 1986, this publication was its first report (and its last, since it was the only one that ever appeared). The reason given for the seven-year delay? When DiPietro and Molenaar had first published their findings on the Face photos, the scientific community had been so aggressive and negative toward their findings that these two men had "gone to ground." Really? It's best to keep a large pot of sea salt on hand because from now onward it's going to be needed quite frequently. A bit of archive archaeology revealed some rather more salient facts about how this enterprise came into being.

The story goes that Dick (as Richard Hoagland is called by his contemporaries) *as an individual* decided to set up research into the Cydonia images. Hoagland, as a journalist and consultant to NASA, had been intimately connected with JPL, the space program, and its attendant media circus since the days of Apollo. At the time of this foray into Mars research, he was involved with SRI and working on a story on Saturn's rings. In the autumn of 1983, he met with SRI's vice president for *corporate affairs,* Paul Shay. The SRI is the same research-and-development (R&D) group that was behind the verification of Uri Geller's psychic talents, and it had many DOD and CIA contracts and connections. Paul Shay, we are told is a "former" intelligence officer. (A pinch of that salt might be required here, since we are also reminded ad infinitum that no one entirely retires from the intel professions.) This meeting about Cydonia did not take place at SRI, but at the Institute for the Study of Consciousness in Berkeley, a foundation run by Arthur Young, a mathematician, engineer, and philosopher. It is not officially clear as to why this venue was preferred, but the name of the organization offers a clue: the stated main purpose for this foundation was "to build a comprehensive theory in which ESP could be integrated into existing scientific knowledge." (It's also true that a foundation is a preferred conduit for government-funded research projects flying under the radar.)

Arthur Young designed Bell's first helicopter. Job done, he resigned from Bell in 1947 to return to his philosophical pursuits. His first foundation had been set up in Philadelphia in 1952. Its aim had been to find out where to put consciousness in the scheme of science. In the pursuit of this aim, he had met many of the movers and shakers of post-war PSI, both civilian and military researchers. And in 1952 and 1955, he had also come across ETI communications.[9]

Ostensibly wishing to get nearer to academia, Young moved to Berkeley, founding his institute in 1973. And ten years later, within its walls, the deal was done on the next phase of Cydonia research. Shay took warmly to the idea and suggested that Hoagland connect with Dolphin, the physicist who had worked on SRI exploration of the Giza plateau between 1973 and 1982. So it looks as if it was already accepted that there is a connection between Earth and the Mars pyramids, the Cydonian complex, Giza, and the Great Sphinx. In fact, during a 1988 interview for their book *The Stargate Conspiracy,* Uri Geller told authors Lynn Picknett and Clive Prince that *the Face on Mars had been discovered by remote viewing in the 1970s, well before the Viking mission.*[10] Essentially remote viewing is the military term for the practice of producing clairvoyance to order.) However, the military also uses the term remote sensing for the actions of an orbiting probe equipped with a camera. So Geller could have been name-checking *Mariner 9.* Its camera certainly fulfilled the definition of remote viewing (RV). Or his statement could have been another sort of mind game, aimed at maintaining funding for consciousness research. SRI was running PSI-based human RV programs for the US military, but that was not known to the general public at the time of Geller's interview with Picknett and Prince. But then, as social anthropologist Pozos surely knows, one can say what one likes and steer the public in any direction when there is no outside source that can provide tangible proof of any claim.

Until we can actually go to Mars with an independent pair of boots on the ground, not just as a NASA employee, the military's PSI-Ops remote viewers can state whatever suits the ethos of the moment. Parallel universes, torn dimensions of time—both have allegedly been "seen" by military personnel remote viewing Mars. Whether these PSI efforts were a fabrication, a political manipulation, or the reality of a particular remote viewer, the implication is that, the US intelligence agencies and therefore NASA already knew about Cydonia—possibly prior to *Mariner 9.* The military's attempts at psychic mind control are a subject to which we will return. Here, let's note that in the 1970s, the Cold War was still officially in play, Soviet PSI research was said to be well in advance of the US efforts, and the news that the Soviets were doing their own PSI research apparently stimulated the US government to set up a PSI program "in the interests of self defense."[11] However, both nations were also running Mars space programs, and their space agencies were sharing the Martian launch windows in a remarkably well-orchestrated and even-handed way, so once again, actions were not quite the same as the war of words.[12]

Back at the Pozos project, within a mere eight weeks of applying, Hoagland and his colleagues received the necessary funding from SRI, and by December 5, 1983, they got going on their conference. Some two months later, an email from Hoagland for February 21, 1984, announced that an artist was joining the group. This person was going to analyze the photos of the Face on Mars in terms of the facial proportions, the supporting mesa structure and the perceived expression. Hoagland introduced him as an artist contributor who was a former colonel in the US Army, assigned to the Pentagon:

> His current occupation is consultant, mainly in the communications field. He is an accomplished artist, using this talent in furthering communications, particularly in multimedia presentations and corporate affairs.[13]

The artist's name was Jim Channon, and his background, as presented by Hoagland, was the understatement of the decade. Channon was a lieutenant colonel in the US Army and had served in Vietnam. In 1977, Channon had embarked on what journalist and author Jon Ronson, in *The Men Who Stare at Goats,* describes as "a Pentagon funded Odyssey"—that word again. Arthur Young had called the period from 1948–1952 his "Gee-Whizz years " and in the same spirit of research into applications for PSI, Channon visited 150 New Age organizations in order to see how their methods could be adapted to combat situations. He never informed these organizations about his ultimate motives, and in 1979, he submitted his report to the Pentagon.

Channon's key belief was that after the demoralizing thrashing it had received in Vietnam, the US Army needed a massive rehab: its soldiers needed to learn cunning. Channon proposed that the Army should become the moral foundation on which politics could then create harmony in the name of . . . not a person, a tribe, or a nation, but the Earth. He envisioned the First Earth Battalion, whose primary allegiance would be to the planet itself, not its inhabitants.[14] Which sounds an awful lot like a declaration of war toward those who are not of Earth.

Channon also stated that making the planet whole would require the ethical use of force, based on the collective consciousness. This last sentence is not mumbo jumbo, it is chilling because Channon's ideas were immediately turned into a variety of PSI-Ops warfare techniques. Some were lethal, and all of them were tantamount to torture of the human spirit via the physical abuse of the body and the mind. Torture by sound was one of the nonlethal weapons developed from Channon's research. The use of aggressively loud music played in reverberating circumstances to prisoners held in large metal containers with the intention of shattering their psyche was much in evidence during the Iraq War. News released in 2014 of other practices considered acceptable by these warmongers (exposure to low-frequency infrasound will certainly render a human physically and unpleasantly useless) underscored the point that despite the Army's mantra of "be all you can be," this view of the world is another example of "the ends justifying the means." Its practitioners were and are both unaccountable and entirely disinterested in the effects that their so-called ethical activities have on the human beings standing in their way. Here's what Ronson has to say on the matter.

In 1979 a secret unit was established by the most gifted minds within the US Army. Defying all known accepted military practice—and indeed, the laws of physics—they believed that a soldier could adopt the cloak of invisibility, pass cleanly through walls and, perhaps most chillingly, kill goats just by staring at them. Entrusted with defending America from all known adversaries, they were the First Earth Battalion. And they really weren't joking.

What's more, they're back and fighting the War on Terror. *The Men Who Stare at Goats* reveals extraordinary—and very nutty—national secrets at the core of George W. Bush's War on Terror.[15]

Manifest within the pages of the Channon manual for the First Earth Battalion is that avowed intention to render the average soldier *more cunning,* that is, deceitful. First, the velvet glove metaphor—soldiers arrive in a combat zone carrying a lamb in their arms, smile at the enemy, and so on—then if that doesn't work, literally produce the iron fist.

With such a handbook to guide them, it isn't difficult to see how Abu Ghraib, Guantanamo, the special rendition of all and sundry including pregnant women, and all other abuses perpetrated by the military on perceived enemies have come about. Ronson might choose the fairly anodyne word *nutty* to describe Channon's philosophy but the sheer hypocrisy of extolling the virtues of life, as does the First Earth Battalion manifesto, while using methods that are abusive in the extreme

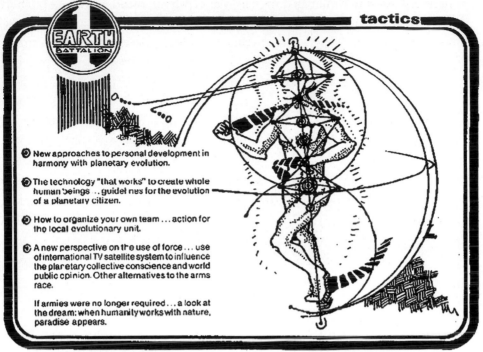

Illustration: Jim Channon.

Figure 6.10. First Earth Battalion.

is appalling. All this might seem off topic, but Channon's involvement with the Pozos report, albeit briefly, means something. The attitudes of a man whose notions have clearly influenced the US military's covert operations is important because if the military PSI-Ops are influenced by these doctrines, then the interface with the public (dubbed "Earth Dwellers" by Channon) concerning ETI is not likely to be open and honest.

Speaking of honesty, we find that Channon was not entirely alone on his fact-finding mission to the New Agers. Tellingly, the report into nonlethal technologies was co-authored by Colonel John Alexander, special adviser to the Pentagon, LANL, and NATO. Talking to Ronson, Alexander speaks fondly of working with Channon on the First Earth Battalion material and adds that he is one of Vice President Al Gore's oldest friends.[16] It is therefore of interest that Gore's climate change documentary, *An Inconvenient Truth,* much lauded at its opening, has subsequently been shown to be factually incorrect in some key respects.[17] While we cannot deny the physical manifestations of our own abuse of the climate, the climate change discussion is now one of the drivers for funding space

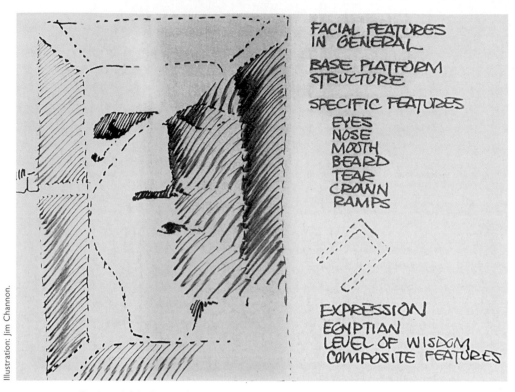

Illustration: Jim Channon.

Figure 6.11. One of Jim Channon's sketches of the Face on Mars.

programs. The premise being that we need to prepare another planet—Mars—for our species.

Meanwhile Channon, having livened up the computer conference report with several nifty designer drawings of Cydonia and the Elysium pyramid group, told everyone on the Mars project what they already knew—that the Face is obviously an artificial construct—and then he ostensibly retired from the Cydonian research scene and never did join the group, as was initially promised by Hoagland.

Despite Channon's absence, the First Earth Battalion ethos was symbolically present nonetheless, since Dolphin, the fundamentalist Christian SRI physicist and explorer of ancient Egyptian artifacts, uses the symbol of a lamb combined with the long cross of the missionary on his personal website. Although supposed to represent the symbol of John the Baptist, the lamb is an unpleasant reminder of the First Earth Battalion's image of a soldier holding a lamb in his arms: *all* these threads tie into Channon's aspiration to form a contemporary group of warrior monks.

Figure 6.12. The Lamb of God.

THE RIGHT HAND NOT KNOWING
WHAT THE LEFT IS DOING

After this opening salvo from the Pozos group, Hoagland followed up with a book of his own. *The Monuments of Mars: A City on the Edge of Forever* was published in 1987, and the data in this book did not quite fit with that sense of official rejection he had extolled in the Pozos report.[18] After the big reveal of 1976, eleven years is quite a long wait before publishing a book specifically about Cydonia Mensae, but checking out the history of space research reveals that in 1987, just a year after the *Challenger* disaster, the future of the Space Shuttle program was in the balance, and the funding and preparation of Reagan's so-called Star Wars initiative was very much to the fore. Cynically, one might note that finding something so obvious as a human-looking face on a planet way out of reach to astronauts would be the perfect stimulus for investing in the crewed space program as well as the DOD arm of the industry. After all, the American people are no different from the rest of humanity when it comes down to the big questions mentioned earlier: Who are we? Where do we come from? And are we alone in the solar system/galaxy/universe?

The Monuments of Mars appeared only a year after the 1986 academic exercise/PSI-Op that was the Pozos report, which means that it was in preparation during the period of the Pozos conferencing. Furthermore, if the very nature of the Pozos report implied a target audience of academics, the Hoagland oeuvre, in which he presented himself as a man of the people challenging the establishment on matters of Mars, implied taking Cydonia to a wider audience. Given the spin that Hoagland had already put on some of his statements, a little more archive archaeology was required, and it turned out to be most instructive.

Hoagland's book was partially financed by a grant from a federal agency: the National Endowment for the Arts (NEA). As the NEA website states:

> A federal grant is an award of financial assistance from a federal agency to a recipient *to carry out a public purpose of support or stimulation authorized by a law of the United States.* Federal grants are not federal assistance or *loans to individuals.*[19] (emphasis added)

In the normal way of an NEA grant, the project requiring the funds would go through its review process, which involves three levels of scrutiny. First, a staff review, then scrutiny by a panel of "experts with knowledge and experience in the area under review" (for Mars, who might they have been?), and finally, submission for approval by the National Council on the Arts/Chairman Review. The NEA states that it rarely lends to individuals, only to organizations, which is true in this case. The publisher of *The Monuments of Mars* is North Atlantic Books, and as its website puts it, this publisher is "within the organizational framework" of the Society for the Study of Native Arts & Sciences. During its early years, this publisher states that it was receiving most of its funding from the NEA under the banner of the society. Even if the society had some arrangement whereby it submitted merely a list of works to be funded rather than the detail of each manuscript, it would seem that this title didn't raise any concerns with the US government representatives. So it must be assumed that the Cydonia region is replete with native artifacts for this grant to have been awarded. And the fact that the book upholds opinions contrary to the official government position on Mars cannot be entirely the case: it has been continually in print, with updates and revisions, since 1987. So presumably, its original funders, the federal government, are entirely pleased with Hoagland's presentation of space-related controversies, including matters Cydonian.

FIT FOR PURPOSE

As well as many works of fiction and the technical publications on the red planet and its environment, and thinking of that "climate change/get to Mars funding driver," (although, ironically, in order to be able to live there, we would have to apply climate change to Mars), there has been much speculation as to why Mars is seemingly no longer fit for life. Authors such as Dick Hoagland and Graham Hancock incline towards the doom-laden view that any intelligent beings that were ever on Mars were either wiped out by some external event or managed to destroy the planet before dying in that self-inflicted catastrophe. This speculation would harden into policy some years down the line, and we'll deal with it when it does, later in the book.[20]

As already mentioned, in 1994, Arthur C. Clarke wrote *The Snows of Olympus: A Garden on Mars,* which dealt with the issues of terraforming (manipulating the climate of Mars so that human beings could eventually live there). Sharing in this rather more positive attitude toward the red planet, authors David P. Myers and David S Percy, in *Two-Thirds: A History of Our Galaxy,* published in 1993, produced a detailed work replete with information about the galaxy, the solar system, ourselves, and the many Earth-Mars connections. Their book concludes that the extraterrestrial presence on Mars was based entirely on beneficial intentions toward the emerging beings here on Earth. It would also confirm that Brian Cox and his colleagues are quite right when they say that we are ourselves Martians. But that is rather a lot for many people take in, and the authors are most often asked these two questions: If that is so, how did we get from there to here, hundreds of thousands of years ago, when we can't as yet build a spacecraft that will do the job? After all if we're descendants of Mars, extraterrestrials, as you suggest, wouldn't some of us have retained the glimmerings as to how to travel such distances safely?

For the first part, NASA is still trying to answer the primary problem: how to actually get there. For the second, this book is attempting to reveal how we might possibly regain our lost knowledge.

In 1952, building on the work of earlier rocket engineers, von Braun had published the aforementioned plan for a journey to Mars. He expected it to be achieved step-by-step over decades or even centuries. Comparing such a plan to the ancient sea voyages of our explorers, who had the same transport technology from start to finish, is interesting. For von Braun understood that if human biology and the stages he planned were a constant, over time the hardware to actually achieve the journey would change as the levels of technology matured and morphed. But von Braun had miscalculated because crewed space travel is the exception to this rule. Even if scientific research and computer technology have improved and probes and rovers are technologically more advanced, in terms of the launch vehicles and crewed spaceraft the hardware is little different from that of the 1960s. Astronauts are still expected to sit on top of a very large "fire-breathing tube" and, upon ignition, cross their fingers that all will go well.

The technology of the 1960s, designed to take men to the Moon, cannot be significantly changed for the trip to Mars, except for size, for to do so would put the viability of the lunar missions into question. Yet scaling up is another recognized technical problem for rocket powered launchers and spacecraft and just launching four people from Earth to Mars requires such an incredible amount of mass that it's way beyond current engineering skills.

The current state of space technology has another fundamental problem—communication. For all the advanced technologies to date, the Mars-Earth time gap remains, and although lasers are being studied as a means of carrying information, publicly at least, the distance achievable thus far is minimal. When traveling to Mars, there can be up to a twenty-four-minute *one-way* delay between astronaut and base. Imagine a scenario as the astronauts arrive at Mars, descend from orbit, attempt to land, and suffer a major systems failure. No communications assistance could possibly be provided in a timely manner; the astronauts would be entirely on their own.

Notwithstanding the public reasons for doing PSI research, in the interests of bypassing such nightmare scenarios the study of PSI communications was ramped up a few notches during the development of human space travel. The premise being that if people could use virtually instantaneous mind-to-mind communication, this would transcend time and space and be more than useful for space missions carrying human beings. If telepathy could be made to work consistently for anyone—no innate talent required—it would be a huge advantage and another item ticked off the list of things to achieve before humans travel to Mars.

It wouldn't be for want of trying: research into human consciousness was being discreetly pursued not only by the likes of Young in the late 1940s but also by US Army officials. In the 1950s, it was dressed up as research into the brainwashing techniques practiced on US soldiers during the Korean war of 1950–1953. From that war came the 1962 film *The Manchurian Candidate*, which certainly educated the public as to the dangers of the use and abuse of drugs to affect the mind and the behavior of soldiers—or anyone else.

Then there was the Apollo 14 telepathy experiment in 1971, but back then, the public was unable to put this event into its proper context. At the time, we were informed that astronaut Edgar Mitchell had owned up after the flight to having conducted a secret experiment with a few friends back on Earth to see if telepathy could work. Everyone laughed and dismissed the affair as somewhat weird. Later, Mitchell would write that having observed Earth from space, he had experienced a profound change of viewpoint.

For me, seeing our planet from space was an event with some of the qualities traditionally ascribed to religious experience. It triggered a deep insight into the nature of existence—the sort of insight that radically changes the inner person. My thinking—indeed, my consciousness—was altered profoundly. I came to feel a moral responsibility to pass on the transformative experience of seeing Earth from the larger perspective.

Mitchell did not go the usual route of post-Apollo career paths, car dealerships and board memberships in established institutions being two favorites, but instead elected to pursue his thoughts on the matter of consciousness. By 1973, Mitchell was heading up the newly created Noetic Institute and was very much in line with Young's thinking:

Looking to provide information and experiences in a way that brings objective reason closer to subjective intuition and thereby help to lessen the unfortunate gulf between these two modes of knowing. We can do this because inner and outer space research are converging.[21]

Whether Mitchell's aspirations have actually helped in space communications is questionable. His experiment in telepathy might have had other reasons for taking place. Post–Apollo 14, Mitchell was effective in bringing Geller to the United States in order to test Geller's PSI abilities. A government-funded think tank—SRI once again—organized and supervised these tests. Some of which were designed to see how a focused mind could affect other materials from a distance. Computer hard drives and magnetic tapes were subjected to Geller's mind, and the ability to influence another human being's willpower and decision-making processes were explored. Geller says he was asked to be present at political conferences and focus on the mind of the person with whom they were negotiating. In other tests held elsewhere, the military were also exploring the idea of using the mind to kill. Geller says that he was asked but declined to do any such test and that he left the program as a result of that request.[22] It is also a fact that the US scientific and military communities were researching mind matters for even more pragmatic reasons. Only slightly less secretly, methods were developed ranging from exploiting designer drugs such as LSD to the intensive brain training programs such as those set up by the Monroe Institute located in Virginia's Blue Ridge Mountains. The mind was being harnessed to harden soldiers to battle, to spy (among other things) on their perceived enemies here on Earth and relative to the exploration and dominance of near space, to search for traces of ETI.

SRI's contracts from the government might infer that he who pays the piper is entitled to call the tune, but not all PSI research or researchers should necessarily be tarred with the same brush. There was apparently a divergence of intention and action that continues today, and the Young school of thought and the Channon school of thought seem to represent these two polarities.

The records (now made public) concerning the US military's RV programs enable the public to see what it is the military want us to see, and what they *want* us to see is that within the official listings of remote viewing targets for defense purposes, there were plenty of reports on the remote viewing of Giza, other ancient sites here on Earth, and off-planet locations on Mars.

For all these exercises, the protocols of RV stipulated the need for very good maps.

Cydonia Lost and Found

How the Mars Complex Was
Lost but Found at Avebury

*We've done the same thing for Mars as the guys of the survey did for
the American West. No one really knew what was out there, and so
they went out and mapped the West. Those maps were the tools for
exploring the West. These maps of ours are the tools for exploring Mars.*

HAL MASURSKY, CHIEF SCIENTIST,
USGS's ASTROGEOLOGY SCIENCE CENTER

THE PHOTO FITS

Once the images from Mars were received, the task of translating them into
traditional maps that reflected the geology of Mars was undertaken by the
Astrogeology Science Center of the USGS in Flagstaff, Arizona. In line with a
desire to forget all about Cydonia, the USGS adopted a "now you see it, now you
don't" approach when mapping the region of Mars in which the Face is located.
Assiduous searching and a deep pocket will ensure some success, but for those just
looking online, it's not always easy to access the appropriate maps for study.

Science writer Oliver Morton's *Mapping Mars* is an informative read on the
subject.[1] He states that NASA has a good portion of people "who just like space
travel for the hell of it," although he adds that "there are those within the organisation who really do want to travel to a destination and explore it properly." This
reads as if those hell-bent for leather are currently in the majority. Morton doesn't
dedicate many pages of his book to the saga that mapping Cydonia has turned out

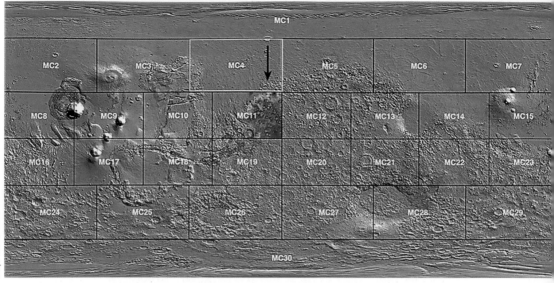

Figure 7.1. Location of Cydonia in the Mars quadrant scheme.

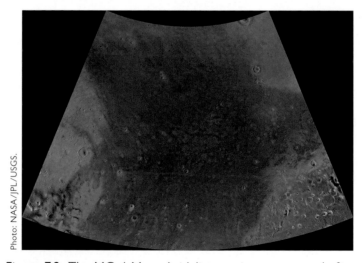

Figure 7.2. The MC-4 Mare Acidalium region was named after the Acidalia Fountain in Boeotia, where the Graces bathed. In Greek and Roman mythology, Acidalia was another name for Aphrodite. Aphrodite was also associated with love and with the planet Venus. Ares (Mars) and Aphrodite (Venus) were the parents of twin boys, Deimos and Phobos, after whom the moons of Mars were named. Thus, the "parents" also designate the boundaries of the astronomers' Goldilocks Zone in this solar system.

to be, and he joins the consensus view that there is no actual face-sculptured mesa on Mars. He ignores the attendant Cydonian pyramidal and polygonal landforms as well as the equally intriguing Elysium pyramidal landforms while happily looking forward to finding a real city one day on Mars. A city he describes as "Inca."

But wait! The *Mariner 9* team at JPL had already named a ground feature in the Martian southern hemisphere Inca City. Morton also recounts the story of Japanese-American sculptor Isamu Noguchi's wish to construct a huge landscape feature of a face, miles long, and as Noguchi had worked with the sculptor of Mount Rushmore, he knew what he was talking about. Some of his impetus to create this new feature was to do with his feelings in the aftermath of the 1945 Hiroshima bombing: Noguchi feared the eventual planetary death of Earth and its inhabitants. He called his project *Memorial to Man* and the design was modeled in sand by 1947. It was later destroyed. Subsequently it is said that the title of this work was changed to *Sculpture to Be Seen from Mars*. But the Isamu Noguchi Foundation and Garden Museum in New York hold a 15-by-34 foot photo of the model and still refer to it as the *Memorial to Man*. So one wonders who put this rumored change of name about. It could be that this occurred after 1976 and a deliberate connection to the Face on Mars was being attempted. From the knowledge that the Noguchi's sculpture was to have a nose a mile in length, it is possible to calculate that the overall sculpted features of forehead, nose and mouth would require a length of just over two miles. According to analysts of the Martian Face, it is about a mile in length. Given that Mars is just over half the diameter of Earth, these two Faces do share remarkable proportions.

Did NASA fake the early images of the Face to match Noguchi's earlier concept? The behavior of those attached to this exploration of Mars suggests not, but if NASA has been inspired by Noguchi, then its choice of that particular mesa, taken together with the anomalous geology of the spiral mound and the overall geometry extant at this anomalous Cydonian complex, has landed them in a whole heap of troublesome consequences. Rather more sensibly, considering the previous PSI discussions, let's consider that the artist Noguchi got his creative inspiration from "out there" in the collective consciousness.

CITY LIGHTS

Morton finds Noguchi's desire to mold the landscape as both a signpost of our presence and an everlasting monument to our existence quite understandable, but he doesn't consider that the reverse scenario of a sculpture on the Martian landscape conveys any sort of intelligent planning.

If that's so, why then name a block of so-called real estate on Mars Inca City?

As far as the public is concerned, the only resources available for making any decisions about Martian matters are the data returned from the various mapping probes and the goodwill of NASA to keep the public accurately informed. However, notwithstanding the professional qualifications required, the individual's perception of reality is going to color, albeit ever so lightly, the ultimate analysis of the data. Indeed, it's interesting to see the different parts of the Cydonian complex on which different researchers have focused. For instance, there is Walter Hain's map from his 1978 book *Wir, vom Mars,* which he worked up from the available Viking data.[2]

At the top of this map, Hain has three different measurement systems spanning the city to the crater. Looking at the mile system, one can see that the city itself has a diameter of six miles; the western edge of the city to the face has a

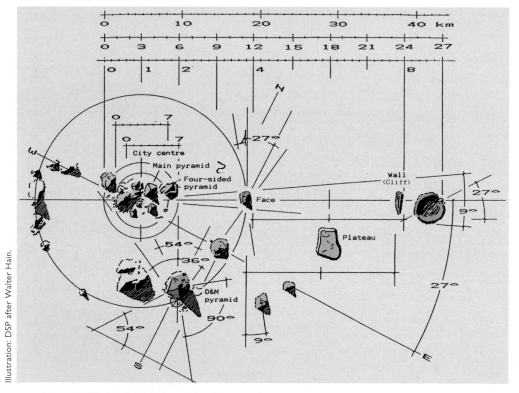

Illustration: DSP after Walter Hain.

Figure 7.3. Cydonia map. Hain's view demonstrates an interest in the western half of the anomalous region, with the Face as the midpoint. One might say that he is defining the hominid section with his circles. The eastern lion half is apparently of little interest since he has left out altogether the rough mesa and the Spiral mound, both of which were available from the Viking data.

radius of twelve miles, double that distance; and from the western edge of the city to the wall the total of twenty-four miles doubles that distance. The crater is not included in this number sequence, but the distance of twenty-seven miles from the western edge of the city hints at a nine-base system, and working eastward from the crater: the eighteen mile marker picks up the center line of the flat topped plateau (the Bastion) and nine marker picks up the landform just to the north of the D&M Tor. When it comes to the kilometer markers, appropriately that huge five-sider links to the ten kilometer mark. The thirty kilometer mark tangents the lower eastern edge of the flat top plateau and would also tangent the western edge of the rough mesa, had Hain decided to include that item. The forty-kilometer mark tangents the western crater rim. The twenty kilometer mark seemingly finds nothing of record but likely merits further research.

Walter Hain created his selective view of the region from the Viking data then added the metric data. The only distance marker that links his thinking to other Cydonian researchers is the lower line numbering 0-1-2-4-8. This line was the very first sightline to be noticed by the Pozos group, and therefore NASA. However Hoagland and the Pozos group, ostensibly ignoring the zero to one

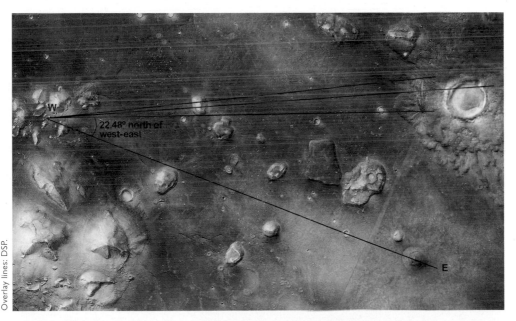

Photo: *Mars Express*, ESA/DLR/FU Berlin (G. Neukum). CC BY-SA 3.0 IGO.
Overlay lines: DSP.

Figure 7.4. Hoagland and his colleagues also extrapolated other sightlines emanating from the City center, noting how their sightlines cut the Face at the top of the nose, bottom of the nose, and under the chin. They also noted that the spiral mound connects to the city center with an angle of 22.48°; on Earth this angle is familiar to astronomers as relating to axial tilt periodicity.

distance, focused on the city center, as their sightline starts at that point.

I saw these Hoagland sightlines and the significant landforms along its length before I found Walter Hain's book and discovered his map. When I looked at his map I realized that these three numbering systems effectively encoded ten-based latitudes and kilometers, and that the statute mile inferred a twelve-based measuring system. And twelve immediately brought to my mind notions of time and space—travel. NASA uses both nautical miles and kilometers when plotting distance in space, and since nautical miles incorporate time, that Cydonia mile measured sightline could equally well consist of nautical miles of 6076.12 feet rather than the 5280 feet of the statute mile. That left the line of distance increases of 2-4-8. I was immediately led to think of musical octaves and harmonics. When a single note or tone is sounded it creates fifteen other harmonics. Four of which are octaves of that fundamental tone.

The distance from the western edge of the City to the Wall on the flank of the crater at the eastern end can be divided into 16 equal parts split into groups of 2-2-4-8, thereby accommodating the fifteen harmonics that ensue from a single tone. Publicly no one seems to have made the link with the sound of music (or frequency in general). Privately, I think that this line has been noticed for what it is and is another reason why Cydonia is a challenge.

The fact is that intelligence in abundance is staring at us from the anomalous region of Cydonia. Measuring systems related to our own cartographic practices,

Figure 7.5. Cydonia relationship modeling the octave markers along the sightline. In musical terms (or any other frequency) the zero point sets the fundamental tone and its octaves occur on the 1-2-4-8 markers.

space travel, and human voices (or sound in general) are marked out by sightlines linking a collection of extremely odd and disparate landforms at Cydonia—the central item being the Face on Mars. Denying that the Face is in any way a mesa that shows signs of being engineered and then describing the western group of Cydonian landforms as a city, which immediately implies civil engineering and intelligent planning, is a paradox. Or a sign of cognitive dissonance.

Whether fantastical or not, thanks to all the hard work over the decades by Hollywood and the media, the public is already primed to be interested in the civilian arm of the space program in general and Mars in particular. As Sagan demonstrated, serious scientists play on this fact whenever addressing the public. Morton does the same thing when addressing the Cydonian landforms. Evoking images of Petra in Jordan and cribbing shamelessly from Percy Bysshe Shelley he writes:

> Mars is a dry and dusty desert; as such we expect it to be old. Deserts are antique lands in which we expect the ancient to be preserved, places where we find rose red cities half as old as time, vast and trunkless legs of stone and other relics of the most distant past.[3]

As such he claims that part of the attraction of the Cydonian complex is that it fulfills these criteria for those who imagine that there is a lost civilization related to the Egyptian-styled Face on Mars. We expect to see ancient ruins because, in some respects, Mars itself is a ruin. All very poetic, but rather sidestepping the problem of Cydonia. As we know, there is no current plate tectonic activity on Mars, so if, as posited by Morton, Mars is of such age and climate that it can preserve potential ruins, then any landforms of geometric relationship, the one to the other, should be worthy of a closer inspection.

Any item showing sculptural tendencies, especially bearing in mind Noguchi's earlier thoughts on the matter, deserves rather more than selective misfiling and collective amnesia from NASA's PR machine.

Morton also relates that Hal Masursky, chief scientist of the USGS's Astrogeology Science Center, head of the Viking project, and a much-respected authority on lunar and planetary exploration, had been asked for his opinion on the Martian Face mesa during a conference held by scientists studying the case for getting crews to Mars. Masursky had "slowly and carefully studied the computer enhancements of the 'Face on Mars' before delivering his damning verdict: a drawn out 'Naah!'"

As an evaluation of a single anomalous landform among the many in the vicinity, it is an opinion that cannot be taken seriously from a scientific point of view. As a guide to public perception, it no doubt worked wonders. Masursky apparently dismissed the fact that Malin, the expert in Mars photo analysis, considers that photos themselves are not valid as a single proof of anything since they are liable to manipulation and subjective interpretation.

Borrowing from sociologist Marcello Truzzi's statement that "an extraordinary claim requires extraordinary proof," Sagan has said that extraordinary claims require extraordinary evidence. Well, the evidence is extraordinary by virtue of its placement, and those who demand proof of intelligent placement might have the courtesy to apply their brains to the problem for somewhat longer than it took Masursky to pronounce that long drawn out "Naah."

Hoagland's personal assessment of the photography led him to conclude that the angles relative to the shadows decreed that the site had been deliberately engineered to provide a sightline from the western City across to the large eastern crater and therefore the whole complex was intended to mark the summer solstice. With some justification, therefore, the Mars digital image processing expert Mark Carlotto averred that Hoagland's assumptions concerning the solar rise and set at Cydonia were based on incorrect data, therefore his dating of the Cydonian location was also erroneous. However, this matter of dating and lighting does not remove the very obvious data we have seen in this west to east sightline across the Cydonian anomalous complex. Until the Cydonian complex is examined from the perspective of cartographic mapping and geology, it is always possible to try arguing away the imagery as mere tricks of the light.

From the point of view of attracting the attention of a passing probe, if you were thinking of how to modify the landscape in order to incorporate simple mathematics that would unarguably denote forethought or intent, you would also know that a single straight line would best be proof against distortion when imaged from above. Sculpting the top surface of a mesa located at the halfway point along that same line only adds to the intention. The skeptic will say, "What you are finding is a matter of subjective opinion. You cannot tell what the relationships are from a photographic image that may have distortions. Therefore, there are no mathematically reliable ratios or any demonstrable geometry to find." There is some truth in that statement as a generality, but it doesn't work for the original Viking images. These were ortho-rectified, meaning that image perspective and relief effects had been removed, resulting in a constant scale wherein features are represented in their true positions. From the point of view of the probe's operator, two prominent Mars researchers publicly arguing the point about *when* the Cydonian complex might have been constructed conveniently removes the focus from those other questions raised by such visible a sign of intelligent planning: By whom? Why there? What for? Or indeed: Who for?

The answers to these questions are far from simple, and because the geological data relative to this region compounded the problem of Cydonian anomalies, so it came to be with the mapping of this region by the USGS wizards in the Astrogeology Science Center. Over time ways of dealing with the Cydonia problem were found and they too were far from simple to discover.

NUMBER CRUNCHING

Before NASA ever reached Mars, the regions of this planet had been named by astronomers using their telescopes and designating an albedo (light reflecting) feature with a classical name. The specific albedo feature originally referenced as Cydonia had its point of origin on the prime meridian at 40° N and was a circle with a diameter of 531 statute miles.

Named after the ancient city of Kydonia on Crete (now Chania), this albedo feature was chosen long before anyone had the faintest inkling of the anomalous complex farther west. Yet the circumference of this old albedo feature *neatly bisects the Face* into its eastern and western components. Which is quite possibly why, when consulting the current USGS data, one finds that the original albedo feature is shifted to a locus point of 59° N, 355° W.

From the time of Schiaparelli, astronomers have designated a prime meridian for Mars. The original Martian prime meridian locus was on Sinus Meridiani, and on the very early maps *with south to the top* (a leftover from the optics of telescope lens inversion) it was most easily discerned by the name Cydonia printed adjacent to it, some 40° below the equator. Indeed, astronomer Sir Patrick Moore was still listing Cydonia as being on the Martian prime meridian as late as 1998.[4]

This south-to-the-top, planetographic longitude system was measured westward from the prime meridian and used latitude and longitude coordinates mapped onto the surface. For nearly a hundred years, this is how things remained. Then, once NASA probes had arrived at Mars, everything changed.

In 1973, the year after *Mariner 9*'s deactivation, the International Astronomical Union (IAU) approved the relocation of the Mars prime meridian locus to a new position—at Airy crater, 5.2° south of the actual equator of Mars, but still on the Sinus Meridiani. This Airy locus was defined as having a diameter of forty-one kilometers (which is a little ironic since 41° N borders the northern edge of the anomalous Cydonian complex).

Acidalium M. (30°, +45°)
Æolis (215°, –5°)
Aeria (310°, +10°)
Aetheria (230°, +40°)
Aethiopis (230°, +10°)
Amazonis (140°, 0°)
Amenthes (250°, +5°)
Aonius S. (105°, –45°)
Arabia (330°, +20°)
Araxes (115°, –25°)
Arcadia (100°, +45°)
Argyre (25°, –45°)
Arnon (335°, +48°)
Auroræ S. (50°, –15°)
Ausonia (250°, –40°)
Australe M. (40°, –60°)
Baltia (50°, +60°)
Boreum M. (90°, +50°)
Boreosyrtis (290°, +55°)
Candor (75°, +3°)
Casius (260°, +40°)
Cebrenia (210°, +50°)
Cecropia (320°, +60°)
Ceraunius (95°, +20°)
Cerberus (205°, +15°)
Chalce (0°, –50°)
Chersonesus (260°, –50°)
Chronium M. (210°, –58°)
Chryse (30°, +10°)
Chrysokeras (110°, –50°)
Cimmerium M. (220°, 20°)
Claritas (110°, –35°)

Copaïs Palus (280°, +55°)
Coprates (65°, –15°)
Cyclopia (230°, –5°)
Cydonia 0°, +40°
Deltonon S. (305°, –4°)
Deucalionis R. (340°, –15°)
Eridania (220°, –45°)
Erythræum M. (40°, –25°)
Eunostos (220°, +22°)
Euphrates (335°, +20°)
Gehon (0°, +15°)
Hadriacum M. (270°, –40°)
Hellas (290°, –40°)
Hellespontia Depressio (340°, –60°)
Hellespontus (325°, –50°)
Hesperia (240°, –20°)
Hiddekel (345°, +15°)
Hyperboreus L. (60°, +75°)
Iapigia (295°, –20°)
Icaria (130°, –40°)
Isidis R. (275°, +20°)
Ismenius L. (330°, +40°)
Jamuna (40°, +10°)
Juventæ Fons (63°, –5°)
Læstrygon (200°, 0°)
Lemuria (200°, +70°)

Libya (270°, 0°)
Lunæ Palus (65°, +15°)
Margaritifer S. (25°, –10°)
Memnonia (150°, –20°)
Meroe (285°, +35°)
Meridianii S. (0°, –5°)
Mare (350°, +20°)
Nereidum L. (270°, +8°)
Nereidum (72°, –28°)
Niliacus R. (270°, +35°)
Nilus Amenthes (260°, +20°)
Nilokeras dum Fr. (55°, –45°)
Nilokeras L. (30°, +30°)
Nilokeras (55°, +30°)
Nilosyrtis (290°, +42°)
Nix Olympica (130°, +20°)
Noachis (330°, –45°)
Ogygis R. (65°, –45°)
Olympia (200°, +80°)
Ophir (65°, –10°)
Ortygia (0°, +60°)
Oxia Palus (18°, +8°)
Oxus (10°, +20°)
Panchaia (200°, +60°)
Pandoræ Fretum (340°, –25°)
Phæthontis (155°, –50°)
Phison (320°, +20°)
Phlegra (190°, +30°)
Phœnicis L. (110°, –12°)
Phrixi R. (70°, –40°)
Promethei S. (280°, 65°)
Propontis (185°, +45°)

Cyclopia (230°, –5°)
Cydonia 0°, +40°
Deltonon S. (305°, –4°)

Figure 7.6. Reproduction of Moore's listing with "Cydonia 0°" highlighted as the prime meridian.

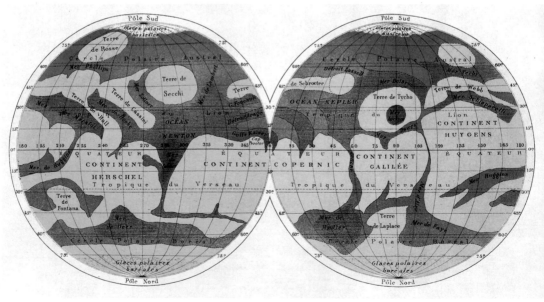

Figure 7.7. Astronomer Giovanni Schiaparelli's Mars map, 1862.

Figure 7.8. Airy Crater.

Since such administrative decisions generally do not happen overnight, it is reasonable to suppose that this decision, too, was taken after the 1971–1972 *Mariner 9* mapping expedition but this diddling about with the Mars meridian locus seems not to have been enough for those in charge of Mars mapping.

ADJUSTMENTS

Three years after first setting about rearranging the Mars coordinates, the IAU managed to lose the anomalous area within Cydonia altogether by doing what colonizers always do when they want run the show: they broke up the region and changed the boundaries. At some point, with the return of the *Mariner 9* probe images the most likely reason for this change, Cydonia Mensae was created. With its locus at 37° N, 12.8° W (13° farther south and 17.8° farther west than the original Cydonia locus), it was described as an area full of flat-topped geological features, many with cliff edges. And with a diameter of 854 kilometers, its circumference largely circumscribed the Face on Mars mesa. Patrick Moore's 1998 book references this locus as official data from the IAU, but they no longer acknowledge it. That is because, in 1976, the Viking period, the Cydonia Mensae locus was again moved—even farther south. This new Mensae had its

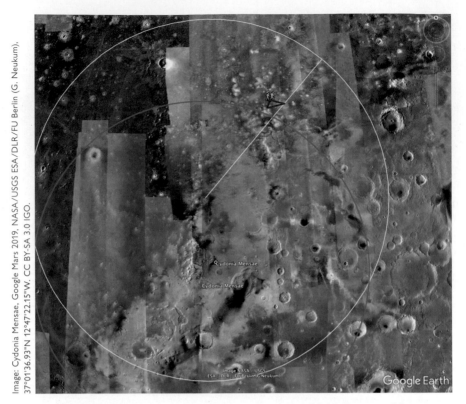

Figure 7.9. Cydonia Mensae: the yellow circle is the 854 kilometer Mensae. The green circle is the 800 kilometer post-Viking new Mensae, The D&M Tor, Face, and spiral mound special right triangle is in red. The two little green men (LGM; upper center) hiking the circumference of this circle are also walking over the Face on Mars.

locus point at 34° N, 13.2° W, and a shrunken diameter of 800 kilometers.

In 2001, it was back to diddling with the meridian. Much to the dismay of the old guard (Moore was especially peeved), the astronomical custom of "south to the top" was discarded. North was reinstalled at the top of the map, and the prime meridian was further refined by repositioning the meridian locus within the Airy crater. This new locus was designated Airy-0. It was also decreed that the new plan-etocentric longitude would now be measured from 0° to 360° eastward from Airy-0, and a new Mars "net" was mapped from a point of origin on Olympus Mons. Some historians state that the meridian locus was changed to honor Englishman Sir George Biddell Airy, who had established the Greenwich prime meridian. Some noted that Airy sounds like Ares (a god of war) and looks like Aries the ram. Others remembered that this constellation had been designated by ancient Earth astronomers as the 0°/360° beginning/ending point for tracking the Sun's 25,920 Great Year cycle known as the precession of the equinox. It is retained as such to this day, and it occurs on March 21, a month also associated with Mars (see figure 7.9).

That was still not enough: with ever-improving results from the mapping experts, much later in 2003 (during the *Surveyor* period, 1997–2006), two more Cydonian zones were created. Cydonia Colles, meaning a region of hills and knobs, had its locus point of origin at 39.4° N and 12.2° W, and this area was defined as having a diameter of 365 kilometers. How poetic! This diameter is nearly, but not quite, the number of days in an *Earth solar year.* The Colles circumference neatly touches the huge eastern rim of the crater at Cydonia.

So where would their third area turn up? Surely not in the same location? Well, yes! Unbelievably, the locus point is at 41.5° N and 12° W. This third area *overlies* Colles, but describes entirely *different* geological features. Named Cydonia Labyrinthus, this third circle encompasses a region of valleys and ridges and has a diameter of 356 kilometers. A number that is just over a *lunar*

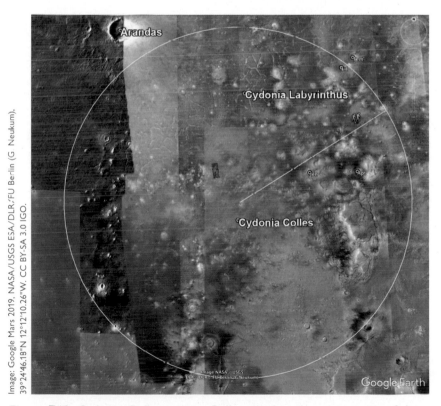

Figure 7.10. Cydonia Colles region; Colles touches the eastern rim of the huge crater, whose analog is at Avebury. And the Sun's motion as seen from Earth is associated with the southeastern quadrant of the Avebury henge.

synodic year of 354.36 days. The name Labyrinthus brings to mind the Cretan tale of the Minotaur. And the circumference of Labyrinthus tangents the southeastern corner of an area included in the anomalous complex of Cydonia—the *Two-Thirds* storyline attributed these two small craters to cattle feed storage and sheep shearing.

Surely, then, that post-Viking Cydonia Mensae diameter of 800 kilometers must also mean something in terms of Earth days. Digging into Moore's trusty tome, one finds that it is remarkably close to the 789-day synodic period of Mars, when Mars is opposite the Sun and nearest to Earth and when launches to Mars can be made with greater economy of fuel.

Looking at Labyrinthus a bit more closely, it's hard to see how the Cydonia Labyrinthus justifies its title compared with the geology of the only other Labyrinthus on Mars, Noctis Labyrinthus (located in the southern hemisphere), yet the addition of its defining circle has resulted in even more geometry and the making of a symbol of great esoteric significance for many people. Its interface

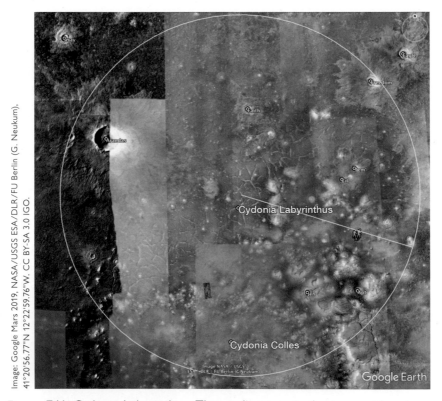

Image: Google Mars 2019, NASA/USGS ESA/DLR/FU Berlin (G. Neukum). 41°20'56.77"N 12°22'59.76"W. CC BY-SA 3.0 IGO.

Figure 7.11. Cydonia Labyrinthus. This circle tangents the two southeastern craters; the analog location at Avebury is the Sanctuary located by the Old Ridgeway path just to the east of Avebury.

Image: Google Mars 2019, NASA/USGS ESA/DLR/FU Berlin (G. Neukum), 39°16'09.16"N 12°34'59.91"W. CC BY-SA 3.0 IGO. Overlay lines: DSP.

Figure 7.12. Vesica piscis formed by Colles and Labyrinthus. The original Mensae is the dark green ring, and the new Mensae is pale green with the square center-mark. Colles is pale blue overlaid in yellow. Labyrinthus is purple. The vesica piscis generated by two interlocking circles (here, Colles and Labyrinthus) also creates three square roots: the square root of two, generated from Colles's center tangents the feet of the LGM to terminate on Labyrinthus's circumference; the square root of five, generated from the same point on Labyrinthus tangents the heads of the LGM and terminates on the circumference of Colles; and the square root of three divides the bladder of the fish in two and is denoted by the turquoise line. To demonstrate how closely the USGS circles match the ideal geometry, a 2-D vesica piscis is overlaid in black, its circles based on the diameter of Colles.

with the Cydonia Colles results in a very close modeling of a vesica piscis, the geometric figure of two interlaced circles that produces the square roots of two, three, and five. And all of this USGS mapping subtly interfaces with the anomalous Cydonian complex and finds correspondence with the original special right triangle that is marked by the D&M Tor, the anomalous spiral mound, and the Face on Mars.

Figure 7.13. Close-up of the vesica piscis. Note that the alignment of the inaccurately placed Google buttons (relative to the USGS coordinates) forms a line that divides the vesica piscis's square root of three into two. Also note how the hidden geometry of an ideal vesica piscis interacts with the Cydonian anomalous complex: the square root of two diagonal interacts with the Face. The post-Viking USGS Cydonia Mensae circumference interacts with the huge five-sided D&M Tor, leaving that imported spiral mound out of this particular coding.

It difficult to imagine that the USGS, driven by geology alone, has accidentally created such significant geometry on the surface of Mars. And, of course, matching 2-D drawings to 3-D spheres is not perfect, but the correspondences are such that whether the USGS staffers were inspired by the already extant geometry at Cydonia or deliberately encoded geometric principles onto Mars for their own purposes, the results merit reflection. Because what happened next reveals something of the intentions of the Earth-based mappers.

Not content with all of this adjusting over the decades, the mappers then managed to make the presentation of their work to the public amazingly convoluted. Today, depending on which Mars map you are looking at, you can find either the older south-to-the-top westward planeto*graphic grid system* or (on maps made later than 2001) the newly installed north-to-the-top eastward planeto*centric net system*, and sometimes these two systems are marked on the same map. Furthermore, many of the USGS areology (Martian geology) maps are now based on 10° grid systems relative to the digital age and satellite mapping. Retrieving all this information requires the cross-referencing of several sources belonging to NASA and the USGS, whose listings are not designed for the layman's ease of navigation.

SHRINKING THE DATA

As it turns out, I had written all of the above chapter before 2005. When Google Earth v5 launched Google Mars in 2009, I found its data to be marginally at odds with the figures stated above. Given that this company has electronic interference all across the anomalous Cydonian complex, I was not surprised and left the data as I had written it. It wasn't until 2015, when Google Pro was available free to the public, that I could actually draw up these Cydonia regions as described all those years ago by the IAU. Checking the data against that of Google and the IAU once more, I found out that the IAU had updated Mensae, Colles, and Labyrinthus on October 1, 2006. The *Mars Reconnaissance Orbiter* had been in orbit around Mars since March 10 of that year, and this update occurred, coincidentally, on the day after the *Mars Reconnaissance Orbiter,* from its science orbit, had taken its first high-resolution image. (The orbiter had a capacity to resolve objects as small as thirty-five inches [90 centimeters] in diameter.)

I was also to discover that the data on Google Earth does not match up to the coordinates supplied by the USGS, but the alignment of Google's Cydonian buttons do lead the eye to the vesica piscis geometry. Finally, with the creation of that shrunken New Mensae still niggling at me, I checked the distance from its locus and found a significant tangent to the D&M. The distance to the massive five-sided pyramid is 366 miles. In Earth days, that is our leap year period. Is this yet another hidden message of our aspiration to leap from the lion's mouth (LEO) and get to Mars. Or a challenge to Ares, the god of war? This is highly significant, given what we will discover about the Pentagon.

LITTLE GREEN MAN TIMES TWO

Coming back to the potential obfuscation of detail through various overlays, yet another item has been added to the Google Mars mapping of the Cydonian anomalous complex. No doubt causing much mirth within Google, two LGM

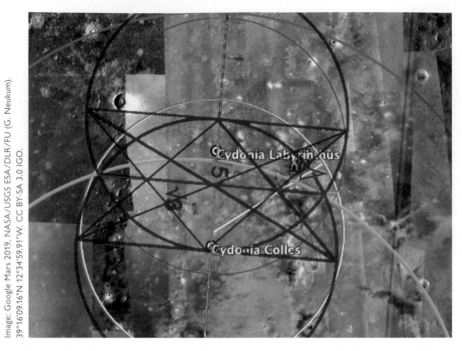

Figure 7.14. Cydonian complex. This closer view shows the location of the special right triangle relative to the vesica piscis. The yellow line is the radius from Colles to the large crater east of the Face.

virtually mask the black-and-white photo already overlaid onto this Martian mesa. Given the association of that particular mesa with the Egyptian Sphinx, it's interesting that these two LGM are reminiscent of the three-legged old man, as in the riddle of the Sphinx. Bowed over their walking sticks, they constitute an electronic link, and here is an extract from the lengthy diatribe that ensues when you activate these LGM:

> The Cydonia area has been the subject of a time-wasting, deliberately orches-trated, and pseudoscientific frenzy about one of the mesas. Above all, the inci-dent illustrates the foibles of twentieth-century life in the presence of powerful but mediocre news media. It all started in the 1970s, when Viking cameras photographed one of the more eroded Cydonian mesas at a time of day when shadows on the rough summit produced a crude face-like pattern, with two small shadows making eyes, and another longer valley shadow making a mouth. *Wags on the Viking team, following planetary scientists' penchant for whimsical nicknames, named this hill the Face on Mars.*

The problem was that a ragtag band of tabloid editors, promoters, and amateur naturalists took this nickname seriously . . . or pretended to.[5] (emphasis added)

It continues with a lecture on pareidolia and all in all confirms the very negative bias of Google and its associates when it comes to the anomalous Cydonian complex, free speech, and thinking differently. This electronic overlay can be seen simply as a multiplexed in-house joke about LGM, Yoda, and *Star Wars* (in both film and defense contexts), or even a snide allusion to the very slow progress we are making along the pathway to Mars. However, its criticism of the media is beyond ironic, since it was JPL itself that initially promoted the Face. When JPL did so, either the agency was asking for trouble or, knowing what would happen, they were making sure everyone was paying attention while laying the blame for all this frenzy on the tabloid press, promoters, and amateurs (for which read scoop merchants, attention seekers, and the ill-informed, respectively). From a digital mapping perspective, the net result of Google Mars's efforts is to be told, "You will get little help from us, and even if there was anything to look at in this area of Mars, you can't!" This schizoid approach is reflected in a final irony: just entering "Cydonia" into Google Mars won't get you anywhere at all. It is necessary to add Mensae, Colles, or Labyrinthus before Google Mars will allow you in via a name check. And even then, none of these locus points will take you to the anomalous complex.

On the other hand, entering "Apt Crater," then, once there, zooming out and scrolling up gently will land you right where it's at. Named after a French Provencal town in the Luberon mountains, in the local Occitan dialect this crater's name would be pronounced "at," as in the email symbol. It was approved for use on Mars by the IAU in 1976, the year of the Face. How apt!

CONTROLLING THE DATA

With the breakdown of Cydonia into three overlapping regions, it looked as if these maps were being presented openly for one and all, but in fact they were not. Not really. The aims of the mapmaker dictate how much information is going to be provided for public access, and even though Mars is some forty-eight million miles from Earth, this vast distance does not automatically mean that all the information is going to be given away willy-nilly to anyone who cares to ask for it. As spacecraft technology improves, Mars will one day become the planet next door. So this mapping data game becomes rather more of an issue than the armchair explorer might expect. Especially, should it already be understood that the anomalous Cydonian complex has the potential for serious military or economic advantages to those human beings who reach it first. If getting crews to Mars is going to take longer than expected, that is more than enough reason to deflect attention from the location by various means. Altering the mapping data, then generally controlling the media output, not least by scorning those interested in this particular anomalous complex, diverting the attention elsewhere on the red planet, using the current understanding of the geology and history of Mars to create dystopian back histories for Mars—it all turns out to be less schizoid than first imagined and more of a pragmatic choice.

It's clear that even when using the kilometer for the encoding of orbital data rather than the statute mile preferred by traditional metrologists and the sacred geometers, the mapping of this region has symbolic and esoteric relevance to those in charge. These Cydonian mapping circles encoding data about Mars, Earth, and the Moon reference the very planets that NASA intends visiting with human beings. The initials of Mensae, Mensae, Colles, Labyrinthus sum in Roman numerals to MMCL (2150), a number perhaps with inherent meaning for NASA or a likely date for the agency to achieve its aims. And that being so, here is an even more fantastic idea: through its actions and mapping on Mars and then on Earth, perhaps NASA is myth-messaging its own cohorts and ETI. Something along the lines of "Look, we know you are around our planet and our Moon; here's our message to you, to show we have understood that you were here as well."

REVELATIONS

Having dealt with the wider areas of the Cydonian mapping, it was still necessary to take a closer look at the anomalous area and for that I needed a good geological map. During 2004, while I was crawling around the NASA and USGS sites for the *nth* time, I came across a geological map of the anomalous region of Cydonia with a straight-on view and north to the top. Dated 2003, it had been made from Viking data and additional new orbital data made available only to these map-makers from the Surveyor and and Odyssey missions.

Finding this map really was a stroke of luck—and a revelation. Without being an expert on geology I could easily see what went together and what did not; and its straight-on view enabled the verification of all the geometry relating to this region. Luckily, I retained a copy of this map because as it turned out, when looking for it again later, in line with the "refining to a better standard" mantra, the anomalous Cydonian complex had become the southeastern quadrant MTM40007 of a map of the Acidalia Planitia, and the Tanaka 2003 map as a stand alone item had disappeared from the web. As the USGS put it: "Because

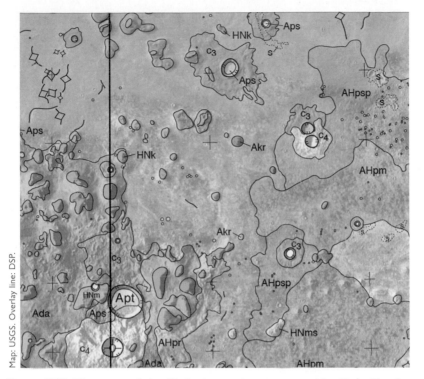

Map: USGS. Overlay line: DSP.

Figure 7.15. The map of the Cydonian region generated by K. L. Tanaka,
J. A. Skinner, T. M. Hare, T. J. Joyal, and A. M. Wenker, 2003.
The black line denotes north-south and goes straight through
the center of the large five-sider pyramid.

this unit is essentially a conglomeration of various materials (Tanaka and others, 2003), it would be counterproductive to retain it in a mapping study intended to provide more detailed spatial information"[6] But then Oliver Morton on discussing the mapping of Mars did say that "The point of geological mapping is to tell a story—to turn landscape into history—and the gaps are the story's articulation."

This reads like more of "Project Lose That Anomalous Cydonia Complex." Having stated all this, the library angel intervened in 2016, just before finalizing the images for this chapter. Looking for a map with a better resolution of this area, David Percy went digging into the USGS archives and serendipitously discovered the very same Tanaka 2003 map—but in a much higher resolution than before.

Now you see it, now you don't. Was this map trying to draw attention to itself? It is after all very special, because it shows that a couple of items within the anomalous Cydonian complex map are also geologically anomalous to the region. One of

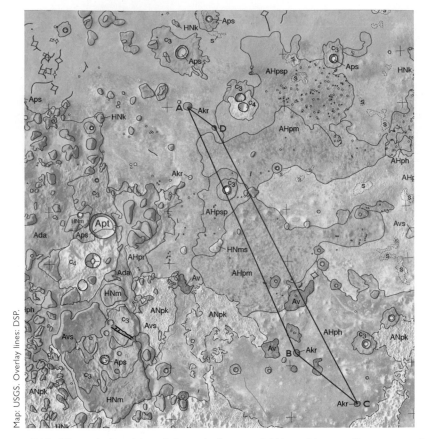

Figure 7.16. The relationship of the spiral mound (A) and its smaller companion (D) to the same geological material 101 km (over sixty-two miles) to the south of the Cydonia complex (B and C). The lines A-D and B-C are virtually parallel to each other. The USGS describes this volcanic material as: "Ringed knob material—Materials of circular knobs in quadrangle MTM 40007 with contour-like, bright, concentric rings on their flanks; associated with putative lava channels and small volcanic deposits. Interpretation: Volcanic constructs."

these, the spiral mound, lies 101 kilometers farther north than another mound of the same material. Labeled AKR on the map, this denotes volcanic material.

AREOLOGY 101

The so-called conical or spiral mound and a small companion conical mound to the southeast (colored dark pink) are not only of a *completely different material, there is nothing like them on the northern* plain on which they stand. They are just as anomalous to their surroundings as the imported red granite in the Great Pyramid or the imported Welsh bluestones at Stonehenge.

Now that it is possible to see that the spiral mound is imported volcanic material, the placement and styling of this particular artifact is obviously intentional. The hand of ETI is apparent here, so it's no wonder this map got "lost." The wonder is that it has actually reappeared. Why? Is it now all right to talk about ETI in relation to Mars?

MANY MILES DISTANT

The closest deposits of the material forming the spiral mound occur over sixty miles to the south.

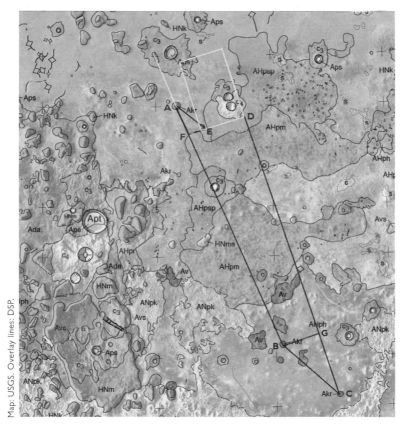

Figure 7.17. The triangle A-E-F formed by placing the two northern mounds with the extended sightline from the southern deposit B is a 30-60-90 right triangle. The extended western Akr line (yellow) locates the wall and the extended eastern Akr line defines boundaries and locations only mentioned in *Two-Thirds*. The triangle B-C-G formed by the volcanic deposits in the south is also a 30-60-90 right triangle. Akr designates volcanic material.

Apart from the parallels and triangles formed by the placing of the northern deposits relative to their southern cousins, another sign of intelligent placement can be found by following the line B-A northward to discover yet another revelation from this geological map. It hits the straight north-south extrusion on the western flank of the large crater and it too is of a *completely different material* from that of its crater and could not therefore be simply lava flow/ejecta from this crater, as has been suggested by many. The crater material is described by the USGS as "moderately degraded crater material—Rim, floor, and ejecta materials of craters with complete but somewhat modified rims and ejecta blankets; commonly partly filled by younger material. Interpretation: Hypervelocity impact materials." While the extrusion on its western flank is described as: "Mesa material—Generally smooth, featureless material with albedo similar to homogeneous plains but forming mesa and butte landforms. Interpretation: Origin enigmatic, but probably residual inliers of a once more widespread deposit." As this mesa material is considered to be geologically older than the crater ejecta that surrounds it, the cliff component of this mesa is therefore of a considerable height, but nothing in the USGS description allows for the formation of a straight wall or ridge down the middle of this mesa, except perhaps the words "Origin enigmatic."

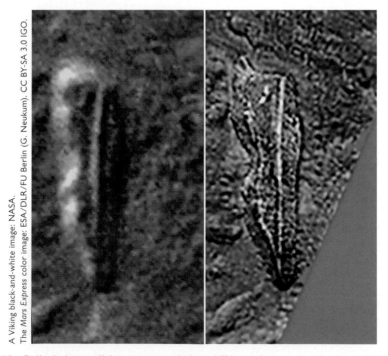

A Viking black-and-white image: NASA.
The *Mars Express* color image: ESA/DLR/FU Berlin (G. Neukum). CC BY-SA 3.0 IGO.

Figure 7.18. Called the wall by some and the cliff by others, visually both ideas are present. The cliff extruding from the ejecta blanket supports the long straight wall.

On the assumption that the Mars data has been accurately translated into its USGS map form, it is hard to see how material belonging to the rough southern uplands occurred accidently on the smoother plains to the north. The contrast between the north-smooth and south-rough terrains recalls the smooth mesa located halfway between the Face and this spiral mound and its rough companion. All of this geological referencing and manipulation is uniquely in the *only location* on maps and photos of this region that manifests an intelligent mathematical form recognizable to us as right triangles and the special right triangle.

DIVERSITY AND PATTERNS OF ORGANIZATION

In the second edition of his book *The Martian Enigmas,* Mark Carlotto took note of Carl Sagan's dictum and wrote this about the anomalous Cydonian complex:

> Claims do require extraordinary evidence, but it does not have to come from a single source. Weak evidence from multiple independent sources will do just as well. As demonstrated here it is the quantity and diversity of all of the evidence rather than any one piece, that makes the evidence in support of our hypothesis so strong. The alternative hypothesis is, of course, that the Face and other nearby objects are simply naturally-occurring geological formations. However, *no convincing geological mechanisms have to date been put forth that are capable of explaining the diversity of forms, the patterns of organization, and the subtlety in design exhibited by this collection of objects.*[7] (emphasis added)

Carlotto concludes that the odds are between 129:1 and 763,000:1 in favor of the hypothesis for intelligent design.[8]

THE CYDONIA-AVEBURY RELATIONSHIP

The American government agencies were not the only minds focusing on the Martian topography; across the pond, David Percy, in London, had found similarities between the Cydonian complex topography and a location in southern England. Having a growing interest in crop glyphs had led Percy to study closely the area of southern England where they were most prevalent. At one point during that time he experienced a spontaneous vision or remote viewing of this area of Mars, and Percy realized that the entire anomalous complex of Cydonia appeared to be replicated on the Avebury landscape.[9] This insight was further confirmed after studying a crop glyph in Avebury that was activated in July 1991.

Illustration: DSP.

Photo: DSP.

Figure 7.19. Representation of the 19.47° crop glyph, Avebury, Wiltshire, England, July 27, 1991.

Figure 7.20. Silbury Hill, Avebury, analog of the spiral mound on Cydonia.

Percy concluded that the upper circle could represent the Avebury Circle, that this, in turn, with its enormous earthen rampart and ditch, also was representative of the crater on Cydonia, and that, consequently, the lower circle in the crop glyph represented Silbury Hill and/or the spiral mound on Mars. And to his mind, the anomalous cliff and wall on the Martian crater's flank was represented by the extension to the left of the large area of flattened crop.

Positioning an ordnance survey map of the area over an image of the Cydonian complex, the match was found to be stunningly proportionate.

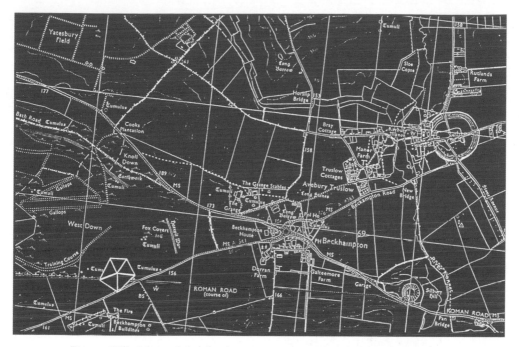

Figure 7.21. Map of the Avebury area, reproduced from *Two-Thirds: A History of our Galaxy,* Aulis Publishers, 1993.

Image: ESA/DLR/FU Berlin (G. Neukum). CC BY-SA 3.0 IGO.

Figure 7.22. An exact match of the key features seen in figure 7.21 updated onto the *Mars Express,* 2006 image of the Cydonia complex.

Photo: DSP.

Figure 7.23. Avebury Circle, analog of the crater on Cydonia,
with the tetrahedral rim pyramid (arrow).

Percy also noted a small triangular mound on the map of Avebury Circle, exactly where the tetrahedral pyramid occurs on the crater rim in Cydonia. This mark was located above a gap in the Avebury rampart, just as the Mars pyramid is above a gap in the Cydonia rim. It is the only mound indicated on the Avebury rampart by the map surveyors, and it is clearly an analog of the tetrahedron located on the northeastern quadrant of the Martian crater rim.

It is important to stress the virtual impossibility, first, of a mound plus a crater rim in Cydonia and a mound plus an earthen rampart and ditch in Avebury having by chance the same relative size and being the same relative distance from each other on both Mars and Earth. Added to that, these two mounds have similar cone form and spiral paths. All these facts strongly suggest that somehow the ancient builders of Avebury circle and Silbury Hill knew about the layout of the Cydonia complex.

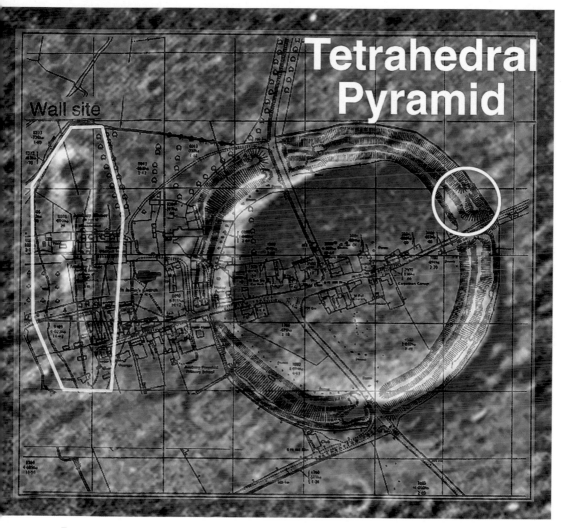

Figure 7.24. Avebury map superimposed over the rim pyramid on Cydonia. This tetrahedral pyramid positioned on the Avebury rampart, north of a gap, is *precisely mirrored* by the tetrahedral pyramid on the Cydonia rim, as originally matched in 1992 to a NASA Viking image and reproduced from *Dark Moon: Apollo and the Whistle-Blowers,* Aulis Publishers, 1999.

SLIDING DOORS

After Percy reached his conclusions in 1991, he thought he should pass on this information to someone for whom this information might be relevant. Having read the Pozos report and seen Richard Hoagland's NASA-Cydonia briefings, he decided that Hoagland was the only obvious "civilian" connected to Cydonia studies. A colleague of Percy's, the film producer Robert Watts, had the connections that made it possible for Percy to recount the full details of this significant Mars-Earth connection in person, and in September 1991, a trip to the United States and a meeting in Sausalito, California, hosted by Garry Gunderson, was organized. Percy, expecting to meet simply with Hoagland was astonished to be confronted by several other people involved in Mars research, including David Myers.

After delivering his detailed presentation to the assembled group, they discussed it amongst themselves while Percy went out onto the balcony for a breath of fresh air. I have seen that presentation, and it's pretty intense for both the lecturer and the listener. When he returned to the room, he inadvertently walked slap-bang into the glass sliding doors that had been closed while he was outside. With hindsight, this closing of the doors was a portent of things to come, but at the time all that was apparent was the gash across the bridge of his nose; traces of the scar remain to this day.

As it turned out, noses then became something of a thing. At a following meeting, Hoagland and Percy were experimenting with a small frameless mirror placed vertically over a close-up of the Martian Face, splitting the nose to reflect each side of the image. What they observed led to the two men rushing off to the local photocopy shop, whereupon Percy assembled the very first rough-and-ready mirroring

Photo: DSP Collection.

Figure 7.25. David Percy (left) and Richard Hoagland (right), California 1991.

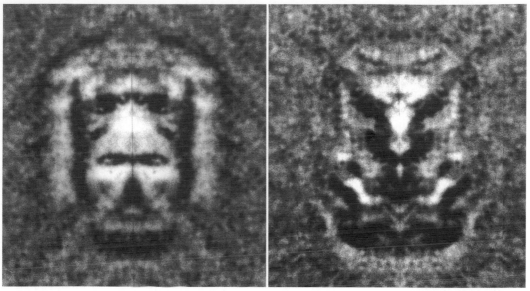

Derived from NASA's Viking photo 35A72.

Figure 7.26. Original display images as created by David Percy in London, 1991, of monkey/hominid image (left) and a lion/feline image (right).

of the Face mentioned in chapter 4. The Face has been often been described as the Martian Sphinx, and this resulting mirror imagery certainly reflected the two aspects of that hominid/lion sculpture we have here on Earth, at Giza.

Following the September 1991 meeting in Sausalito, Percy returned to London but liaised with Hoagland's Cydonian research team, which Hoagland had dubbed the Mars Mission. This group published *Martian Horizons,* a quarterly journal at that time edited by David Myers, and in Volume 2 it was announced that Percy had been appointed director of European Operations. Percy then went on to produce all the visual material, including high-quality 35 mm slides of images and graphics for *The Terrestrial Connection,* a major presentation on Cydonia and its exciting connections with Avebury. This was presented in February 1992 by Hoagland to delegates, the media, and special guests at the headquarters of the United Nations, New York, in the Dag Hammarskjöld Auditorium. Percy also prepared all the photographs, maps, and graphics for an exhibition to be staged in the run up to the presentation. These were shipped to the United States and one of the US Mars Mission team, artist and psychic researcher Colette Dowell, arranged for the mounting and hanging of the display in the south lobby of the United Nations building.

After the event, Hoagland updated his book *The Monuments of Mars* and asked Percy to provide him with some images for the 1992 edition. However, in

Photo: DSP.

Figure 7.27. Part of the exhibition in the south lobby of the United Nations building, New York, February 1992.

the analysis of a crop glyph illustration, Hoagland was prepared to tweak some angles to validate his own hypothesis. This was not an acceptable practice to either Myers or Percy, and this led to a parting of ways. Thereafter, they would continue to explore the ramifications of the Wiltshire scale model analogs on their own.

Figure 7.28. David Myers exploring the Avebury connection to Cydonia in the summer of 1992.

Photo: DSP.

Myers and Percy published the detailed results of their research a year later on August 20, 1993, just two days before the *Mars Observer* went silent, and their book, *Two-Thirds: A History of Our Galaxy,* turned out to contain far more information than they realized at the time. One principal issue concerning their Avebury-Cydonia connection was that while the spiral mound and the crater were fully present, when it came to the other components of the special right triangle, the analogs of the Face and the massive five-sided pyramid were only *implied* on the Avebury landscape.

Finding the analogs of these components and understanding the other representations of the Cydonia complex had to wait until research for this present book was underway, and it involved traveling to Washington, DC, the capital of the country whose space agency is obsessed with Mars—the United States.

The Washington Connection

Significant Analogs between Mars and Earth and US City Planning with Connections to Scotland

A WASHINGTON, DC, CONNECTION
WITH CYDONIA

Granted, many people have found various hidden geometric forms at work in the layout of the city of Washington, DC, and have written of the care taken by the original designers of the city. Those authors have found relevance in the layout to the cultural, political, and spiritual aspects of the American psyche and the history of the nation. But what about the future? As incredible as it might at first seem, in the layout of the city and the location of its principal buildings there is also a representation of the special right triangle we found on Cydonia. This new finding leads to several questions: Is human consciousness affected more than is currently appreciated by matters such as location and intention? Is it a question of mind over matter? Has the geometric plan underlying this administrative capital culminated in the expressed desire of the United States to be the first nation to get human beings to Mars within the first four decades of the twenty-first century? While these questions are well worth asking, to find answers we have to delve back into the past.

The heart of the US capital city was laid out according to criteria set out by the Founding Fathers, and it is well documented that most of these founders were Masons, immersed in the ancient Egyptian religious and sacred arts. However, if the land of the pharaohs influenced the spiritual aspirations of these founders, the republican model adopted by the Romans and later the French informed the body politic, and the notions of Greek civilization that of the academic mind.

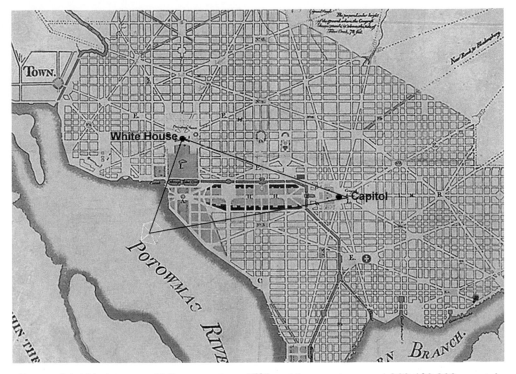

Figure 8.1. Washington, DC, map, circa 1792, with superimposed 30° 60°-90° special right triangle. The 90°-30° diagonal corresponds with Pennsylvania Avenue, extending from the White House to the United States Capitol.

These men rejoiced in incorporating their philosophies into the symbolism of their city plans. They created a grid matrix within which were hidden two-dimensional geometric forms. Certain streets and parks defined these forms, their angles and intersections punctuated by significant buildings or sculptures, the whole animated by those who lived there.

Although it is doubtful that they had thoughts of Mars when they laid out Washington, as you might remember with the case of Isamu Noguchi, it might be that great minds really do think alike. In other words, deploying geometrical exposition to convey philosophical notions produces similar forms across time and space.

When it comes to the parallels between the Martian special right triangle and its unwitting analog at Washington, DC, the Capitol (as the spiral mound) forms the pivotal 30° angle and the White House occupies the 90° Face angle. The analog of the five-sided pyramid should be located in the waters of the Potomac—an impossibility, of course. Yet when land was reclaimed on which to build the city

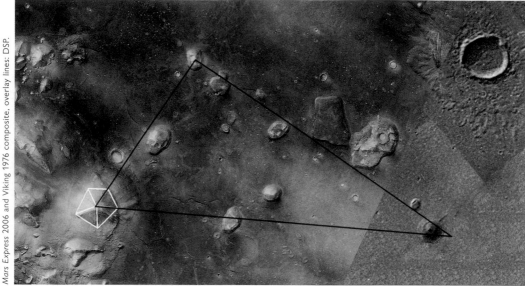

Figure 8.2. Cydonia complex with overlaid special right triangle, incorporating the five-sided pyramid (left), the Face (top center), and the spiral mound (lower right).

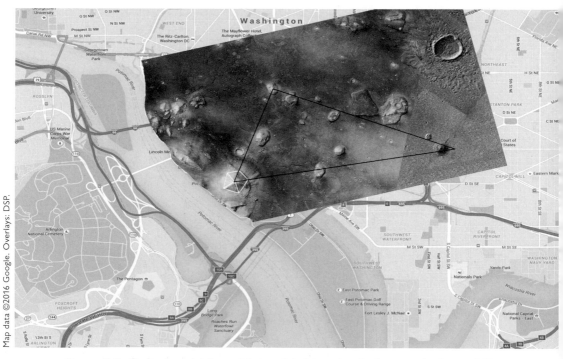

Figure 8.3. Cydonia complex with special right triangle (as seen in figure 8.2) overlaid onto present-day Washington, DC.

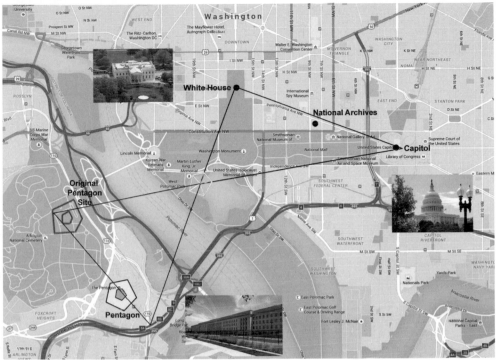

Figure 8.4. The 90°-30° diagonal of the special right triangle parallel to Pennsylvania Avenue, with the White House equaling the Cydonia Face, the Capitol as the spiral mound, and the five sided pyramid extending to a right-angled triangle, locating the originally planned (pre-1941) Pentagon site and the Pentagon location as finalized in 1943.

and a tidal basin was created, this 60° analog angle remained within its waters, conveying the notion of a mythical primeval mound ready to rise from the watery depths and a subconscious connection to Mars. This invisible analog location turns out to be an essential component of the Washington city plan, as will become clear.

STAR GROUPS

Author David Ovason has written extensively on Washington, DC, and the incorporation of zodiacs into its architecture, and he stated that its construction "marked one of those rare events in history when a city was planned and built for a specific purpose."[1] Inferring that the founders of Washington, DC, like the ancients, considered the zodiac motif to be a practical magic tool. The thinking went that linking heaven to earth through design would not only symbolize the transfer of power from above onto the city below, it would actually reinforce and sustain those incorporating the role of governance. Noting the founders' interest in ancient Egypt's buildings and its gods and NASA's determination to associate itself with the Orion constellation to the extent of incorporating the mythology surrounding this constellation into the design of its space projects, one might ask if NASA has not perpetuated customs set out by the administrators and elites hundreds of years earlier in Washington, DC—and thousands of years ago in Egypt.

Ovason considers that the constellation of Virgo primes in the overall starry schemes laid out across the city. His research raises interesting points (particularly for Virgo's association with other goddess figures, including the Egyptian goddess Isis), but the proliferation of zodiacs and various astrology motifs all over the capital's administrative buildings belie the fact that any one constellation was of prime importance.

A zodiac motif may be aesthetically pleasing simply as an art form, but it also affects the emotions of those who observe it according to their personal philosophies and their reason for being in the building or room that it adorns. This motif can also be used as a convenient means of hiding in plain sight esoteric knowledge or memorializing within its associated building a date or an event. Are there any hints of such practices in the city's recent memorials that would confirm such matters? There certainly are.

MEMORIALS

It would appear from the available historical data that important US national ceremonies were held at astronomically significant times for the celebrants, and that is certainly holding true for the space agency's exploits. Just to make the point, the most recent installation of another zodiac in Washington fulfills all these criteria. The Einstein Memorial has been constructed within the grounds of the National Academy of Sciences with a zodiac laid out under Einstein's right foot. It's surrounded by a grove of elm and holly trees to the southwest of the National Academy of Sciences building, itself a neighbor to the Federal Reserve and the Department of State. Einstein holds a paper inscribed with the famous equation from his theory of relativity ($E = mc^2$) in his left hand.

This twelve-foot statue was created in 1979 to commemorate the centennial year of Einstein's March 14 birthday. But the astronomical positions set out on the zodiac and calculated by the US Naval Observatory mark quite another date. A date that seemingly has nothing very ceremonial about it at all, as it merely marks the National Academy of Sciences' *annual meeting,* held on April 22 each year.

There is probably rather more to this choice of date.

Photo: DSP.

Figure 8.5. Einstein Memorial, Washington, DC. Ironically, for a tribute to the stars, Einstein's statue, with its zodiac data, is located in Foggy Bottom.

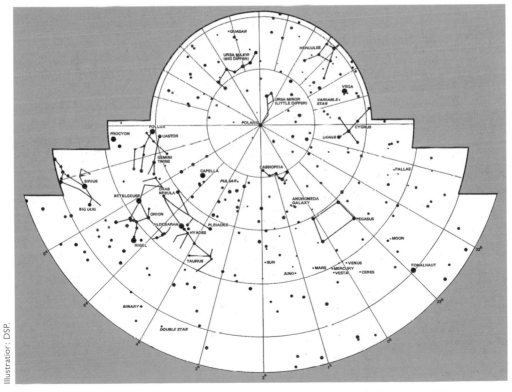

Illustration: DSP.

Figure 8.6. Einstein Memorial Zodiac (simplified version). Re-created from *The Celestial Map at the Einstein Memorial* pamphlet, National Academy of Sciences.

Einstein's theories associated with the concept of space-time are strongly supported by the scientific community, and any challenges to his assertions as to the values of the force of gravity or the speed of light arising from ongoing space exploration are generally explained away. The unity of the National Academy with its preferred scientist is thus encoded in this memorial.

However, the head of the Manhattan Project, Robert Oppenheimer, also celebrated his birthday on April 22. Over the decades, this particular day has been favored for carrying out nuclear tests by the military of many nations, including the Russians (then Soviets), the Americans, and the British, sharing Australian facilities and also US facilities in Nevada.

So that's the past taken care of; what about marking future intentions?

STARGATES

Should the triangular geometry on Mars have been too subtle for many to notice, as we shall discover in the next few chapters, developments in Washington indicate that individual components of the Cydonian layout have indeed been quietly noticed in some quarters and by certain agencies. *Quietly* being the operative word. It is quite remarkable that there has been little comment from anyone, whether NASA, scientists in general, or researchers such as Hoagland, concerning the highly anomalous flat-topped mesa that sits halfway along the Martian triangle's adjacent side. Nor should its shape relative to the Egyptian hieroglyph for a nose have escaped notice, given the city's Masonic roots and NASA's preference for Egyptian references.

Photo: *Mars Express*, 2006. Image: ESA/DLR/FU Berlin (G. Neukum). CC BY-SA 3.0 IGO. Overlay lines: DSP.

Figure 8.7. Flattened "nose" Egyptian hieroglyph superimposed over the flat-topped mesa on the Cydonia complex.

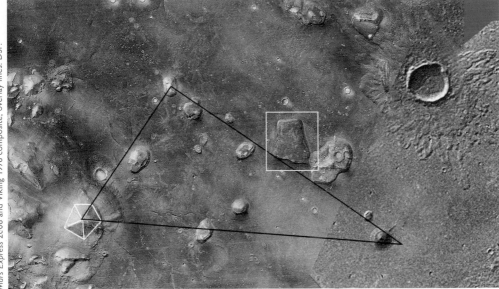

Figure 8.8. Location of the large flat-topped mesa on the Cydonia complex.

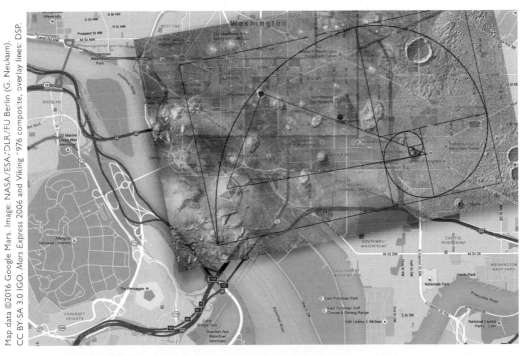

Figure 8.9. The US Naval Heritage Center, on Pennsylvania Avenue, is the analog of the flat-topped mesa on Cydonia. The correspondences between the Cydonia landform layout and Washington, DC, are further supported by the positioning relative to the Fibonacci spiral.

This link with Egypt makes the analog location at Washington a good place to look at next. Was there any activity in Washington after the return of the Cydonian images from the Mariner and Viking photo missions that would qualify as a hidden acknowledgment of intelligent landscaping out on Mars? The adjacent side to the Cydonia/Washington triangle is Pennsylvania Avenue— "America's main street." And fittingly for that Mariner connection, the analog halfway point is entirely associated with US history and the Navy, as is shown in figure 8.9 on page 191.

This location on Pennsylvania Avenue houses the Naval Heritage Center and the US Navy Memorial Plaza in Market Square opposite the National Archives. The center's website describes this location thusly:

> The Naval Heritage Center is located in the heart of the Penn Quarter and offers visitors the chance to learn about the history and heritage of the men and women of the United States Navy—past, present and future.[2]

It is difficult to learn about the history and heritage of the *future* navy personnel—unless one has psychic or ETI capacities. Then again, the US Navy was the first branch of the US military tasked with UFO research. This nautical linking of past, present, and future also emerged in chapter 4 with the 2017 updating of the UFO story, again led by footage from the US Navy. All this

Photo: 2010 AgnosticPreachersKid. CC BY-SA 3.0.

Figure 8.10. US Navy Memorial Plaza.

time-shifting fits with the space and RV programs, along with the beliefs and actions of NASA astronaut Ed Mitchell, who ticks off all the boxes, having been a naval officer and aviator, test pilot, aeronautical engineer, researcher into consciousness and PSI, and a ufologist.

Seen in this context, that word *future* is less of a grammatical error and more of a witting reference to contemporary planning for future space matters. To verify this assertion, it is necessary to go back again in time and dig down into the historical record.

Pierre Charles L'Enfant, the original architect for the Washington city layout, had originally envisioned a memorial in the nation's capital "to celebrate the first rise of the Navy and consecrate its progress and achievements." His aspiration of 1775 did not make it any further than the dream stage for another two hundred years. In 1977, *the year after* the revelations concerning the Viking images of the Face on Mars, the plan was back on the drawing board, reactivated by Admiral Arleigh Burke. Groundbreaking for the memorial eventually took place in December 1985. Ironically, that same month in Nevada, the theoretical physicist Edward Teller was doing some groundbreaking of another sort: he was conducting an underground nuclear test connected with the conversion of the resulting x-rays into antiballistic missile laser weaponry. Two years later, the Zenith Star feasibility study for space-based chemical laser defense systems was set up.*

Whether all these actions and timings were intended for passing ETI or for the secret satisfaction of those indulging in these events is a moot point at this stage, but it will become clearer further down the line.

Back above ground, the Navy Memorial was finally dedicated on

---

*Teller's December 28, 1985, experiment was ultimately intended to supply "X-ray battle stations in space." It was equipped with numerous lasing rods capable of destroying incoming missiles on command before itself being destroyed by its own fireball. Albeit this was said to defend against ballistic missiles launched *from* Earth, one laser eventually emerging from Teller's idea was Zenith Star's alpha hydrogen fluoride laser. It had its start-up on October 5, 1987, with details released in November 1987. The weight of this laser would have been one hundred thousand pounds at launch; by April 7, 1989, it was deemed too expensive, too difficult to operate, *too heavy for the launch capabilities of all extant rockets.* But the program continued. It was slimmed down to ninety-six thousand pounds by May 24–26, 1989, but budget restrictions and shuttle reliability thwarted the DOD desire to use the shuttle to launch the prototype "Zenith Star Laser Battle Station," as David Baker describes it in *Spaceflight and Rocketry.*[3] Yet on June 22, 1991, the prototype was unveiled, then promptly replaced by Starlite, essentially a forty-thousand-pound version of Zenith and capable of launch by a Titan rocket. After which, silence. The Berlin Wall came down in November 1989 and the Cold War officially ended December 1, 1991. One might wonder if the actual direction these laser weapons were supposed to be pointing is not necessarily inward, toward Earth.

October 13, 1987, the 212th birthday of the US Navy. Rear Admiral William Thompson, one of the founding fathers of this particular monument, was sticking with the past, and he had this to say:

> The Navy Memorial is new, but it is rich in tradition and heritage that parallel the history of the Navy and the history of the United States.[4]

While on the subject of the Navy and history, and recalling the Apollo 11 astronauts, NASA has always asserted that Neil Armstrong was a civilian astronaut. Two facts emerge that contradict this story. During his lifetime, the Apollo 11 commander was described as a *Navy* astronaut on the national Navy Memorial. When he died of complications following heart surgery, CNN described him as a former US aviator. His cremated remains were *buried at sea* on Friday, September 14, 2012, from the deck of the guided missile cruiser USS *Philippine Sea*. Interestingly, given that the designated splashdown ocean for the lunar missions was the Pacific Ocean, Armstrong's burial took place in the Atlantic, an ocean reserved by NASA for re-entries from LEO and any failures or accidents occurring shortly after launch.

Returning to Thompson and his reference to rich traditions, he was right. Historically, the October 13 dedication date was the day in 1792 that George Washington laid the cornerstone of the Executive Mansion, now known as the White House and an analog to the location of the Cydonian Face on Mars. However, this October date also marks an even older, more esoteric, foreign historical event of great importance to the Founding Fathers and one that still carries weight today. It was on October 13, 1307, that King Philippe IV of France finally turned on the Knights Templar, accusing them of heresy and bringing about the demise of their organization. The decimated Templars are reputed to have fled France in their ships, many escaping to Scotland, where their memory remains well curated by the Sinclair family in the Rosslyn Chapel near Edinburgh.[5]

This is not simply a question of a French king versus a group of powerful knights. Pertinent here is the fact that the form of Masonry that prevails in Washington, DC, is the *Scottish Rite*. There is also an area called Rosslyn located due west of the White House in Arlington, just across the Potomac. Overtly personified at NASA, the Scottish Rite Masons' flag was incorporated into the Apollo 11 mission, and Freemason Buzz Aldrin made sure that his Masonic ring was visible in the official Apollo 11 crew photograph.

All Freemasons describe their notion of God as the "Great Architect of the Universe" and use the compass and set square to symbolize the creation of the universe. Without necessarily joining with the Masonic worldview, it is certainly

Figure 8.11. Rosslyn Chapel, Scotland, 1852. Located seven miles south of Edinburgh, this fifteenth-century chapel has now become popular as central to the "sacred feminine" plot of the movie made from Dan Brown's book *The DaVinci Code*.

true that learning to draw geometric shapes using a straight edge, compass, and occasionally a protractor and set square, brings its own rewards. Concentrating the brain and contemplating on the drawing in progress can lead to ever deeper insights and questions such as these:

- Does the geometry itself create a form of energy that interacts with the natural elements and the rotational and orbital physics of a planet?
- Does this energetic transfer occur in such a way that essential information encoded in its form is potentially available to any intelligence making that same geometric pattern anywhere at all, or only within a particular region—such as the Goldilocks zone of the inner solar system?
- Does the act of drawing a geometric shape focus our brain waves so that they become not only emissaries of our own thoughts but also potential receivers of other thoughts?
- Does the geometric form used (for example, a circle, triangle, cube, pentagon, hexagon, or octagon) define the nature of the information transmitted or available through contemplation?

- Does the geometric form result from a material response to an audible or inaudible stimulus?

At Rosslyn Chapel over two hundred cubes of stone carved with a series of geometric patterns decorate the arches of its Lady Chapel. These arches are supported by pillars decorated with stone carvings of musicians playing various instruments. Edinburgh musician Stuart Mitchell noted these two facts and made a list of the different geometric patterns carved on these cubes. Mitchell was also familiar with the work of Swiss scientist Hans Jenny. From 1958 to 1972 Jenny had experimented with matters pertaining to wave phenomena and vibration (ta Kymatika or Cymatics). Essentially, Jenny found that the effect of vibration on matter resulted in the material substance arranging itself into recognizable geometrical patterns, and he held that the principle underlying Cymatics was periodicity. Mitchell had found a repeating sequence in the arrangement of the Rosslyn cubes and with his knowledge of Cymatics he turned the Rosslyn stone cubes into a musical compostion that was performed on October 11, 2013, two days before the anniversary of the fall of the Knights Templar.

Image: Austrian National Library Archive.

Figure 8.12. The Great Architect, circa 1220–1230. The compass, indeed, circles in general, was related to the designs of the heavenly realm. The set square, squares, and cubes in general inferred the earthly realm. This knowledge is reflected in Leonardo da Vinci's *Vitruvius Man,* where the center of the circle is the navel, or solar plexus, and the center of the square is the adult genitalia, or root.

Returning to Washington, DC, the analog location of the Egyptian nose hiero-glyph already houses the United State's most important national documents in the National Archives building. The addition of this Navy Memorial decades later is a perfect example of adding to an original plan across time and is reflective of the encoding of different criteria according to the knowledge of those responsible for the planning. Most historians consider that any Masonic influence on the capital's buildings to be irrelevant beyond the eighteenth century. To position and construct two important buildings containing the political and the nautical memory of the nation at such a powerful geometrical hub would suggest this is not the case.

SitRep (Situation Report):

- A special right triangle can be deduced as existing on a planet millions of miles away by dint of its 30° angle marker and the anomalous presentation of the other two angle markers, one of which—the Face on Mars—has acted as a beacon to human beings on Earth and stimulated the desire to get a crewed spacecraft to Mars.
- Mars has been uninhabitable for millennia, and here on Earth the tools required for deep scanning under the surface are only now coming to the fore—at least in the world of archaeology. Within the military, it might be another matter.
- Once the spiral mound's imported geological material is revealed, the subtlety of this particular triangle's placement is immediately obvious. Set between the attractive Martian city and the huge crater with its straight wall, it has not received any official comment from anyone, and that might be precisely because once it is seen, there is only one conclusion to be drawn. The fact is that selecting to create a special right triangle rather than any other sort of triangle demonstrates intelligence and choice. Creating its inferred form through the intelligent placement of an intrusive geological feature and its relationship to two other anomalous structures at Cydonia demonstrates intelligent planning.
- Finding the same hidden geometry in landscaping on Earth as on Mars suggests that there is a connectivity—the thought processes of the creators must operate on similar lines.

THINKING ON

Exactly how the transfer of data and information via thought processes actually works across vast distances of space and time periods might be a reason for this special right triangle's placement on Mars. Once we accept the very idea that ETIs have left us an attractor to get us thinking and wanting to travel to Mars, we will be in a position to expand our own thinking processes in order to do just that. Comparing and contrasting the geometry of Cydonia with the analogs we find on Earth can bring about new ways of thinking about everything, including space travel and ETI

THE COLOR WHITE

It is certainly true that architects design buildings according to strict geometrical criteria and forms and that we illustrate our thoughts and emotions when we use color and textures to decorate them. The White House was so named because the color of the stone used to build it differed visually from the red brick used at the time on other buildings along Pennsylvania Avenue and elsewhere in the city. After the British had set fire to it during the War of 1812, it was restored and the smoke-damaged pale stones were covered with white paint.

However, there might have been an esoteric reason why pale stone was initially selected for this building. In the organization of ancient cities, the buildings were colored according to their status. The houses of priests and administrators were of unpainted baked clay, and only the ruler's house was painted white.

Ancient cities, such as those built at the time civilization was emerging in Sumeria, were built around the genius loci, the place considered the most energetically powerful and thus the precise spot that anchored the heavens above to the land below. This point of origin was generally associated with a naturally emerging mountain or hill, but if it was discovered on flatter land, an artificially constructed mound, ziggurat, or pyramid would be erected on the designated sacred spot. Whether natural or artificial, it was at this perceived power point that the temple was placed. However, the temple in earlier times was not merely a place of worship dedicated to the local deities, it also functioned as a part of the administration and even as the treasury.

The highest mound or building wasn't only the most appropriate place locally for viewing the heavens. The knowledge acquired from learning to scan the skies accurately and their position literally at the top of the mountain made the astronomer-mathematicians the most powerful of all the temple administrators. Even above the ruler. Over time, in some cultures, these two roles were held by one person—the Priest-King, who acted as the link between heaven and earth in the eyes of the people. In other cultures the high priests of the skies became the governors of the spiritual while the political ruler became

Painting by Allyn Cox, "Architect of the Capitol" project.

Figure **8.13**. George Washington laying the cornerstone at US Capitol in 1793, displayed in the Cox Corridors in the House Wing of the U.S. Capitol Building.

the governor of the land, and with the ruler attempting to gain equal status with the priest an eternal rivalry between sacred and secular was established. However, when it came to extraordinary events such as meteorite sightings or eclipses, these sparring partners were united in doing or saying whatever was necessary in order to keep control or manipulate the populace for their own ends. Nothing much has changed in that regard; the parallels with modern governments' policies are obvious, as we saw in chapter 4.[6] And these ancient ways of organizing society, from the astronomer-priests to the ruler and to the people, is another good reason for the visual display of zodiacs throughout the administrative buildings of Washington, DC.

STAR PEOPLE

When archaeologists first discovered the ancient Sumerian sites, they considered that a god had been dedicated to a specific city. Whether this interpretation of the record is accurate or distorted by modern perceptions is open to discussion.

The ziggurat at Ur (now in Iraq) was attributed to the Chaldeans, but the word *Chaldean* means "astronomer," so Ur of the Chaldeans was interpreted as "the place of origin of the astronomers," but this would translate better as the "observatory" or the "lookout."

Today, instead of saying that they don't know, archaeologists attribute a ritual or religious function to any site or artifact not completely understood at the time of its cataloguing. Which is perhaps rather narrow-minded. Also today, as in the past, the high priests of astronomy occupy sacred mountaintops, and in Hawaii, the indigenous population is still upset by the violation of their sacred Mauna Kea by modern-day astronomers. Therefore, it is also possible that those former Sumerian astronomers, in building their own observatories, took over the sacred locations of an indigenous population and, in so doing, engendered the enduring myth (recounted by Zecharia Sitchin) of an invading "star people" who dominated the population through their stellar knowledge and superiority.[7]

OLYMPIAN RACES

At the time of the selection of the Washington city site, not many people were in the know, as few understood why this new capital city was to be set on the swampy, mosquito-infested Foggy Bottom beside the Potomac. Thinking in more esoteric terms and based on the foregoing information, it could have something to do with ancient stellar/lunar/solar/terrestrial energy games—putting into context the obsession of the Founding Fathers with ancient knowledge, mythology, and geometry, and their affinity with zodiacs leading to the prolific use of astronomical symbols in the nation's capital city. All of which would later fit in with the NASA habit of cherry-picking its way through the gods of ancient cultures.

LATITUDES AND ATTITUDES

Washington, DC, is at the same latitude as Delphi, home of the Pythe, the other half of the word Pythagoras. For the Greeks, the agora was the meeting place for the exchange of both ideas and goods. Over time, the market predominated, eventually leading to the covered bazaars, malls, or shopping centers of today. All of which lends deeper meaning to the naming and usage of the capital city's National Mall.[8] Especially because in the domain of ideas it was the combination of Pythe and agora, under the guise of the ancient Greek philosopher Pythagoras, who is credited with passing down ancient knowledge, perhaps acquired during his stay in Egypt, which ordered matter and quantity as four aspects of number:

Mathematics: number as pure abstraction outside of space and time
Astronomy: number in both space and time
Geometry: number in space
Music and harmonic theory: number in time.

Subsequently considered the essential component of the university curriculum in the middle ages, these four disciplines were called the Quadrivium. Of course, as previously mentioned, the Pythagoreans were not above suppressing any newly acquired information when it didn't fit with their philosophy.

The ancient Greeks had also adopted the Egyptian tale of Orion (mirrored in the Greek tales of Alexander's travels to India), and Orion, often conflated with Osiris, had as his consort Isis. The Sun and the Moon were attributed to their persona, as were the Apis bull and the cow. Isis was also portrayed as a woman holding a globe and ears of wheat, symbolizing the fertility of the land and again fitting the iconography of Virgo and of the Earth triad: Taurus, Virgo, and Capricorn.

This circuitous Egyptian heritage finds another anchor, and makes the point that the important connections between America and France go deeper than their common republican values, when the ancestry of the Founding Fathers' Scottish Rite masonry is taken into account. The Virgo/Isis/fertility motif can be seen to relate to the seeding, growth, and harvesting of the ideas and knowledge brought to America via the Scottish Rite Masons but originally acquired from the Templars of France.

Thanks to knowledge acquired in the East, these French Templars are reputed to have influenced the architectural style and the building of the great gothic cathedrals throughout France and England, and author David Ovason asserts that the French cathedrals are laid out according to the star positions of Virgo.[9] This might be to miss a particularly important point.

The greatest of the French gothic cathedrals, Chartres, has a crypt dedicated to the Dark Virgin; that is, to all of those Earth goddesses, especially Virgo and Isis. However, the whole cathedral was deliberately built over a mound sacred to the Celtic peoples. Even more significantly, Chartres is not aligned due east, but instead mimics the layout of Stonehenge, with its northeastern avenue aligned to the summer solstice sunrise. This great cathedral was wittingly built with connections to an ancient sacred site connected to the Earth and to a stone circle reputed throughout the ancient world for the observation of solar and lunar

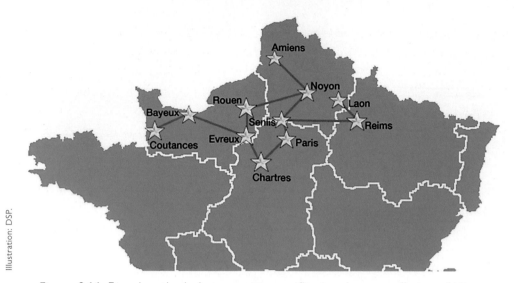

Figure 8.14. French cathedral-star positions reflecting the constellation of Virgo.

Illustration: DSP.

phenomena. Chartres is a jewel in stone and glass set among the wheat-growing fields of the Beauce region, and its two western towers representing the Sun and the Moon are slightly offset to the southwest-northeast axis, thereby producing some interesting dynamics and harmonics within the building. This visual bringing together of earth and sky, along with the hidden resonances and the symbolic and secret connections to the sacred feminine and, as it would later prove, crop glyphs,[10] suggests that the Templars were fully exploring the potential of transferring aspects of their acquired knowledge and philosophy via geology, location, art, and architecture.

THE WATERWORKS

As would the future builders of Washington, DC. The city's geometry and architecture are not simply expressions of local cultural symbolism, they are also astronomical, organizational, and navigational aids. Navigational, how so? Returning to the watery aspects mentioned earlier in this chapter and that missing analog of the Martian five-sider in the waters of the Potomac, this watery location cleverly links itself to the seaside branch of the military by encoding the nautical mileage system in this angle of 60°. According to geomorphologist Erol Torun, the D&M Tor pyramid on Mars has a specific geometry that renders it an imperfect pentagonal form. How fitting; water distorts objects seen through it. The sea is also associated with the subconscious creativity, our inner knowing. And according to the American psychic Edgar Cayce, the submerged lands of Atlantis are strongly associated with the waters off Bimini, within which there allegedly lurks another five-sided pyramid. The terms used for the navigation of space with spacecraft and crew are also directly inherited from the Navy, and deep space beyond our atmosphere is considered in the same terms as the depths of our oceans. The myth analysts at NASA apparently keep a close eye on all these matters; indeed, the naming of a NASA shuttle *Atlantis* says as much.

In fact, bearing in mind the dual nature of Mars as a god of war and a god of agriculture, finding a Cydonian pyramidal form virtually mirroring the architectural form of the US war offices and anchoring the notions of agriculture expressed via the Virgoan symbols scattered across their city, the NASA culture vultures must have recognized all their favorite myths and taken an even closer look. Or action.

Notwithstanding the Washington/Cydonian special right triangle layout, the fact is that a five-sided pyramidal mound has been located in the very region of Mars that has raised the hackles of the US space agency. The fact that it mirrors the very form chosen for the US DOD headquarters way back in 1941—when no-one on Earth had any conscious knowledge of the anomalous region of Cydonia—is nothing short of astonishing.

As such, it's a natural progression to leave the ancient past and come closer to the present. It's time to find out if the seemingly disparate matters discussed in this chapter inform us as to how the original Pentagon site was selected and whether the current Pentagon site also evolved after the imaging of Cydonia on Mars.

9

A Real French Connection

*The Pentagon and the D&M Tor Link to City
Planning in France and on Mars*

PENTAGONS PLACED

In 1976, as far as the public was aware, the five-sided object on Mars was a geophysical anomaly, its purpose, if any, a mystery, whereas the history of the Pentagon is well documented. So, bearing in mind the heavy hand of esoteric symbolism that threads throughout Washington, DC, the process of selecting and building such an odd shape for the US DOD administration building might elucidate useful areas of thought applicable to the contemplation of the Cydonian five-sider (recall figure 5.12, page 113). Thereby, let's further test the hypothesis that one of the purposes of Cydonia is to get us thinking about the transmission of thoughts, data, information, and knowledge across vast distances.

Until we humans can actually get there and see Mars for ourselves, Cydonia will have to come to us, and it does when we explore such obvious parallels as those between two locations some forty-eight million miles apart: the five-sided D&M Tor on Cydonia and the US Pentagon, along with links with their political colleagues across the pond—the French Republic.

PATTERNS ACROSS TIME

The War Department's initial pentagonal design of 1941 is said to derive from the awkward shape and nature of the originally selected site at Arlington Farms, just to the east of Arlington National Cemetery.

Photo: USGS featured on the historical section of Google Earth Pro.

Figure 9.1. An aerial view of the Pentagon taken in April 1988.

Photo: USGS featured on Google Earth Pro. Overlay lines: DSP.

Figure 9.2. Arlington Farms area, 1949.

Map circa 1792. Overlay lines: DSP.

Figure 9.3. Original site for the US DOD building, located on a continuation of the Capitol line. Washington, DC.

The original location selected for the US Pentagon can be found by extending the special right triangle's hypotenuse from the Capitol westward. An extension of side one, from the White House/Face analog southward creates the opposite side of an equilateral triangle. The analog of the Martian five-sider anchors the geometric crossing point between these two different triangles. Astonishing.

Then it was decided that the building would be so bulky that it would harm the views from Arlington across to Washington's nerve centers. The decision was made to move the building to the southeast. Its swampy new location, adjacent to Hoover Field, the former airport for Washington, DC, encompassed the appropriately named Hells' Bottom. There was no longer any site-specific requirement to conform to a pentagonal shape, but by the time the site decision was made, it was "too late and too costly" to alter the construction blueprints. So the original five-sided design was retained. The Pentagon's new position turns out to be three-quarters of the way along the remaining side of the equilateral triangle.

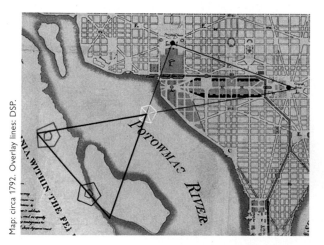

Map: circa 1792. Overlay lines: DSP.

Figure 9.4. Final site for the US DOD building—the Pentagon. Washington, DC.

Photo: DSP.

Figure 9.5. Side view of the Pentagon from the memorial garden, 2018.

In contradiction to the earlier "time and money" excuse for not changing the original form, it was now considered neither too late nor too costly to scale up the existing blueprints of this already bulky building. The resulting mass was to be so much bigger that two levels of this massive concrete building were to be set underground in order to lower the overall skyline.[1]

Once all the blueprints had been finalized, the Pentagon's construction was to be overseen by Lieutenant General Leslie Groves, who was also directing the Manhattan Project. Indeed, in another case of one operation masking another, the overt operation of the Pentagon's construction managed to conceal much of the logistical organization required for the very secret Manhattan Project.

That is the official story of how the Pentagon came to be at its location and how it gained its five-sided form, but it would be naïve to imagine that time and money were the only reasons for retaining a pentagonal structure or, indeed, that the shape of the Arlington Farm site had ever truly dictated the design. For the American people, there is much esoteric symbolism in a pentagram or pentagonal form, which reinforces the understanding that the presence of a huge five-sided landform southwest of the Face on Mars would have struck a chord with those US officials studying the Mariner and Viking imagery.

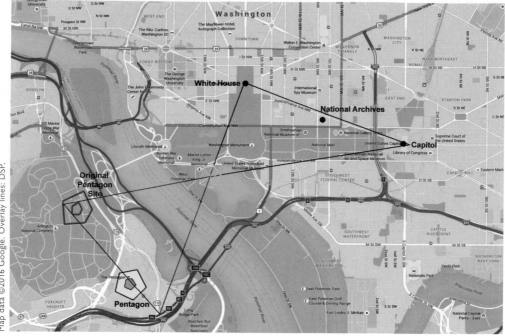

Figure 9.6. Locations of the Pentagon building, the implied Mars location, the original site, and the final site along with the landfill alterations carried out over the last 240 years.

The US military has adopted the five-pointed star for insignia across all of its branches. The most senior military rank is represented by a five-pointed star arranged as a pentagonal group, while the top rank of both the Marines and the Coast Guard is represented by the same five-pointed star arranged as a linear group of four.

Regarding pentagonal or five-sided star-shaped crop glyphs in grain fields, the five-sider crop glyph shown on the next page had a single point set on the diagonal. That's 1/30th of the total number of points in this glyph and the spiral mound. Silbury Hill also has a slope angle of 30°.

Figure 9.7. Five-pointed stars grouped in fours: upper, insignia of an admiral in the US Coast Guard, and lower, the insignia of the US Navy and US Marine Corps, set on the diagonal.

Figure 9.8. The pentagram glyph. Five whole pentagrams and a sixth with a point set out of alignment. Found on July 24, 2000.

PATTERNS IN EUROPE

If pentagonal architecture was especially favored by the US military and its engineers for their administrative building, it was not only for esoteric reasons. Building the national war office in a pentagonal form echoed the military architecture of the European Renaissance of the fourteenth to the seventeenth centuries. Those forts were often called bastilles or bastions.[2] Their five-sided form together with their fortified towers attached to the ramparts had proven the most effective for building defensive castles. Interestingly, the defensive work at the forts' five corners was usually four-sided (tetrahedral), with a sharp point, enabling active defensive and flanking fire. The location of one of these European forts in the western Parisian suburb of Suresnes has very particular links with the administrators of America. The fort itself was only built in 1841, relatively recently in the history of this site. But the elevated ground on which it sits is another story. It has always been recognized as a point of origin. No doubt the genius loci was harnessed by all who used this site, including, sadly, the Nazis during World War II. In the 1970s, I lived nearby and frequently traveled through Suresnes on my way home. The sadness from those times was still tangible, and I found that most Parisians avoided referring to the fort itself, and instead referred simply to the place itself: Mont Valérien.

Figure 9.9. Fort at Mont Valérien, France.

In the Arlington National Cemetery, the Tomb of the Unknown Soldier contains the remains of an American soldier sent over from the American Cemetery for the fallen of World War II, which is located just outside the walls of the fort at Mont Valérien. He lies in his Arlington grave surrounded by the French soil within which he was originally interred. As a rule, the repatriation of the fallen does not include soil from the land in which they died. This is a highly ritualized gesture and even more significant when it is known that President Thomas Jefferson used to take retreats at the monastery that had been on Mont Valérien since the fifteenth century. However, the monastery had only been built in that location because—as at Chartres—in even earlier times the very ground itself had been considered sacred. As already stated, it was a point of origin, the site of the genius loci. The perceived power and/or sacredness of this location puts that very rare transfer of soil from Suresnes to Arlington into context.

It can be construed that the Pentagon's original location also adhered to such hidden philosophies. And comparing the Washington five-sider with the Martian five-sider, one might justifiably conclude that these two, just like Mont Valérien, occupy powerful local energy spots. And as such, any desired symbolic form would also be subject to site-specific constraints, whether geological or merely spatial. This certainly offers a justification for altering the plans at the new location.

Figure 9.10. Aerial view of the five-pointed Fort at Mont Valérien, France.

Moved from its original Arlington site, the Pentagon lost the benefits of a natural power spot allied with its desired geometry. Seeking to recapture and compensate for this loss, the building's five-sided geometry was anchored deeper into the ground at its new location, thereby connecting more strongly with any perceived genius loci while its overall increase in mass provided an imposing visual presence to the onlooker.

This deliberate linking of military forts in France with those in the United States has not been particularly remarked on. Despite the fact that Pierre Charles L'Enfant, a French-born American, had been chosen as the original designer of Washington, DC, any links with France are generally considered to be due to the historical fact that both nations were emerging republics.

Someone must think otherwise, because the Franco-American associations do not end there. Links with the Pentagon, French city planning, and amazingly, Mars continue to this day.

AN ALTERNATIVE CITY

In France, an entire new city and park complex has been built northwest of Paris. Author Philip Coppens penned an interesting article on the process of constructing this city, considering it one of the more secret components of President François Mitterrand's schedule of Grands Projets (Great Works).[3]*

These public architectural projects were so grandiose and such was Mitterrand's interest in Egyptian iconography that he was variously dubbed "the Pharaoh" or "the Sphinx" by the French media. The new city was ostensibly designed to accommodate a burgeoning population while linking it to Paris via the engineering of an enormous terraced park oriented toward but not quite in line with, the principal east–west Parisian axis.[4] Starting from the Louvre (I. M. Pei's glass pyramid is another of Mitterrand's Great Works), this Parisian axis proceeds via the Place de la Concorde (with its obelisk) up the Champs Elysees to the technical and administrative district of La Défense. It then continues straight on to the island of the Impressionists in the middle of the Seine. From there it turns through 30° to align itself with the new city of Cergy and its new leisure spaces and park. The whole enterprise was to be known as Cergy l'Axe Majeur. As it is principally the layout of the new town's architectural and landscaped park that is of interest, for the sake of the reader this project will be referred to here as the Axe Majeur park.

Back at La Défense (the towers of which can be seen from the higher altitudes of the new city), and set slightly askew in order to frame the sunset for the Parisians, La Grande Arche (yet another of Mitterand's Great Works) is the ultimate land monument on the Louvre sightline and its penultimate end point. It also has a hidden zodiacal connection to the new city.

This Grande Arche de La Défense is of such a size that it could accommodate the cathedral of Notre Dame, a cathedral dedicated to the virgin Mary that sits on the island that was the point of origin of the city, Par Isis one might say, since the Christian image of mother and child is also attributed to Isis and Horus. Interestingly this monumental building also mirrors NASA's *Saturn V* large Vehicle Assembly Building in Florida.

---

*Mitterrand's Great Works officially consisted of eight monumental building projects that transformed the city skyline. These were the Bastille Opéra, the Musée d'Orsay, the Arab World Institute, the Ministry of Finance, the Bibliothèque Nationale de France, the Louvre Pyramid, the Grande Arche de La Défense, and the Parc de la Villette.

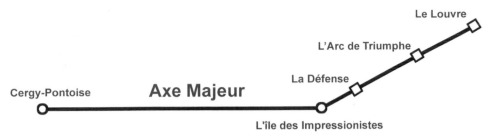

Figure 9.11. Linear layout and sight lines for the Axe Majeur park at Cergy, France. Note that this illustration reveals that these two principal sightlines are in the phi ratio 1.618. Wherein going from east to west, the Parisian Louvre to the Island of the Impressionists is the smaller part. Then going west to east, Cergy to the Island of the Impressionists is the smaller part of the sum of both lines.

Figure 9.12. La Grande Arche, Paris, France.

Figure 9.13. Vehicle Assembly Building, Kennedy Space Center.

The Axe Majeur project at Cergy matches the Cydonian anomalous complex at vastly different scales. The twelve-component landscaped park is located sufficiently far from the capital to become of itself the analog of the whole Cydonian city. When it's geometrically linked to the principal monuments of the Parisian capital, significant monuments match locations on the west–east Cydonian sightline, so that the Axe Majeur park, the Grande Arche gateway at La Defense, and the glass pyramid at the Louvre act as components of the west–east Cydonian sightline. And at this particular scaling, relative to the Tower component of this new city, the Fort at Mont Valérian is at an approximation of the D&M Tor location. Then the Grande Arche (or perhaps it should really be "the great architect") becomes the analog position of the Cydonian Face. Although it's true that *architecture* is the buzzword (no pun intended) when describing space hardware design programs, that's some new town planning committee.

- No surprise then, that the new French park is known as the Axe Majeur.
- No surprise that the roof of Grande Arche is laid out as a twelve-constellation zodiac. It was initially opened to the public but very quickly made inaccessible.

- No surprise, either, that the new city's park has a total of twelve elements along its west–east sightline and that the top of the tower marking the starting point has a compass inscribed into the roof's edging stones. This rooftop was also briefly opened to the public but then made inaccessible.

Plus ça change, plus c'est la même chose.

All things considered, this new city build looks to be a combination of matters practical and philosophical, cleverly incorporated into the layout and into the timetable of construction. Although Coppens didn't make the link to Cydonia, he did think that there had to be a reason, other than financing, behind the stop-and-go construction of the twelve elements composing the park and that there was some process waiting to happen before the final elements of this park could be completed.

Looking at this development and conscious of Mars exploration, it is glaringly obvious that some sort of messaging is going on here. Whether between the European space agencies and NASA or whether intended as a message to ETI, the twelve components of the Axe Majeur park echo those twelve components of the northwest pyramidal complex known as the City at Cydonia. And this Axe Majeur park layout becomes a miniature version of the Cydonian sightline from the city center in the west to the large crater in the east.

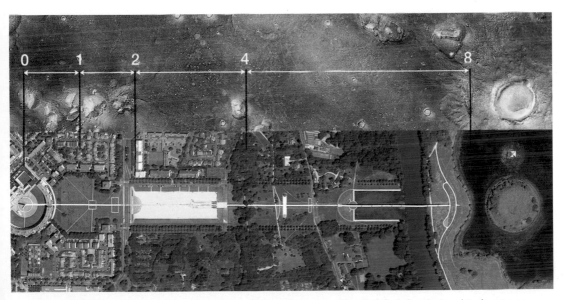

Figure 9.14. As above: Mars. So below: Cergy Pontoise. The 0-1-2-4-8 relationship between Cydonia structures and the Axe Majeur park, near Paris, France.

Above: *Mars Express*/ESA/DLR/FU Berlin (G. Neukum). CC BY-SA 3.0 IGO. Below: France: 49°02'54"N 2°02'08"E, Google Earth, 2020. Overlay lines: DSP.

The hypothesis that there is more to this project than the mere expansion of the French infrastructure might seem a little far-fetched for some, but the links between the exploration of Mars, the layout of Cydonia, and this new city and its park are substantiated by the time lines. The French project was initiated in 1975, while the Viking mission was traveling to Mars. And as space travel is the issue, the name of the principal community in the region where the new town was to be sited becomes more than apposite: it is St. Christophe. The principal legend of this third-century saint recounts his act of kindness in carrying a stranger—a small child—across a river. The child then revealed himself as Christ. The origins of this legend are uncertain. Nonetheless, St. Christophe has become the patron saint of travelers for many people of various faiths. With the construction of the new town of Cergy and the Axe Majeur, the ancient community of St. Christophe became the hub around the newly constructed Giant Clock (Place de l'Horloge).

At the time of writing, this giant clock has the largest clock face in Europe; with a diameter of 10 meters, it links space to time, and a circumference of 31.41592654 meters makes it a memorial to pi. It is really two transparent clocks supporting each end of a transparent roof that protects the entryways to the region's major transport link. This design is aesthetically charmless, but it much beloved by the local pigeon population. (It cost some 25 million Euros to clean up in 2016.) The clock faces, designed with twenty-four radials overlaid with twelve Roman numeral hour markers, also operates as the start of the minor axis connecting the community to the first element of the Axe Majeur park. In Roman terms, this minor axis is the north–south Cardo relative to the east–west Decumanus of the Axe Majeur. Notwithstanding the interesting usage of Roman city (and fort) design practices, adding this element to the twelve stations of Cergy Park brings the twelve around the thirteenth. From biology through astronomy, religion, mythos, and politics, this is a theme that resonates throughout our cultures. The Arthurian legends are a particular relevant example of the twelve around the thirteen. Modern authors Stan Gooch and John Michell also examine the roots and ramifications of this theme. Relative to this discussion, the placing of this clock at a transit hub links time and motion to this thirteenth element of the Cergy project. While the central number of 13 also recalls the European Union's thirteenth administrative floor in the Brussels headquarters, the thirteenth constellation of the zodiac, the thirteen original US states, and so on. Adding another layer to this number game, the twelve elements of the clock face are repeated in the twelve elements along the sightline, and having already got 12 + 12 = 24 on the clock face of the Cardo axis, multiplying that (12 × 12) produces 144. In esoteric symbolism, 144 is associated with light. By now it should come as no sur-

Figure 9.15. L'Axe Majeur park, the red footbridge looking back toward Cergy, with the long flight of steps leading up to the park's plateau.

prise that a laser projected eastward from the Tower toward the eleventh element is designated as the twelfth element of Cergy Park. Currently it is a green light.

The layout of the park itself began in 1980, and big surprise, in 2020 it is still not finished. In fact, a visit in 2018 revealed that much of it is now falling into disrepair while construction is currently halted at the end of the eighth element, the Passerelle, a red footbridge/flyover that from the air looks like the shaft of an arrow– one that has not yet reached its target.

Directly opposite this metaphorical flight path, in the lake sits the ninth element of the park and it is a round empty island. Located at the halfway point between the first and the eleventh element, this island is eventually intended to be filled with astronomical instruments, but in the meanwhile its artificially flattened surface makes it look a lot like a small version of England's Silbury Hill and therefore a clone of the Cydonian spiral mound. Silbury Hill's flat top at 30 meters (98.42 feet) is the same diameter as the sarsen ring at Stonehenge and at 100 meters (328.08 feet), this island's flat top is the same diameter as the Stonehenge earthworks. Parallel to this island and to the left, and accessible only by boat, there is a four-sided stepped pyramid. It has been designed so that one can navigate into the blue-painted interior and note some very interesting architectural detailing. Photos on the Cergy website of the pyramid under construction

Photo: DSP.

Figure 9.16. The Lake Pyramid is the tenth element of this Axe Majeur park and offset from the sightline. It is composed of 177 stacked concrete slabs and measures half as high as its base side of twenty meters. Expressed in feet, that base number of 65.616 is a number that recalls cartographer Carl Munck's grid point for the Face on Mars.

reveal that items placed on its stepped sides were all symbolic of Masonic, alchemical, and philosophical principles.*

The eleventh component of this complex (situated on land just east of the lake) was originally symbolized by the astronomical symbol for the planet Mercury. This has since been removed. The actual stop sign at the end of the complex now consists of a metal plate looking like an *H*, and *H* can stand for Helicopter pad, if you're at the Pentagon; here, at Cergy, it can signify Hermes, the messenger—another version of Mercury. Yet even this eleventh component of the park is considered to be unfinished, and the *H* plate also a temporary feature. As it stands, this *H* plate is designed to receive the twelfth element, that green

---

*At the time of writing this chapter, there was an official French website with photos of all stages of construction for this Axe Majeur development at Cergy. Access was checked several times prior to publication, these images seem to have been taken down now. Nevertheless, the reader is encouraged to try entering keywords such as "Cergy l'Axe Majeur" or simply "Axe Majeur" because, as with NASA sites, things can pop up in unexpected places.

laser. Should the red footbridge infer the road to Mars, currently inaccessible to space agencies, it will be interesting to see when, if ever, the island is completed and connected to the footbridge.

On the other hand, should this park reflect the desire to communicate with ETI, then in Eastern understandings, red is the color of cinnabar. Esoterically, cinnabar represents the lower dantian, translating as "the field of the elixir of life" or "sea of *qi* [energy]." This dantian, located just below and behind the navel, is considered to be one of three significant energy hot spots in the human body. In the past in China, those seeking immortality sought to reinforce their natural reserve of qi by drinking small doses of cinnabar. As cinnabar is also a toxic form of mercury, this had the opposite effect, generally hastening the death of the imbiber. The rivers of mercury found under the Chinese Emperor Qin's pyramid and that of Teotihuacán's Quetzalcoatl pyramid symbolize this notion of immortality. Which makes the red-painted Passerelle at Cergy even more significant. As does the fact that cinnabar is often found near hot springs, also present at the Cergy site.*

On the color spectrum, red is opposite green, which in 2018 was the color of the Cergy laser. Of course, both red and green are connected with the red planet and again that hint of ETI, since the term "little green men" (LGM) is a common epithet for Martians. Remember the LGM that Google Mars placed on the image of the Face on Mars. Note that Russell Targ, one of the first RV researchers, is also a laser expert. Consider the green laser light across a red suspended bridge to a far-distant Mercury sign. The message could not be clearer.

The fact that terrestrial or military history links between these two nations have been quietly encoded through multilayered symbols since the founding of their respective republics is one thing. Planning a landscape in 1975 that unavoidably invites comparisons to major aspects of an ancient Martian landscape imaged by *Mariner 9* only three years earlier is surely a massive subconscious coincidence on the part of these two spacefaring nations. Or it is all entirely intentional?

---

*The other two significant dantian spots that further refine the energy of the being are located at the heart level and between the eyebrows. In the Taoist system, the lower (qi) dantian is considered to be the root. In the Indian chakra system, the root is at the base of the spine. The middle (*jing*) dantian, "the Crimson Palace," relates to the thymus gland, not to the heart itself, while the upper (*shen*) dantian, "the Muddy Pellet," corresponds to the pineal gland, and that is the chakra system's third eye. These notions of energetic connections within the human body pertain across all cultures. The Muddy Pellet brings to mind the Egyptian images of Horus the Hawk meditating while sitting on a lotus. The lotus itself rose from muddy waters, and that cross-references to the lotus position of Indian culture used for meditation purposes.

Furthermore, since one could conclude that these visual messages are intended for observation from above, the attitude of the space agencies and other authorities toward ETI might be radically different from any pronouncements made in public on the matter of UFOs and their putative occupants. Whether or not the predictions of the *Brookings Report* on the consequences of finding signs of intelligent design on other planets have played their part, it is quite likely that responses to Cydonia have been affected by covert policy decisons regarding ETI.

MATCHMAKING

NASA's Viking project scientist Gerry Soffen might have been obliged to quip away any visible evidence of ETI based on the images returned from Cydonia, and SETI director Seth Shosak's abrupt dismissal of the Cydonia landforms might also have been a required response designed to deflect attention from any contact experiments that might be underway. Remembering the Osiris-Orion connection beloved of NASA and that Isis was the consort of Osiris, it certainly makes mythical-historical sense to establish this version of Cydonia near Paris.

This French park could fulfill several agendas: It could serve as yet another link in the partnership between these two republics, this time via their respective space agencies: NASA and the ESA. When viewed from above or on a plan, the park could be a silent visual acknowledgement of the Cydonian decumanus to observers, ETI or otherwise. Last but not least, it could be considered a useful tool in getting the public to absorb the opinions of these agencies merely by walking and experiencing the Axe Majeur park and its environs. But has this triple approach worked?

MIRRORING

Upon reading the official website, there is the sneaking suspicion that the whole park has been designed to create an experiment—with people, time, and space. One gets teasers about the symbolism of this artificial landscape, so clearly, there is a hidden depth to this design for those in the know. The public is given heavy-handed hints and then told to decide for itself what it all might mean. The promotional photographs of the location that accompany these teasers are very impressive, and the park looks like a wonderful place to explore.

Sadly, on the ground, it's a different matter. When actually walking it, moving through this grandiose architecture is a sterile experience. And that's because despite all the careful planning, behind any project the intention is everything.

Here it feels as if people truly are of marginal importance to this project: the last shall not be first. This place has provided no heart for the people who inhabit it. First you lose heart, then you die. Which is perhaps the reason why this park's elements suffer from graffiti, weeds, and malfunctioning components.

Those visiting the Axe Majeur at Cergy are asked to make their own minds up on this mostly esoteric site, so here are further observations on this architectural endeavor. As a part of Mitterrand's Great Works, there are other interpretations of this park's twelve elements (for example, Masonic symbolism is also in evidence across these Great Works), and these can be usefully studied alongside this space agency interpretation. I find the links between the five-sided Fort at Mont Valérian to the Pentagon and a newly built city imitating a layout on Mars significant. That this new city's park is equipped with a light that is also a weapon is also important. A green laser targeting a symbol attributed to the Hermetic messenger god Mercury leads to questions as to whether this display is supposed to be peaceful or whether aggression and defensiveness are the message. The manipulation of light also brings up the observation that in the eastbound Axe Majeur sightline, an astronomical island is the penultimate physical marker. A hidden notation for Mercury the end point of both the laser and the name of NASA's first human piloted space program. For the Parisians, the Grande Arche is the penultimate physical marker and the Island of the Impressionists is the end point, the switching point. Both astronomers and artists are concerned with the understanding of light, especially as it concerns time. The astronomer's island is empty while the Grande Arche has the zodiacal constellations hidden on its rooftop. The switching point of these two sightlines is associated with the French Impressionists, who were trying to capture light's changing qualities and to convey the passage of time by using thin brushstrokes to give literally an impression of a fleeting moment seen and captured on canvas. The analogies between these two locations and the apparent lack of desire to furnish the astronomical island and connect it to its park leads one to ask some questions: Is this is a reflection on our own inability to manage crewed space travel to the red planet? Do we even have the right technology to achieve our aims? Or is it all about silent messaging, ETI Communications and the perceived limitations of the speed of light?

It's time to head back across the pond to see what exactly *was* NASA's official attitude toward ETI contact.

SETI, CETI, or DETI:
That Is the Question

*Searching for Evidence of Intelligent
Life beyond Earth*

Officially, NASA started funding SETI in the months just prior to the August 20, 1975, departure of *Viking I* to the red planet. Financial planning is not generally undertaken overnight or on a whim, so no doubt it is a happy coincidence that the overt funding of an organization specifically looking for ETI occurred after the successful 1971 arrival of *Mariner 9* and its 1972 photo-fest of Mars—especially of Cydonia.

SETI states that in 1975, NASA was at last able to fund the organization because there had been a fundamental change of opinion within both the scientific and academic communities as to the validity of the hypothesis that ETI might exist.[1] No mention was made as to whether this change of opinion had occurred overnight or was the result of events dating from years earlier. Or even decades, since this 1975 start-up date conveniently excludes troublesome events such as: The inadequately explained the 1908 Siberian Tunguska crash or unexplained airburst with its legacy of radiation burns, the aerial light phenomena of the so-called foo fighters seen by all nations' air forces during World War II,

Figure 10.1. SETI Institute logo.

the observations and conclusions of Fermi and his colleagues at Los Alamos, the 1947 Kenneth Arnold UFO sighting in Washington State, and the New Mexico Roswell event of that same year. These receive no mention at all.

In order to ascertain how the official opinion that ETI might actually exist had radically changed from "highly unlikely," the view held in the early decades of the last century, to "highly likely" by 1975, we have to scroll even further back in time—to the year that NASA historians mark as the moment that the hunt for intelligent extraterrestrial life began.

The origin story for the searching for signs of ETI goes back to before the founding of the SETI Institute, as it is known today. The story goes that in 1956, a twenty-five-year-old astronomer at the University of California, Santa Cruz, pointed a twenty-five-foot radio telescope at the Pleiades, 440 light years from Earth, and saw two spikes on the read-out that should not have been there. It was a false alarm, but apparently Frank Drake had been alerted to the possibilities held by the emerging radio telescope technology.

Three years later, in 1959, Giuseppe Cocconi and Philip Morrison of Cornell University were visiting CERN near Geneva, Switzerland, during which time they wrote their seminal paper "Searching for Interstellar Communications" and proposed that the radio frequency of 1420 megahertz (corresponding to the twenty-one centimeter line of neutral hydrogen) was the best frequency to search for signals from extraterrestrials. Their article was published in *Nature* on September 19, 1959, and the sound and light show had begun.[2]

As Morrison said of the proposed Search for ETI:

> The probability of success is difficult to estimate but if we never search, the chance of success is zero.[3]

Morrison had been at Los Alamos on the Manhattan Project, and he later contributed to the secret wartime project for the recuperation of all nuclear research documents held by some fifty physicists and scientists in Germany, France, and Italy. Set up by the very busy General Groves in 1943 this task was accomplished by May 1945.[4] Wittily, but to the displeasure of Groves, the project had been called "Alsos," the Greek name for a grove of trees. Equally wittily, thirty years or so later, during the Viking years of 1975 and 1976, Morrison's co-author, Cocconi, was again working at CERN when, coincidentally, a Swiss farmer, Eduard "Billy" Meier (a man who would become equally concerned about nuclear warfare), was discovering flattened circles of grass in a forest clearing (surrounded by a grove of trees, one might say).[5]

The Cocconi and Morrison paper didn't please everyone. Over at the National Radio Astronomy Observatory (NRAO) in Green Bank, Virginia, Drake (by then an alumnus of both Cornell and Harvard) was preparing another ETI experiment. According to Drake, the genesis of this new search had taken place in the winter of 1958, the construction of the necessary hardware was going to take some thirteen months, and they were only six months into this project when the Cocconi and Morrison paper appeared. Although most scientists are naturally competitive, at Green Bank, they were keen to embrace the great minds think alike approach, but director Otto Struve was said to be agitated and frustrated at the appearance of this paper suspecting it would take credit (and therefore future funding) away from the NRAO. Struve went into major PR mode, and in October 1959 announced the future Green Bank ETI experiment to a conference at the Massachusetts Institute of Technology.

With hindsight, given that the NRAO was the country's national observatory, Struve's funding worries were largely unnecessary. The intention to discretely alert other scientists to the instigation of SETI, first via the media and then directly via a conference platform was a far more plausible reason for this preemptive strike.

OZ MAN

Drake and his colleagues conducted the next search for ETI from April to July 1960. Sticking with the tradition of attaching witty but relevant fantasy titles to serious space projects, Drake called it Project Ozma. For those unfamiliar with American fiction, this was in homage to another Frank—author Frank Baum and his Princess Ozma of the Land of Oz, a place "very far away, difficult to reach, and populated by strange and exotic beings," as Drake put it. Project Ozma apparently didn't produce any results, although the notes by Drake on the matter make it even clearer that PR for SETI was one of the objectives of the experiment. As for that name, he might have been making an even wittier in-joke, referring to the new telescope facilities for NASA's Deep Space Instrumentation Facility (DSIF-41), which were under development in Australia during the very same months Drake's experiment was taking place. This facility in Oz was at Island Lagoon, just outside the British missile test site at Woomera (310 miles northwest of Adelaide). DSIF-41 functioned from 1960 to 1963 and was renamed as DSS-41 from 1963 to 1972. After which the eighty-five-foot telescope was dismantled and sold for scrap.

However, something is still operating at that location because Google Earth in 2015 showed the facility hosting a "golf ball" geodesic dome similar to those at the US facility of Pine Gap, near Alice Springs, Australia.

Photo: DOD Australia.

Figure 10.2. Domes at the Pine Gap facility southwest of Alice Springs, Australia, 2016.

Established during the 1960s after secret talks between the United States and the Australian government and following geological surveys by the USGS, the site was selected by late 1964, early 1965. The treaty itself was only signed officially in 1966. The facility was partly operational by 1969 and fully operational by 1970. Run by the CIA and other alphabet agencies as a US satellite surveillance base titled the Joint Defence Space Research Facility, to the public it was simply involved with space research. Its first code name was Merino.

Back at Green Bank, Drake's Project Ozma was pointing the eight-five-foot telescope at the star Tau CETI (another play on words from the space boys when choosing acronyms) in the constellation of Cetus, the whale. Result: nothing. That afternoon, the telescope was pointed at Epsilon Eridani, the thirteenth star of the constellation Eridanus, the river. Bingo! An anomalous signal. Then this:

> Suddenly I realized there had been a flaw in our planning. We had thought the detection of a signal so unlikely that we had never planned what to do if a clear signal was actually received.[6]

Drake then recounts the discussions that followed as to whether the signal was from Earth or from the star. However, two paragraphs previously, Drake

had informed the reader that the gear installed for this exercise included hardware *specifically intended to separate out the incoming signal from space and any
Earth-based interference.*

The signal did not return for a week or so. In the meantime, to the chagrin of
Drake, news of this potential ETI signal had leaked to the media via contacts in
Ohio. (Those familiar with NASA-speak will see a pattern here.) The press went
wild; the observatory, suddenly coy, refused to comment. Much drama. An additional filter was installed. The signal returned, and lo, it was coming from much
nearer Earth than Epsilon Eridani. Why the existing filters had not earlier made
the distinction was not explained. The anomalous signal was put down to radio
communications emanating from a secret military project. Whether the NRAO
was in the loop is another matter, but it reads more like a DOD exercise to see
how both the military and the NRAO performed. Although flying a secret operation over the NRAO, whose activities were announced well in advance, was not
going to be a very secret operation as far as the military went.

As to the amount of chagrin that the "leak" really caused, take a pinch of salt
from that pot, because Struve's earlier efforts at PR had made Green Bank's search
for ETI perfectly clear, at least to the scientific community. This leak now made
it clear to the public that ETI was potentially out there. The detection and resolution of the mystery signal also clarified to everyone that the military was in charge.

Another year, another declaration. In 1961, the Space Science Board of the
National Academy of Sciences in Washington, DC, sponsored a meeting at the
NRAO. It had four objectives:

> To examine the prospects for the existence of other societies in the galaxy
> with whom communications might be possible; to attempt an estimate of their
> number; to consider some of the technical problems involved in the establish-
> ment of communication; and to examine ways in which our understanding of
> the problem might be improved.[7]

This statement did not exactly define whether the understanding of the problem referred to the hardware or the presence of extraterrestrials in the galaxy,
but in the following year, 1962, a new exobiology laboratory was established at
NASA's Ames Research Center (N-ARC, another fun acronym for the cogno-
scenti), and John Billingham moved across to Ames in 1965. Trained as a medical
doctor, he had previously headed up NASA's Environmental Physiology Branch
in Houston, working on the Mercury, Gemini, and Apollo programs. At Ames,
he first headed up the Biotechnology Division, then as Viking was sending back

its images in 1976, he moved to the Extraterrestrial Research Division and later the Life Science Division. A senior scientist and on the board of trustees of SETI, Billingham has been inducted into NASA's Hall of Fame for his efforts as "the Father of SETI." The media called him "the seeker of ET," which is interesting, because the titles used by NASA cross the boundaries between true extraterrestrial research and the biomedical research for humans and the animal research then being undertaken prior to human space flight. Other NASA documents also refer to an exobiology laboratory at Ames, although whether it's the same as the Extraterrestrial Research Division is not clear. Confusingly, the term *exo* means different things to NASA. Attached to *biology,* it means the presence of living organisms indigenous to any planet other than Earth but including this solar system. Attached to *planetary,* it means an organization of planets orbiting a sun other than our own.

FUNDING CONFUSION

It is acknowledged in the NASA documentation that although SETI (or more prosaically, "interstellar communication") was just a small element of NASA research projects back in 1974, the subject was of intense interest to the general public.[8] This is disingenuous, as interstellar communication with ETI would be the most significant thing to happen to human beings. Therefore, "the general public" actually means "all of humanity" and hopefully, should include the agency itself.

This NASA document, when examined closely, also revealed that notwithstanding the Mercury, Gemini, and Apollo programs, the agency had been involved at some level with ongoing research into ETI since its inception in 1958. However, in yet another confusion, and a good example of the mismatched document modus operandi, chapter 1 of its publication on ETI states that SETI funding officially started in August 1974, whereas the funding chart in the appendix of the very same document commences much later, in the 1975 financial year. Oh dear.

Playing down the matter of ETI research, it is stated that the SETI group at Ames Research Center received what was called a miniscule budget of $140,000, funded from the Office of Aeronautics and Space Technology. Bearing in mind that Ames already had a laboratory working on matters exobiological, that's not quite as sad is it sounds, and if that acronym TOAST is tempting, it wasn't going to be toast. In 1977, JPL joined Ames as an active SETI partner. In 1978, the very person who had dismissed the Face images as a mere trick of the light two years earlier, Gerry Soffen of Langley Research Center, transferred to NASA headquarters as the director of Life Sciences, a broad title that included exobiology.

Chart: Jens Feeley, NASA HQ Space Science, June 1997, HRMS.

SETI Area Funding ($K)	FY75	FY76	FY77	FY78	FY79	FY80	FY81	FY82	FY83	FY84
SETI Microwave Observing Project	140	310	400	130	300	500	1895	0	1800	1500
Definition/R&D	140	310	400	130	300	500	1895	0	1800	1500
Program/ Project C/D	0	0	0	0	0	0	0	0	0	0

SETI Area Funding ($K)	FY85	FY86	FY87	FY88	FY89	FY90	FY91	FY92	FY93	TOTAL
SETI Microwave Observing Project	1505	1574	2175	2403	2260	4233	11500	12250	12000	56875
Definition/R&D	1505	1574	2175	2403	0	0	0	0	0	14632
Program/ Project C/D	0	0	0	0	2260	4233	11500	12250	12000	42243

Figure 10.3. Funding history of the NASA SETI Program.
Note that in the financial years FY92 and FY93 all monies were allocated to
NASA's High Resolution Microwave Survey (HRMS).

This was also known as the astrobiology program, for which read SETI. All of this *officially* NASA-funded SETI program of the 1970s ran parallel to a significant portion of Mars exploration, starting with Viking and ending with the *Mars Observer* fiasco, of which more later. In fact, when it came to agencies accounting for themselves, things were not straightforward in any domain. The establishment of this triad of Ames, JPL, and NASA headquarters engendered some rivalry as to which group did what and when—and for how much.

The High Resolution Microwave Survey program was specifically intended to look for signs of ETI. From October 1992, Ames worked out of Puerto Rico's Arecibo Ionospheric Observatory, and JPL worked out of Goldstone in the Mojave Desert. Just a year later in October 1993, Congress instructed NASA to discontinue these searches. Mostly thanks to the efforts of Nevada Senator Richard Bryan who famously said "The Great Martian Chase, may finally come to an end. As of today millions have been spent and we have yet to bag a single little green fellow. Not a single Martian has said take me to your leader, and not a single flying saucer has applied for FAA approval."

If the funding details were confusing, that was nothing compared with the identification of these SETI programs, which were often deliberately mislabeled. It was recommended that inconvenient funding applications, those liable to be rejected by Congress or other critics as unfeasible, could feature a different project name in order to hide their true nature. This also applied to serious projects likely to acquire *unscientific labels from the media*, such as "seekers after little green men." Never mind the fact, as already noted, that scientists were the first to use the LGM label and that now Google flaunts these same LGM across their Mars maps.

In British jargon from the world of theater, such a project has been "bowler hatted": operating under a different label, the project continues to be funded but escapes the notice of all those who might endanger its existence. In military terms, it's "gone dark." One wonders if this principle applies to just about any "inconvenient" aspect of a space exploration project.

The hiding of a program within a program might explain why even today every article about exploring the galaxy for biomarkers includes only a small mention of SETI. To illustrate the point: as a part of the NASA PR summer efforts, the July 2014 issue of *National Geographic* featured a front cover showing Jupiter and its moons, with a headline shouting "IS ANYBODY OUT THERE?" and subtitled "Life beyond Earth." This is a very attractive cover and fully engaging on the notion of ETI. The twenty-page spread opened with the statement that "one of the oldest questions may be answered within our lifetimes. Are we alone?" So far, so good. Then, surprise. An illustration titled "Listening for Life" included every kind of sensor except those used by SETI, and its text exclusively referred to environmental sensors. Of the 458 lines of text devoted to the search for life, SETI was allocated just 7 percent. And these thirty-two lines of text cited Drake's eponymous equation, which he had designed in 1961 to establish the specific factors inherent in the emergence of ETI civilizations.[9]

$$N = R^* \cdot f_p \cdot n_e \cdot f_l \cdot f_i \cdot f_c \cdot L$$

N – The number of civilizations in The Milky Way Galaxy whose electromagnetic emissions are detectable.

R^*– The rate of formation of stars suitable for the development of intelligent life.

f_p – The fraction of those stars with planetary systems.

n_e – The number of planets, per solar system, with an environment suitable for life.

f_l – The fraction of suitable planets on which life actually appears.

f_i – The fraction of life bearing planets on which intelligent life emerges.

f_c – The fraction of civilizations that develop a technology that releases detectable signs of their existence into space.

L – The length of time such civilizations release detectable signals into space.

Figure 10.4. The Drake Equation as it was presented in 1961.

This July 2014 article further stated that it was the 1961 SETI conference at the NRAO in Green Bank, Virginia, that had kickstarted the current astrobiology research (not entirely accurate), but that by 2014, SETI funding had mostly dried up (also not entirely accurate). However, forget Mars, this article reinforced in the reader's mind that the focus for discovering astrobiology and potentially, ETI lay with finding exoplanetary systems capable of supporting life. *Exoplanetary,* as noted, describes planets outside the solar system.

The same year as this *National Geographic* effort, NASA's Public Outreach Division produced a 330-page document edited by SETI astrobiologist and psychologist Douglas A. Vakoch and titled "Archaeology, Anthropology, and Interstellar Communication." Rumors later abounded on the web stating that NASA had removed this document from its website due to the inappropriate LGM label. On past form, those internet rumors were more likely deliberate PR exercises intended to stimulate the "what can they be hiding" crowd into actually looking for and, hopefully, reading this document.[10] It certainly makes interesting reading, and within its pages lurked the SETI mission statement. As of 2014, this is what that organization had to say:

> What we have actually been searching for, of course, is unassailable evidence of the existence of an extraterrestrial technological civilization, born of cognitive intelligence.[11]

At first glance, Shosak's dismissal of Cydonia takes another knock. Or in reality, perhaps not; it would seem that at the time of the Viking project, SETI was not quite ready for the reveal, thus Cydonia was dismissed. This next section of the SETI mission statement explains their disturbance:

> The anatomical and physiological structure of the extraterrestrials is a topic of major *theoretical* interest.[12] (emphasis added)

SHOCKING NEWS

When the Viking craft flew over the Cydonia complex and imaged a five-sided landform looking like the Pentagon in Washington, DC, would that have been a shock? And when it imaged a landform resembling a human face, would that have been another shock? Moreover, if, from above, the structure also looked very much like an Egyptian pharaoh, would it all be too much to cope with?

Yet what could be more wonderful than seeing a huge mesa, now eroded but still retaining enough indications of artificial design to infer both the seat of human intelligence (the head) and the art of an ancient human culture brought to life by the play of sunlight and shadow on its mesa so that there is, quite literally and wittily, evidence of *intelligent* life in the Martian soil. Even though it was officially stated that the Viking project was about looking for signs of life on Mars, present or past, for the masters of the clever acronym, this juxtaposition of theory and the apparent reality returned from Cydonia was not amusing.

To a degree, one can only see what one expects. Working on the assumption that the Martian environment had been hostile to advanced forms of life for millennia, the reality of actually imaging the apparent remains of a civilization so close to home might not have been sufficiently considered. A bit like Drake's sudden realization that they hadn't thought through exactly what to do if they did get a signal back, this is the nub of the problem; to this day no one has actually decided what to do. A 2019 poll revealed that over 50 percent of the population was comfortable about being in contact with ETI, but exactly who should manage that contact was less clear. Back in the day, prior to the 1975 Viking launch, NASA administrator James Fletcher, who believed that life was present in the universe and that Mars was a likely place to find out more on the subject, added:

> Although the discoveries we shall make on our neighboring worlds will revolutionize our knowledge of the universe, and probably transform human society, it is unlikely that we will find intelligent life on the other planets of our Sun. Yet, it is likely we would find it among the stars of the galaxy, and that is reason enough to initiate the quest. . . . We should begin to listen to other civilizations in the galaxy. It must be full of voices, calling from star to star in a myriad of tongues.[13]

THE LISTENERS

All of which is very poetic, although given the deafening silence that Fermi was complaining about, Walter de la Mare's poem "The Listeners" springs to mind. However, Fletcher's statement has little to do with biology markers on Mars. It rather suggests that *Mariner 9*'s discoveries informed his statement and that seeing geometric and architectural traces of a cognitive intelligence in our own backyard was too close for comfort. Pushing the boundary back into interstellar space within our galaxy, Fletcher concluded his 1975 statement thus:

> Though we are separate from this cosmic conversation by light years, we can certainly listen ten million times further than we can travel.[14]

SETI concluded the mission statement set out in NASA's outreach document with:

> What matters most for our search is that these beings will have figured out, almost certainly a long time ago, how to build powerful radio transmitters.[15]

For most ETI seekers, as in the Fermi discussion at Los Alamos, any ETI civilization that we are likely to come across is potentially far more technically advanced than we are. If that were the case, would radio signals truly be the epitome of technological progress? SETI's statement might be better construed as the intention to avoid putting on the public record any advanced technologies in research and development by the military in regard to communications to ETI. Or it is simply a lack of imagination directed at the public, who are not expected to think further on the matter. But let's go that extra mile.

Had ETI civilizations "almost certainly" (science speak for probably/possibly/ perhaps) developed powerful radio transmitters a long time ago, then it is almost certain that such a civilization would have found rather more advanced communication tools in the intervening millennia, and therefore there is no guarantee they would still be using radio transmission for deep-space transcommunication or use any local radio receivers to listen for incoming communications from the likes of us.

Nor is there any guarantee that this was even the hoped for outcome from the ambivalent SETI organization. SETI has stated that it is not necessarily about actively seeking to *communicate with* ETI, for which the acronym CETI is used. Here, acronym-ophiles might like to think about the associations with Tau Ceti, cetaceans in general, and the Order of the Dolphin in particular. This was the title given by Frank Drake to his first group of SETI colleagues, and he was referencing the dolphin communication experiments conducted by the physician and neuroscientist Dr. John C. Lilly, not Lambert Dolphin of SRI's Mars group. Seeing parallels between his work and Lilly's, in that they were both attempting to contact another intelligence, Drake had some influence in getting funding for Lilly's own project: a dolphinarium on the Caribbean island of St. Thomas.* Lilly had a beach house converted into a partially flooded environment for the purpose

---

*Lilly's work resonates with human space travel research. He experimented with humans in isolation tanks. With the dolphins he carried out verbal communication experiments and attempted telepathic connections by administering LSD. However much dolphins are considered by the public to be truly empathic, sensitive creatures, the way his dolphins were treated at the end of the St. Thomas experiment indicates that once the experiment had finished, they were of little interest to Lilly.

of dolphin-human communication experiments, and it was funded by NASA and other government agencies.[16]

SETI also thinks it possible to signal *our* presence to ETI without *necessarily expecting an answer,* and the acronym SETI refers to this passive mode. This notion did not then, nor does it now, resolve the issues within the scientific community as to the wisdom of seeking any sort of contact with ETI in the first place. But it does create great PR, while giving the funders and the naysayers (Stephen Hawking was one) a bone to chew on. It also explains why SETI and the science community might be hardwired against any signs of contact with ETI uninitiated by themselves.

That *Brookings Report* was turning out to be a self-fulfilling prophecy.

PROJECT CYCLOPS

Around the time that *Mariner 9* was leaving for Mars, in 1970 (or 1971, depending on the source), over at Ames, Billingham had begun a campaign to get NASA involved in SETI. He did this in conjunction with Bernard "Barney" Oliver, a former vice president of research at Hewlett-Packard Corporation with a long-standing interest in SETI. As Drake noted, Oliver had already visited Green Bank ten years earlier, during Project Ozma:

> A third visitor was, of all people, the vice-president for research of the Hewlett-Packard Corporation, a company which made a lot of oscilloscopes and meters and other electronic gadgets which we and every other observatory used. Bernard M. Oliver dropped out of the sky one day in a chartered plane, full of enthusiasm, to watch the goings-on in the West Virginia wilderness. Actually, as it turned out, it was not at all surprising that he was there, because he too had thought about the means for detecting other civilizations for many years. A successful inventor, electronics expert, and physicist, he already knew all about it, and was glad that someone had the opportunity at last to do something.[17]

Drake's use of the term *detecting* when referring to ETI signals is not quite the same thing as the active communication mode (CETI) or the passive listening mode (SETI). Here, DETI sounds like a deter-and-detect defensive mode. And using a giant round "eye" to do just that engendered Project Cyclops. Oliver now joined Billingham as the project director of Project Cyclops, and this meant that Oliver would be co-opted into Stanford University for the duration of the project. About which Drake had this to say:

"Barney" Oliver has become the leader in the development of plans for enormously sophisticated systems for the detection of extraterrestrial intelligent radio signals. When he has his way, as he will some day, we will see radio telescopes ten thousand times larger than the 85-foot telescope scanning the sky, on not one, but perhaps billions of frequencies at once. Good.[18]

According to those NASA ETI historians, Project Cyclops was a detailed NASA study proposing an array of one thousand 100-meter telescope dishes that could pick up radio signals from neighboring stars. Rendered unfeasible due to its estimated $10 billion price tag, it was never adopted. The NASA report states:

> An especially unfortunate result of the study was the creation of a widespread misperception that the Cyclops project required an "all-or-nothing" approach thus SETI got nothing for several years.[19]

Another especially unfortunate aspect of this statement is its skewed representation of the facts. The introduction to the report generated by Billingham and Oliver's Project Cyclops states that NASA had been funding educational "summer programs" since 1967. Held at four different universities for the benefit of young members of the faculty, these programs consisted of papers given by eminent scientists on research and engineering aspects relative to the selected theme of that summer. At Stanford in 1971, the theme was SETI. In the light of this understanding, it is unsurprising that the Project Cyclops study was not put into action; it was ostensibly a blueprint for the future.

Indeed, the SETI website describes Project Cyclops rather differently:

> Two-dozen academics spent three months considering what sort of equipment was needed to make a serious, systematic search for signals, and where they should point the antennas. Their conclusions, published as "Project Cyclops," became the bible of SETI research for decades to come, and are still important today.[20]

However, that summer school project in the United States was running parallel to another aspect of Project Cyclops, taking place in the same year but on the other side of the world. From this example, we see another version of hiding a program within a program—by calling two different events by the same project name. Author Stuart Holroyd reports that in 1971, there was an international conference sponsored by the American and Russian Academy of Sciences and that it took place in what was then the Soviet Union. So much for the Cold War.

He records that this conference at Byurakan in Soviet Armenia was the launch of Project Cyclops, an international research project on the practical possibilities and foreseeable consequences of establishing contact with extraterrestrial beings. According to author George Basalla, Carl Sagan, upon giving the audience the salient facts about ETI, asserted that:

- Any civilization we contacted would be vastly superior to our own.
- Alien civilizations have very long lifetimes.
- Alien civilizations have extremely advanced technologies, and by drawing on their superior knowledge, we may solve the technological problems that plague us in the twentieth century.

The attendees were in agreement with Sagan and others who claimed that extraterrestrial solutions existed for terrestrial problems. The absolute certitude of Sagan's statements about, as he puts it, "aliens," begs the question as to exactly how Sagan obtained his facts.

A conference held the following year might provide a clue. "Life beyond Earth and the Mind of Man" was a symposium held at the University of Boston, organized by astronomer Richard Berendzen. Coming a little late to the party, given NASA's thoughts on the matter, Berendzen stated, or rather repeated, the mantra of "strong indications of the highest probability of the existence of extraterrestrial life." But wait, maybe Berendzen was bang on time after all. He supported his statement with two interesting items. Those "strong indications" had emerged from recent developments in the sciences. By 1972, radio telescopes had been operational for decades, so his remark had little to do with Drake's SETI friends. A better candidate for a recent development, quite literally, would have been the technology on *Mariner 9* and its return of the images of Cydonia and Elysium. Berendzen then added this second little nugget: that an effective way of communicating with extraterrestrials might be through psychic channels. Remember, he is speaking to a bunch of scientists and he himself was not a fan of New Age hippy think! Had the Face on Mars inspired such thoughts of brain-to-brain communication? Or more prosaically, had the covert military research into PSI, ongoing since the 1950s at the very least and acquiring a new impetus with the dawn of the space age, produced such certitudes? Berendzen's statement was probably not as radical for his audience as it sounded to the public.

BROOKINGS AGAIN

It appeared that by 1972 scientists were willing to acknowledge the existence of ETI and discuss the possibility of making contact, even allowing some of those discussions to emerge into the public domain. And in that case, the authorities should have been well prepared for the chain of events that next occurred in Switzerland between 1972 and 1990. However, events indicative of ETI contact, coming in the other direction—from out there to down here—were definitely not going to be for public discussion. Which is either an indication that the authorities had an inkling that the Swiss anomalies might be linked to our Martian exploits or that they were not themselves as psychologically prepared for ETI as they imagined.

Either way, Switzerland in the 1970s was going to be something of a shock to the system.

Switzerland:
The Anomalous Years
(1972–1990)

*Puzzling Evidence of Intelligent
Nonhuman Activity in Switzerland
and the Connection to Anomalous
Events in the United Kingdom*

BACK TO THE FUTURE

Much of this chapter is an attempt to see if the events in Switzerland were indeed signs of *incoming* ETI contact. And if so, to what extent the presence of particular human beings affects such events. Whatever one's personal opinion might be, many authorities think that it is worth investigating (often covertly) events like UFO sightings, channeled communications from extraterrestrial entities, and unexplainable geophysical imprints on the Earth's surface. These are generally taken to be isolated incidents relative to a specific region—or in this case, a country. When considered as separate aspects of a same interaction that is taking place over long time periods, at different locations, and via different media, there emerges a more complete window into the possibilities presented by these anomalous events.

Six months into the *Mariner 9* imaging mission, in the Canton of Vaud north of Lake Geneva, a Swiss farmer named Mertinat awoke one June morning in 1972 to find the oddest thing. A hole in the ground had opened up near his cattle sheds, and it bore no signs whatsoever of being a natural event. Measuring

2 meters (6.5616 feet) in diameter, its sides were 1.5 meters (4.9212 feet) deep and straight as a die, while the bottom of the hole was smooth and flat. Most intriguingly of all, the 4.71 cubic meters of earth that had been extracted from Monsieur Mertinat's land were nowhere to be seen. This bizarre event filled the Swiss news media for a day or so and was then was quietly dropped. If Mertinat ever got a plausible explanation from any authority as to the disappearance of the soil from his farmyard, it was not made public.

Over in the United Kingdom, architect, crop circle researcher, and author Michael Glickman embraced numerology when he stated that the numbers 5 and 6 appearing together (and summing to 11) are the first step of the master number sequence. He asserts that the sequence is indicative of first contact by ETI. In his 2009 book *Crop Circles: The Bones of God* he wrote that this sequence from 11 through to 99 was said to refer to "the progressive levels of contact with, and stages of incorporation into, other dimensions."*

It is asserted by those who put credence in this system that the master number 11 is linked to intuitive ability; the number 22 refers to the qualities of a master builder and the number 33 to the attributes of a spiritual teacher. When I discussed the matter privately with Michael Glickman he told me that human beings will not get past the first three of these master numbers. Thinking that architectural training and the Masonic triad of apprentice, journeyman, and master might have some bearing on that interpretation, I disagreed with him, saying that limitation was not particularly helpful when trying to open the mind to thinking differently and that there is likely to be more information within this series that has been forgotten or ignored by modern day numerologists, architects, and Masons. Looking at this sequence differently it is possible to see that the solar cycle of 11 years fits with the first 5:6 set, and that the magnetic equator of the planet is at a difference of some 11° with the equator. Again on a different scale, the whole of this series of 5:6 could be about solar or photonic matters: taken beyond the ninth iteration, we come to 50:60. And it just so happens that our planetary electrical currents operate at 50 Hz and 60 Hz. According to data accumulated by NASA, the chest wall of a human being resonates with 60 Hz and the eyeball resonates with frequencies of 40 Hz, 50 Hz, and 60 Hz.

---

*The sequence is formed from these numbers: 5:6, 10:12, 15:18, 20:24, 25:30, 30:36, 35:42, 40:48, 45:54. These nine sets are then added together within themselves, so that 5:6 sums to 11; 10:12 sums to 22; 15:18 sums to 33; and so on. Numerologists (and Glickman) only consider the first three of the sequence to be important. They take the numbers from 44 onward to be "power numbers."

WHEAT AND BREAD

If introducing crop circles seems irrelevant here, Mertinat's empty cylinder might suggest otherwise. Not only did its diameter in feet recall Glickman's 5:6 notion, the event also recalls data relative to the Face on Mars. Mertinat's farm was situated near to Echallens, the "capital" of a wheat-growing region known as the Bread Basket of Switzerland; its bread museum receives some eleven thousand visitors a year. (For those who did not get the memo, that's 1,000 × the sum of the first group of master numbers: 5 + 6 = 11.) Perhaps Deuteronomy 8:3 is appropriate here; after all, it does contain the well-known saying "Man shall not live by bread alone," and its chapter and verse also add up to 11.

Furthermore, Lake Geneva, with its waterspout at Geneva, looks like a cetacean, so Frank Drake's Project Ozma and Cetus constellation experiments again spring to mind, along with the square granary, that Chinese name for the constellation. A conversation has begun, or is continuing.

Twenty-seven years later, on August 6, 1999, a unique crop glyph would appear in wheat in the village of Bishops Cannings in the county of Wiltshire,

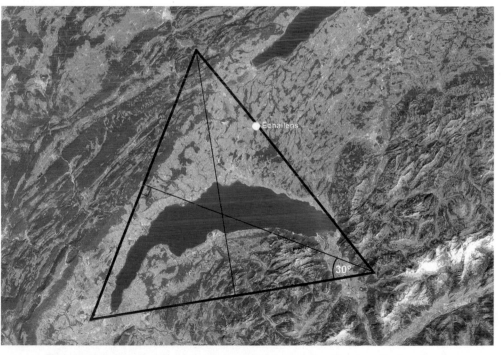

Figure 11.1. Echallens's location in relation to Lake Geneva, Switzerland. The volume of Mertinat's cylinder of missing material is calculated as: pi × radius squared × the depth: 3.14159265 × 1 × 1.5 = 4.71 cubic meters.

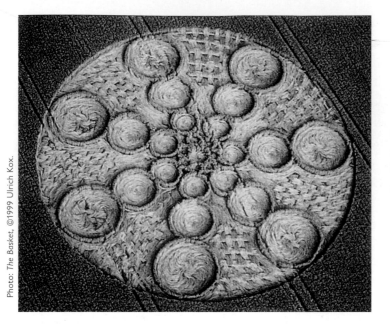

Photo: *The Basket*, ©1999 Ulrich Kox.

Figure 11.2. The Basket crop glyph, Bishops Cannings, England, August 6, 1999.

England. Clearly visible from the busy road between Devizes and Marlborough, it was so complex that it clearly troubled more people than just the farmer, who destroyed it within hours of its arrival.*

Figure 11.2 shows the only aerial image ever taken of the Basket glyph. It was dubbed the Basket due to the complex weaving of the wheat stems within and round the twenty-eight circles set out on seven principle axes, each of 51.42°. This is the latitude at which this glyph was found. It was discovered by crop circle researchers Andreas Müeller and Werner Anderhub, August 6, at 6:15 a.m., 1999. And in the context of this chapter, it is of interest that Anderhub is Swiss and that the glyph's location was just across the road from Glickman's home. Glickman later showed the photo taken by Kox to the farmer, who refused to look at it, turning away and covering his eyes.[1]

---

*In 2002, three years after this event, a known crop circle maker gave a lecture in Devizes, during which he claimed to have made that Basket glyph, as it been named. He had done all the complex weaving and a friend had made the circles. This drew more than a few incredulous gasps from the audience. I reported on this lecture for a crop circle website and a few months later was reproached by the hoaxer for dismissing his claim. He said he did not like my words. I replied that they were his own: he had said that he had made the crop circle (as he called it) because he had been upset about being taken to court over another crop circle he had made. I said that he had omitted to inform his audience that the court appearance took place in November 2000, fifteen months *after* its arrival. He had no reply to that. Nor could he tell me why he had waited until May 2002 to assert ownership of this crop glyph, as I called it.

ETI

The Echallens cylinder's imperial measurements of approximately five feet in diameter and six feet in depth might fit a first footing, but that would need deducing, since the Swiss prefer to use the meter and only the Scots practice first footings. Scrolling forward to the twenty-first century, the Australians have set up an array of thirty-six radio telescopes designed to hunt for anomalous input from space and in particular for fast radio bursts, about which scientists currently have little data. However, these tend to get headlines in the press such as "Have ETI Signals Been Detected?" This array is set up at Murchison, northeast of Perth, and is being run by Australia's national science agency, the Commonwealth Scientific and Industrial Research Organisation (CSIRO), which, in its wisdom, has given this Australian Square Kilometre Array Pathfinder the apparently appropriate acronym ASKAP. This is how CSIRO describes this array:

> The Australian Square Kilometre Array Pathfinder (ASKAP) is the world's fastest survey radio telescope. Designed and engineered by CSIRO, ASKAP is made up of 36 "dish" antennas, spread across a diameter of 6 km, that work together as a single instrument called an interferometer. The key feature of ASKAP is its wide field of view, generated by its unique phased array feed

Photo: CSIRO Radio Astronomy Image Archive.

Figure 11.3. ASKAP's "field of view" is depicted, showing the thirty-six beams as individual circles. As CSIRO states: "We get all of this in one go. By comparison, the field of view of a traditional telescope would be a single slightly smaller circle. The Moon's diameter is half the diameter of one of these circles."

(PAF) receivers. Together with specialised digital systems, the PAFs create 36 separate (simultaneous) beams on the sky which are mosaicked [*sic*]
together into a large single image. . . . The PAFs were assembled in our CSIRO
workshops in Sydney—for 36 PAFs, there are some 6 million parts, including 20,000 printed circuit boards. Each antenna has 216 optical fibres that
transport data to the MRO Control Building—that's a total of 7776 fibres.
The total length of this fibre is approx. 15,500 km—enough to wrap around
the Moon 1.5 times! These fibres transport data at a rate of 1.9 Tb/s from the
antennas to the correlator for on-site processing. After this processing takes
place, the data is sent along high-speed optical fibres (40 Gbit/s) down to the
Pewsey Supercomputing Centre in Perth.[2]

This is undeniably brilliant engineering, but its creaking acronym gives the
game away as to its very real interest in ETI. The CSIRO personnel were active,
of course, at the time of the Apollo space missions; in the 1990s, they were also
on the ground in England inspecting the Wiltshire crop glyphs, whose ground
zero was near a village called Pewsey. So it's another nice irony (or another discreet messaging service) that CSIRO's Murchison ASKAP operation has its
supercomputing center at a locale also called Pewsey. Coca-Colas all round, and a
time travel back to the Swiss anomalies.

If Echallens was literally a first footing, there had been plenty of anomalous
activity in the Swiss skies before 1972. Which takes us to another Swiss farmer,
Eduard "Billy" Meier, then living not far from Zurich. He claimed that from 1953
to 1964 he had received messages from an extraterrestrial entity called Asket.
Although the authorities publicly ignored him, privately, things were different.
Three years after the arrival of the Echallens cylinder and two days after the United
States upgraded the capabilities of its Minuteman ballistic missile program, on
January 28, 1975, Meier took his first photographs of UFOs (later to be described
as originating from the Pleiades), and during the time period that Meier received
the urge to take his spacecraft photos, he allegedly received messages too.

Meier's photos would not get to US laboratories for analysis until 1978, so
that story comes later in this chapter. For now, it's useful to note that the eastern
end of Lake Geneva is overlooked by the Swiss pre-Alps named Les Pleiades. We
have already noted that the Pleiades star system is in the constellation of Taurus
the Bull, itself adjacent to the constellation of Orion. And that in Egypt Orion
is conflated with the Bull and with Osiris, and representations of this god sport
a cap with two horns. The Minoans of Crete had their bull-leaping ceremonies;
the Romans used the same imagery and attributed it to Mithras. The modern-day

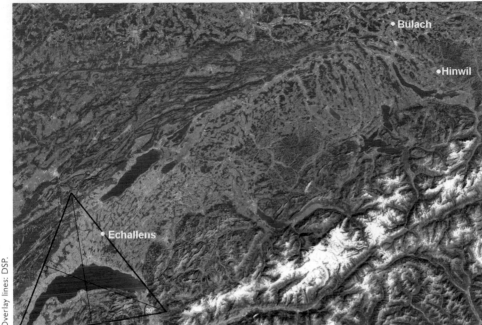

Image: Google GeoBasis-DE/BKG (©2009) Landsat/Copernicus, Google Earth, 2016.
Overlay lines: DSP.

Figure 11.4. Echallens in relation to Meier's home in Hinwil and his birthplace, Bülach.

bullfights and the bull's-eye of the dartboard are all vestiges of these human concepts of the tussle between the physical and the spiritual. Were the Mars Viking team members at JPL in their horned helmets aware of these antecedents?

The fact that two anomalous events occurred to two different Swiss farmers—one relating to earth and fire (wheat is associated with fire) in the French-speaking region and the other relating to the images seen in the sky and to mind contacts in the German-speaking region of a country more renowned for its financial acumen and its particular affinity with watchmaking, clockmaking, and chocolate than its farming skills—is a puzzle for the moment.

Synchronicity was again in evidence when on February 24, 1975, Japan launched a satellite to observe solar X-rays, ultraviolet radiation, and the ionosphere.

The next day, and without being aware of the Japanese event, Meier psychically received a long communication about the environmental damage the military was doing to the planet through nuclear testing. The industrial extraction of fossil fuels was also mentioned.

He passed on the content of these environmental warnings to various authorities in both the United States and in Europe, but replies came there none—at least not to Meier.

The decades since the detonation of the atomic bomb had seen much nuclear testing in the atmosphere—the hugely damaging Argus and Starfish Prime exercises being just two examples. Yet information regarding the subsequent damage to the planet and its environs didn't emerge into the public domain until 1988, and even then it took until 1992 to fully register as more than a slight inconvenience. So back in the 1970s, these messages from Meier had more relevance for the US and Russian authorities than could have been known by a Swiss farmer.

CUCKOO!

It should be noted that "cuckoo!" is used colloquially as "hello!" in France and in French-speaking Switzerland. Remembering that one of the prime instigators of SETI, Philip Morrison, was also a nuclear physicist and also taking into account the mythological role of the planet Mars as a god of agriculture and war imparts connectivity to seemingly unrelated and anomalous events. Add to that the profession of farmers, the guardians of living things and of the soil, and the proximity of the Meier farm to those financial wizards, the "Gnomes of Zürich," and there are connections with ETI, space exploration, the investment it requires, and the fact that the rockets designed for space exploration were also those used to launch and test ballistic missiles. And also in the land of the cuckoo clock, we have the Swiss watchmaker Omega, NASA's go-to company during the space race. When it came to racing around the Earth, six years after the arrival of that first Swiss cylindrical hole, *Briefing for the Landing on Planet Earth* by Stuart Holroyd was published in the United Kingdom. It recounted the intercontinental adventures of a small group of people experiencing ETI communications from 1974 to 1977, most of which were relative to the nuclear threats of that time. Their backgrounds and their intergroup "etiquette" make the whole book an instructive read, especially when taken together with the events going on in Switzerland.[3] All of these connections give even greater significance to that January 26 Minuteman upgrade and Meier's ETI communication of January 28, 1975. Especially when one notes that Janus, the god who looks both backward and forward and after which the month of January is named, is the god of thresholds and doorways.

PHOTOGRAPHY

By 1977, the Meier UFO photographs officially come to the attention of the US authorities, and a team began assembling to examine them. As to the authenticity of these images, the jury is *still* out, and people on both sides of the

Figure 11.5. Typical craft photographed by Eduard "Billy" Meier.

true-or-false argument are equally adamant as to the validity of their opinions. Some have agendas and not all are necessarily interested in ETI, but the fact that really big names in NASA's science teams were involved in investigating the Meier phenomenon ought to give pause for thought. If these UFO images are just faked-up "dustbin lids" (remember that Fermi story and the *New Yorker* cartoon from chapter 4), then the question must be: Why do those in authority to this day, with huge responsibilities in the Mars program, think otherwise?

For six years, from 1978 to 1983, Meier's UFO images (many of very good quality) were closely scrutinized in the United States. One of the organizers of the examination project, Jim Dilettoso, writing in 2000, described the challenge:

> When I was 28 years old. [*sic*] Lt. Col. Wendelle Stevens (USAF Ret.) and Jim Lorenzen, founder and director of APRO (Aerial Phenomena Research Organization), visited me to discuss testing UFO pictures taken by a Swiss man named Billy Meier. . . .
>
> Trying to locate equipment and experts in image processing to assist in testing UFO pictures was a little frustrating. In 1978, computers were mainframes and workstations. State-of-the-art image-processing equipment had 64K of ram and a 5MB hard drive—and the cost was $100,000. Desktop scanners cost $50,000 and up. Even worse, most of the equipment we needed resided in labs owned by or contracted by the US government and defense agencies. Wendelle Stevens and I made the rounds of trade shows (like the Society of Photo-optical Instrumentation Engineers), NASA labs, America's finest companies (like IBM and Northrup) and organizations such as the US Geological

Survey and the US Navy. It was like "Mission Impossible." Penetrating labs like Sandia and the Jet Propulsion Laboratory required special credentials, and sometimes even a masquerade in order to get people to assist us.

Some said the case was a hoax, but lab results differed. We persevered, though, and eventually had found many professionals, who under secrecy and non-disclosure agreements tested these UFO pictures. The secrecy was critical. These labs were not generally authorized to perform personal projects, like testing UFO pictures. So when other UFO researchers, hell bent on getting into the case, made inquiries into some of the places we had been, they would, as agreed, deny any involvement on their part in testing the Billy Meier UFO photographs. *Although not one lab found the pictures to be a hoax, UFO clubs like APRO and the Mutual UFO Network (MUFON) were claiming that it was a hoax.*[4] (emphasis added)

It's mildly interesting that APRO, against all the evidence, declared them fakes when their organization's founder had instigated this picture testing in the first place! However, all of this report is off-kilter. Trundling around governmental organizations that are hard to get into in the first place, looking for top people to work on UFO photos, and then stating that they all agreed to do the work "personally" and thus signed nondisclosure agreements is somewhat exaggerated. What this actually means is that the scientists who looked at the pictures offered a personal opinion and not one representing the organizations or laboratories within which they were working. And if those labs were so difficult to get into and the scientists so hard to convince, then one wonders how the next part of this narrative works:

We decided to go somewhere to perform testing—somewhere like NASA-JPL. I would do the research, get some names, call Wendelle Stevens, talk about it in our own code, and go.

From the front gate onward, we faced little obstruction. We had credentials and access materials that literally gave us the run of the place. Although every building was available to us, we always met our confederate in the cafeteria or the ERC (employee recreation center). We took the photos in their picture form, and showed them to our contact over a hot dog or chipped beef, then proceeded to the confines of a NORAD looking room, where the latest in image-processing equipment could be found. . . . The image-processing experts sat with us in a secure facility at the Deep Space Network, looking at pictures of UFOs. Things seemed too convenient, indeed.

We actually thought we were getting away with something, as we were granted free access of this place, using expensive equipment, and getting opinions from leading folks in the space program. As time went on, though, I increasingly suspected that someone knew our every move and was opening doors, if not outright sanctioning the caper. Test results were always "positive"—no evidence of a hoax. But the situation was a lot deeper than that. These people were as amazed as we were at the quality of the pictures. . . .

[At JPL] not only were we given full-access press passes, we were given one off [sic] the press suites in Von Karman—the media-relations center at NASA. National Geographic and Ted Koppel had to walk through our suite to get to their desks.[5]

According to Dilettoso he got those full access passes because he had he sold the idea to NASA that both JPL and his other client, the prog-rock band The Moody Blues, would get good publicity by using the Voyager space program as their hook. "We'd propose, I thought, to call the tour—and album—Voyager, get photos and video from JPL during the fly-by, and use them in the tour. NASA loved it, and so did the Moodies. At the same time, Junichi (Jim) Yaoi from Nippon TV hired us to acquire images from JPL and get them to Japan. Cameras were the key: We had a second reason to be on site. Soon, we'd be heading to every lab with a video camera." And even though the Billy Meier case was completely under wraps, that freedom enabled Dilettoso to collect as he put it, "all the video we needed. But we still couldn't talk about our experiences." In fact, Dilettoso's unique talents in the digital realm, whether with video, film, or computing, were fully utilized by the emerging space industries and in that regard while sharing a Cray Supercomputer with NASA, he was also creating special effects videos for other acts such as the Pink Floyd. This happy marriage took place in Tempe, Arizona, where Dilettoso's own graphics lab was conveniently sited within the Arizona State University's computer institute.

THE ARIZONA STATE UNIVERSITY FACULTY

Dilettoso was not only the person closely associated with this Meier photographic analysis. A key player in both the Meier photo analysis and the Mars imagery would emerge from the Arizona State University faculty: the previously mentioned Michael Malin. When Dilettoso came knocking at his door in May 1981 bearing Meier's UFO photos, Malin (then thirty-one) had been a professor of geomorphology and lunar and Martian geology at Arizona State University for two years. Nobody's fool, Malin has a degree in physics from Berkeley, a

Ph.D. from Cal Tech in planetary sciences and geology, and his doctoral thesis incorporated the science of image processing and analysis of spacecraft images from Mars. After spending a considerable amount of time on the Meier UFO images, Malin thought that they were the best UFO photographs he had ever seen, but as he was unable to see either Meier's camera or the film stock that actually went through the camera, he could not say much more about them. He did though. Malin added that he could not see any obvious signs of hoaxing and that although he was not convinced that they necessarily showed extraterrestrial spacecraft, the images appeared to represent real phenomena.

Seven months after Dilettoso had knocked on Malin's door, another anomalous event occurred in Switzerland. During the night of December 5, 1981 (nine years after the Echallen event), a new cylindrical hole was discovered. Dilettoso had thought that someone behind the scenes here on Earth was "opening doors" for his research. The question is: Was the door really being opened from the other side?

This *second* cylinder measured five meters in diameter and eight meters deep, with 157 cubic meters of soil completely missing. The location was between Ollon and Villars in the hamlet of Les Combes, just to the north of the ancient salt mines of Bex. In much earlier geological times, the eastern shores of Lake Geneva had extended as far as Bex, so the analogies with other ancient dried lake beds, to be discussed later, can apply to this particular hole. The press ignored the geologi-

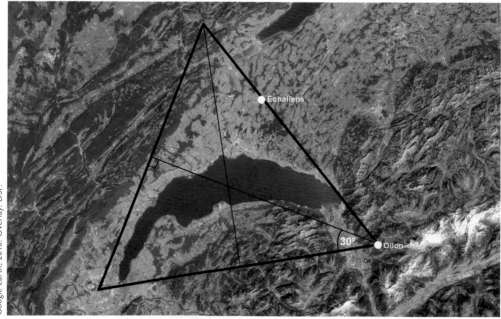

Image: Google GeoBasis-DE/BKG (©2009) Landsat/Copernicus, Google Earth, 2016. Overlay: DSP.

Figure 11.6. Echallens to the northeast and Ollon to the southeast.

cal time line and instead specifically mentioned the fact that the hole was near to an airline pilot's holiday home, oddly informing their readers that he normally lived on the other side of the lake in Nyon. Which might have been something of a clue, but nothing more was said at the time. End of story.

EXPLANATIONS

Twelve months and twelve days later, that odd reporting would become somewhat relevant when, on December 17, 1982, a *third* cylindrical hole appeared diagonally opposite the Ollon location, on the other side of Lake Geneva. Situated near Begnins, not so very far from Nyon, where that airline pilot owned a house. If this detail was meant to imply organized mischief by human beings, the size of the cylinder indicated otherwise. It was 5.5 meters in diameter, 7.5 meters deep, with the now standard formula of straight sides. The amount of missing mass totaled 178.1872083 cubic meters. These were the figures published in the local press at the time and quoted in publications ever since. However, in October 2019 David Percy was able to interview the farmer, Roland Haefeli, who corrected these published figures, stating that in fact the hole had been five meters in diameter and ten meters in depth. (thereby giving 196.35 cubic meters of missing material).

Figure 11.7. Cylindrical hole of missing material ten meters deep
near Begnins, Switzerland, 1982.

Photo: H. Schaffer.

In the thirty-seven years since this event, for which there was no explanation at all, the reasons for its existence have now multiplied. In fact, Farmer Haefeli had far too many explanations for this hole's arrival in his field, when one would have been quite sufficient. None of them really fitted the overnight arrival, except for that of ETI, which he also offered and dismissed in the same breath.[6] Haefeli explained that an expert had assured him that his missing material was due to shifting underground sediments and currents. That none had occurred anywhere at all in that locale before or since, and especially not overnight, was ignored. However, this sounded a lot like the relatively recent explanation of the factors contributing to the tsunami that had occurred in 563 CE in Lake Geneva. The tsunami had created turbidite currents and channels in the sediment of the ancient glacier bed that could facilitate another tsunami in the future. Should another massive rock fall, such as that of Tauredunum, occur in the eastern delta near to Ollon and declench another tsunami, it would reach Begnin in thirty-five minutes and Geneva in sixty minutes.

This new theory was presented to the public in 2012 in a paper by Katrina Kremer published in *Nature Science*. Overturning the notions that tsunamis were a thing of the past, Lake Geneva's geological sediments are now being surveyed for any signs of another tsunami occurring (as is Lake Lucerne). The notion of

Figure 11.8. Analysis of the 563 CE tsunami and location of Nyon (adjacent to Begnins), with times of wave propagation (in minutes) and spot heights of the tsunami at key locations during the Tauredunum event.

increasing energy transfer via water is of interest when thinking about the place-ment of these anomalous holes. Especially since they all occurred around a lake that as already noted, with its waterspout at Geneva, looks like a dolphin. This cetacean is highly sensitive to humanity and is especially good at detecting hidden malaise in the human body. It also relies on echolocation to "see" underwater.[7] I will return to this subject later. Here, it makes the placement of the Begnin hole particularly apposite, as its five-by-ten dimensions in meters are 1:1000th of the five-kilometer-wide by ten-kilometer-long turbidite bed found by Kremer. And Kremer, just like a dolphin, used an instrument called a "pinger" to transmit and then analyze the reflections of sound waves that can penetrate the material of the lake bed. The five-meter-deep turbidite sand and silt sediment is laid down by rapid water movement, as such it is not sorted by grain size; for those who are making connections, the small hole at Echallens in the grain fields of Switzerland springs to mind.

It soon became clear that the arrival of these anomalous Swiss cylinders obliged the authorities to find some sort of answer for the landowners concerned. If that led the authorities to examining their native geology with even more atten-tion than usual and thus drew their attention to hitherto unappreciated dangers within the lake, then notwithstanding one's personal opinion on the matter of ETI, these three events have already been more than helpful to humanity. But these anomalous cylinders were not finished with Switzerland: the best was yet to come.[8]

Eight years later, the *fourth* and the largest anomalous hole to appear in Switzerland was found on February 4, 1990. If it had turned up overnight, as per the general fashion of these events, that was a nice touch because Friday, February 3, was Billy Meier's birthday. Another nice touch, it also arrived within a whisker of the underground facilities of CERN, which straddle the French-Swiss frontier. Which is very appropriate, given CERN's association with SETI scientists Philip Morrison and Giuseppe Cocconi. When this cylinder of missing material appeared, Cocconi, a much-loved and respected particle physicist, had been retired from CERN for eleven years. That number again. Retired he might have been, but Cocconi kept in touch with the developments within his profes-sion. Now it appeared that his subject had kept in touch with the co-author of that original SETI paper in *Nature*.

With all this in mind, it's even more interesting to note that the size of this cylindrical hole, being ten meters in diameter and twelve meters in depth, ful-filled the second of Glickman's master number sequence, but *this time in meters*. It was very precisely positioned in a field of winter planting just above the tunnel of Confignon, part of the Geneva ring road system, at a place where the tunnel runs just eighteen meters below the surface. The tunnel was at that time still being constructed, but this particular cylinder of missing material was not considered to be caused by any excavation works. It went unmentioned by its constructors in their documentation of all such matters.[9] Which is unsurprising, because just like its predecessors, there was no trace of the 942 cubic meters of missing soil in the bottom of this perfectly formed cylindrical hole nor in the immediate vicinity. According to the Swiss crop circle researchers and authors Werner Anderhub and Hans Peter Roth, removing this amount of earth would have required one hundred trucks each loaded with twenty metric tons, yet no such activity was reported by the local residents.[10] This final cylindrical hole was remarkable for another reason. Not only was there a difference of six meters between the bottom of the cylinder and the roof of the tunnel but it was also six kilometers from the Large Hadron Collider ring at CERN, itself running under Geneva Airport at Meyrin.

Over a period of eighteen years, just over 1,300 cubic meters of Swiss soil had been removed. Three out of four of these cylinders had alleged flight or air associations. As did those Pleiadean UFO photographs. Regarding holes in the ground, at that time, CERN itself was running rings under the ground for its particle accelerator experiments and actively searching for the answers to subjects such as dark matter, dark energy, *missing mass* within the galaxy, and for the

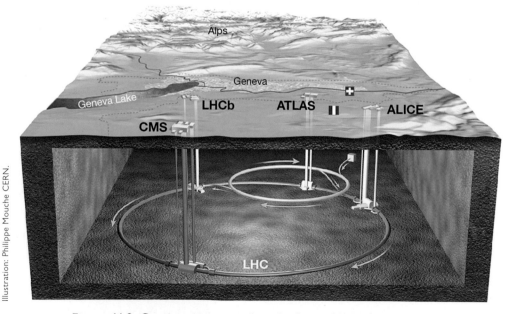

Illustration: Philippe Mouche CERN.

Figure 11.9. Cutaway area revealing the Large Hadron Collider ring under the ground at CERN.

Higgs boson—or the God particle (as some would have it, albeit not Peter Higgs himself). In other words, scientists at CERN were seeking answers to everything, for then (never mind ETI), in the words of Stephen Hawking, "We should know the mind of God."

Less publicly, if past experience is anything to go by, some of the science undertaken at CERN might be aimed at seeking new technology for the benefit of all mankind but only once it has been annexed by (a) the military, (b) the space agencies, and (c) the intelligence agencies, although not necessarily in that order. All in all, the location of this fourth cylindrical hole could not have been more pertinent. And when taken together with the warnings from Meier, the placing of these four holes can also be modeled as a Combinatorial Hierarchy of forces, which may have relevance relative to the protection of self-aware beings (represented by the dolphin-shaped Lake Geneva) from damaging cosmic radiation as well as tsunamis.

DARKER MATTERS

Given that Switzerland is a country associated with the management of time, there was yet more encoding relative to the occupations of CERN physicists that was sitting, or rather, not entirely present, in that field near CERN. The numbers 10 and 12 reference binary and duodecimal systems of counting, and also latitudes and longitudes, and can imply a reference to *space-time,* a term coined by Albert Einstein. He had worked in Switzerland and lived in Bern, not far from the village where Meier would live decades later. Bern is also aligned with both Echallens and Confignon. Could it be that the 942 cubic meters of missing material, removed in the hours of darkness, referred to aspects of the research into the perceived *missing mass* of the universe, dubbed dark matter?

Most scientists discuss dark matter as demonstrable fact rather than an unproven hypothesis. It is just a theory developed to support what is called the standard model as a result of an obvious bad fit between the data acquired by Newton and Einstein and the data emerging from the continued exploration of space by probes. By September 2015, a new theory termed "the stealth dark matter theory" (no useful acronym!) was postulated within the online pages of *Astronomy Now,* and again the article takes the dark matter hypothesis as fact. Here is an extract:

Dark matter makes up 83 percent of all matter in the universe and does not interact directly with electromagnetic or strong and weak nuclear forces. Light does not bounce off of it, and ordinary matter goes through it with only the feeblest of interactions. Essentially invisible, it has been termed dark matter, yet its interactions with gravity produce striking effects on the movement of galaxies and galactic clusters, leaving little doubt of its existence. The key to stealth dark matter's split personality is its compositeness [*sic*] and the miracle of confinement. Like quarks in a neutron, at high temperatures, these electrically charged constituents interact with nearly everything. But at lower temperatures they bind together to form an electrically neutral composite particle. . . . In a particle collider with sufficiently high energy (such as the Large Hadron Collider in Switzerland), these particles can be produced again for the first time since the early universe. They could generate unique signatures in the particle detectors because they could be electrically charged. [Pavlos Vranas of the Laurence Livermore National Laboratory added that] "Underground direct detection experiments or experiments at the Large Hadron Collider may soon find evidence of (or rule out) this new stealth dark matter theory."[11]

"They could do this, they could do that—we may soon find . . ." However, if the data taken to prove this hypothesis is itself faulty, then no amount of experimentation with the Large Hadron Collider is going to produce the results sought. As space probes move out into the farther reaches of our solar system, so does Einstein's notion of space-time look at best incomplete and at worst basically flawed, or, taking this cylinder analogy, basically floored. In 2006, Edward "Rocky" Kolb, chair of the Dark Energy Task Force, a group comprised of people from NASA, the National Science Foundation, and the Department of Energy, said that what they really needed was another Einstein or, failing that, to wind back the clock eighty-five years to the point when, in order to solve Einstein's equations, it was assumed that the universe was isotropic. An assumption that is gathering fewer and fewer enthusiasts as time goes by. Yet when someone does turn up questioning the misconceptions that have arisen within the standard model, apart from the very few discerning physicists, backs are turned and the standard model is held as the paradigm, flaws and all. As for dark matter, by 2018, contrary to the expectation of the scientists, a galaxy with no dark matter has been detected. Which rather points to the cylinders' missing soil. At its very basic level: "Look, no 'dark' matter." The missing soil is so obvious a clue, it is most surprising that this missing matter would be ignored by the science boys next door. But something did resonate in the collective subconscious. When CERN started looking for the Higgs boson particle, worriers in the media queried as to whether it would send us all into some state of "missing matter."[12]

Then again, the understanding that this ETI initiative might be *incoming* rather than *outgoing,* with the attendant inference that we do not in fact control the environs of our planet, may not go down very well. After all, when anyone inside or outside the scientific community offers an alternative viewpoint on the nature of the universe (including unexpected ETI interventions), such a person is considered beyond the pale. If the science is irrefutable, then ad hominem attacks will be targeted on the character and probity of the person upsetting the scientific applecart. Nobody has looked at these anomalous Swiss cylindrical holes from a scientific point of view, and Kolb obviously hasn't taking any notice of the putative hints being sent to us from an ETI source—at least publicly.

GAMES OF SERIOUS INTENT

Bearing in mind that Carl Jung was also Swiss, those not comfortable with this ETI interpretation might prefer to ascribe the anomalies and coincidences of these cylinder events to the manifestation of the collective subconscious of the locals.

Thinking laterally, it is possible that everything about an anomalous event such as the Confignon cylinder contains clues for CERN as to how to proceed. Taking another step sideways, according to one analysis, these four Swiss holes have volumes that recall the Sirius star system.*

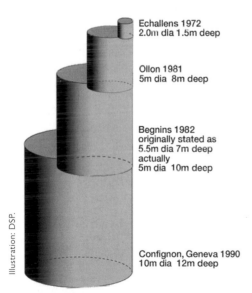

Echallens 1972
2.0m dia 1.5m deep

Ollon 1981
5m dia 8m deep

Begnins 1982
originally stated as
5.5m dia 7m deep
actually
5m dia 10m deep

Confignon, Geneva 1990
10m dia 12m deep

Illustration: DSP.

Figure 11.10. Relative sizes, as given to the press, of all four cylinders of missing material, Switzerland, 1972–1990.

---

*David Percy proposed this connection in a personal communication with author Robert Temple, who despite his own interest in Sirius was not overly enthused. Sirius was better appreciated by NASA's space suit manufacturer, ILC Dover.[13] It was the code name used for all the garments assigned to Apollo astronaut Neil Armstrong. The Sirius analysis is available at Aulis Online.

That a perceived "superior" intelligence could be stretching human intelligence into new ways of thinking might not be appreciated by the scientific intelligentsia, especially when it happens for real outside the laboratory and through methods deemed as "game playing."

If these seemingly disparate anomalous events on Earth are manifesting through the use of the very technologies that scientists today are seeking, game playing seems a wise way to proceed. Especially when dealing with scientists brought up on the *Brookings Report* doctrine. Such technologies in the wrong hands may turn out to be equally lethal, and if that is the case, then these anomalous ETI events inferred by these Swiss cylinder events are very clever indeed. Encoding locations with sufficient answers to alert those paying attention requires a paradigm shift in thinking to "get the rest." Placing such anomalies adjacent to locations or organizations that resonate with or in some way reflect the functions of the technology perhaps facilitates an intuitive and subconscious understanding that will only flower into consciousness when the wider ramifications of this interaction with ETI are fully understood. By which time, deploying such technology for war-like or destructive purposes would be anathema to a brain that fully grasps all the ramifications of these anomalous events, because with great discoveries come great responsibilities and the need for accountability.

These missing matter cylindrical holes have occurred in a country that went through its own inner turmoil in the fifteenth century. Since then, politically, it has generally adopted the view of its citizen Nicholas von Flue: dialogue was a better alternative to schism and war. This is a mindset relatively unknown to those whose rockets served the dual purposes of launching and testing ballistic missiles and were then adapted into space launch vehicles primarily for spy satellites. In fact, all of these cultural references have relevance to those experimenters at CERN—should they wish to consider them.

ENQUIRE WITHIN

In this context of information sharing, it is very interesting that Sir Tim Berners-Lee was at CERN from June through December 1980 designing the very wired project ENQUIRE. This was an in-house development intended to facilitate information sharing among researchers—hint, hint!

From that very early computing network, CERN would become the hub of the largest computing grid extant: the Worldwide LHC Computing Grid.[14]

Project ENQUIRE's 1980s debut coincided with the decade encompassing three of the Swiss cylinder events and the moment when the topic of crop glyphs emerged in England. Luckily for the authorities, the Swiss missing-soil cylinders did not make any impact at all on the British media. And not being easy to copy, they were swept under the magic carpet of officialdom and totally ignored. Crop glyphs were another matter entirely. Meier had seen crop circles in grass in 1975 and 1976 (the Viking years), but his findings were buried within his UFO stories. Again, the British media failed to make a link, although the authorities certainly did when, in the 1980s, the crop glyph phenomenon became more than a farmer's kitchen table topic of discussion. The British authorities, including the Scottish scientist and academic Archie Roy, sat up and took notice. Roy was known for his mirror theorem in celestial mechanics and had such an understanding of orbital motion that in the 1960s NASA had asked him to help calculate the trajectories for their space probes, including Apollo. Of crop glyphs he wrote this:

> We called them crop circles to begin with. They looked like a minor quirk of nature, another of those little local difficulties which farmers get patiently used to. Few of us took much notice at the time. Those that did expected the problem—if problem was not too great a word—to be solved in a month or two. But that was the summer of 1980. Our journey of exploration had only just begun.[15]

Archie Roy was equally interested in matters of human potential, including the so-called psychic abilities latent in human beings, for which he had earned the nickname "Glasgow's Ghostbuster," and being based in Scotland, the fact that he was alerted to a problem predominately occurring in southern England at that time is doubly interesting. Precisely which of his professional tools was sought most is a moot point but that aside, Roy's mission statement is charming but slightly inaccurate. The authorities had already been aware of anomalous cylindrical holes and strange events in crop fields way back in the 1960s, when a field near Charlton in Wiltshire had received a flattened circle placed across the boundary of two different crops, one of them root, one grain, and alongside it there was a small cylindrical hole. With no explanation as to who exactly called for them, the military and a young astronomer called Patrick Moore turned up to inspect this complicated event. After which a silence ensued and the whole affair was consigned to the archives.[16] Fast forward to the 1980s, when a farmer in southern England finally sought someone official to give him an explanation for the circles of flattened crop that had been appearing in his fields over the preceding years. Someone else was available, perhaps not to actually explain what was happening, but at least to provide the sense that things were being officially managed or at the very least monitored. That someone was Pat Delgado, and as was the case with Roy, not only was he an extraordinarily decent human being, but just like Archie Roy, his previous work experience fitted right in with the Swiss scientists, the domain of defense/communication networks, and NASA.

A highly trusted mechanical engineer, Delgado had worked for the British government at the Woomera Rocket Range (as the Australian's missile testing ground was then known). Later he worked for NASA at its first deep-space tracking stations to be built outside the United States—the very Oz location I linked to Frank Drake's Ozma allusion in chapter 10. Relative to his work at NASA's DSS-41, Delgado was one of only fifteen personnel entitled to travel between Island Lagoon (in reality, a site located at the edge of an ancient salt lake bed just outside the Woomera boundary line) and NASA's Goldstone and JPL facilities.[17]

A prize contender for inexplicable phenomena would manifest just next door to the Woomera boundary, and 149 miles from Island Lagoon. In 1998 a giant geoglyph was discovered by a NASA/USGS Landsat 5 satellite. This time the soil was not missing as in Switzerland but it had been furrowed into the form of

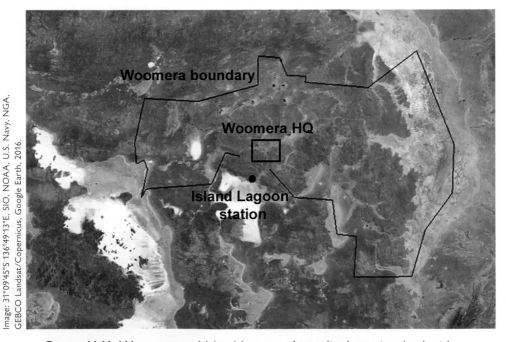

Image: 31°09'45"S 136°49'13"E, SIO, NOAA, U.S. Navy, NGA, GEBCO Landsat/Copernicus, Google Earth, 2016.

Figure 11.11. Woomera and Island Lagoon, Australia. Later involved with nuclear technology and the emerging electrical windmill technology, Pat Delgado says it was his seven years spent at the Australian desert facilities that got him interested in unusual and inexplicable phenomena.

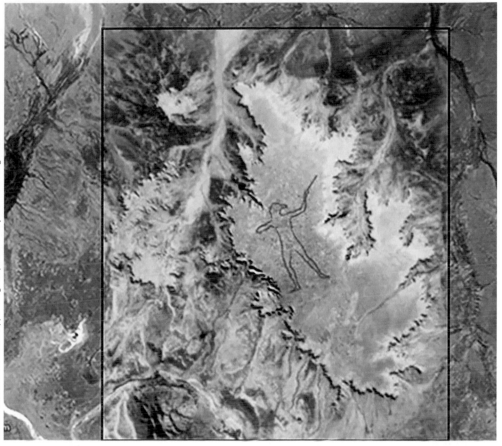

Photo: Landsat 5 Thematic Mapper image acquired on 28 June 1998, using bands 1, 4, and 7.

Figure 11.12. Woomera Marree Man. This figure has two remarkable facets: viewed close up, it looks like the Greek bronze statue of Zeus of Artemisius. The hairline and beard are clearly drawn. Albeit, this earth carving is the reversed image so that it is throwing with the left hand. Yet viewed from a distance, those details are less distinct, so the tied-back hair becomes a beak and turns this figure into the Egyptian god Horus the Hawk, looking over his left shoulder.

a man. It was at first considered to be a representation of (and perhaps made by) the local Aborigines, especially since the figure was thought to be using a throwing stick known as a Woomera (or *atlatl*). However the Aborigine community denied making it and furthermore, disowned it culturally. Not so picky, the Woomera authorities expanded their boundaries to absorb it into their exclusion zone. Astonishingly, the furrowing was similar to that of the US 1990 Sri Yantra event, and it also recalled the boustrophedon plowing practice, and that has commonality with the ancient Mount Tauredunum in Switzerland. The etymology of

Tauro is the Bull and *dunum* is not only an old Celtic word meaning "to come full circle" but also an old Ottoman word for the amount of land that can be ploughed by a team of oxen in a day.

Ancient dried lake beds were associated with the weird events in Australia, Oregon's Sri Yantra in the United States, and the Egyptian El Gouna event. This last was a complicated response by the Egyptians to an event that had all the hallmarks of a geoglyph and which was to affect the croppie community in the UK. (The full story of El Gouna, which means "the lagoon," and its connections to these other anomalous geoglyphs is available at Aulis.com/pathway.)

Given the association of the Woomera region with space exploration and missiles, it is astonishing that this geoglyph found an echo in the first American edition of the Russian novel *Omon Ra*. This important Russian novel concerning the nature of reality and the crewed space programs was written by Victor Pelevin in 1992.

During a meeting in Moscow, while researching *Dark Moon,* David Percy had a meeting with Pelevin and asked him how he came to write his story. Pelevin referred to "the internet without wires." Left unsaid was the precise definition of that concept. In this instance, these two men understood each other perfectly. Elsewhere, the mix of ETI events and military-scientific disciplines apparent throughout this book often leads scientists or pragmatists, open to the weird and wonderful yet mindful of the professional risks in speaking out, to adopt the tactics of creative thinkers such as Pelevin. His readers are left to work out for themselves whether he referred to intuitive connections gained through meditation (or any other means beyond the so-called norm) or, more practically, to inside information concerning the US-USSR space programs. This is a very special book; even in translation it haunts the memory. Reading it in the original Russian text no doubt packs an even greater punch.

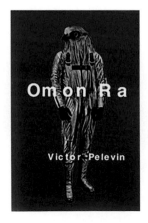

Figure 11.13. *Omon Ra* by Victor Pelevin. The cover of the first US edition, in 1994, features a falcon-headed man in a soft cotton spacesuit.

ASSURANCES

Back in England in the 1980s, notwithstanding his retirement, Delgado took to crop glyphs as a duck to the lagoon. He alerted the British press to the phenomenon in 1981, which conveniently enabled the LGM causative factor to be aired immediately and just as quickly be dismissed as ridiculous. With such heavyweights as Roy and Delgado on board, it would be fair to say that the authorities, potentially on the lookout after Switzerland's Billy Meier case and the Echallens cylinder of missing material, were in on the game from the start of the "crop circle phenomenon," as it was dubbed at the time.

Thanks to the media coverage, the number of people sitting up and taking notice increased, and in 1983, Delgado was joined in his research by an electrical engineer working for the UK local government in the county of Hampshire. Was it sheer chance that this person, Colin Andrews, was also responsible for formulating, implementing, and maintaining the arrangements for any civil emergency incident? Probably not. In 1985, Delgado and Andrews were joined by pilot Busty Taylor. His ability to fly with one hand and take a good photo with the other completed the skill set seemingly required for "a detailed investigation of the flattened swirled crops phenomenon," as they put it. And "sticking entirely with what they knew from looking at the fields and not attempting to 'prove' anything," they published the first book on the subject in 1989.[18] While not imputing connivance on the part of these three researchers, it is remarkable that whether entirely fortuitous or merely designed to look innocent, theirs was an excellent setup from the point of view of covert monitoring. NASA's technical knowledge allied to British local knowledge and emergency organizational procedures combined with a skilled navigator-photographer could create opportunities to learn as much as possible about this disruption in normal agricultural proceedings, while revealing very little.

Certainly, the varied responses of those reading the media, visiting the fields, and importantly, seeing the images of crop patterns from above would enable the formulation of policies that could (and did) control the public response to this phenomenon. Remembering the *Brookings Report,* one might ask if Delgado and Andrews, while collating information from the farmers, were also intended to reflect back to the inquiring public those two aspects of the Brookings persona likely to be most affected by ETI: the scientist and the mystic. With the benefit of hindsight, it becomes clear how, over the decades, this left brain/right brain approach to these anomalous events, with no middle ground, has succeeded in creating adversarial impediments to genuine research. As for Delgado and Andrews, not long after their immersion into the world of crop, something interesting occurred. Delgado, the fully-fledged scientist, became more attuned with the mystical aspects of the phenomenon while Andrews, the engineer, increasingly embraced the role of scientist.

Back at CERN, Berners-Lee had unwittingly kept pace with the emergence of the crop glyph phenomenon and the cylinders closest to Lake Geneva by working at his prototype software project throughout the 1980s, and by 1989, ENQUIRE had become the World Wide Web. In 1990, Robert Caillau, the Belgian computer scientist and creator of CERN's first hypertext system, joined Berners-Lee in finalizing his web design, and the Confignon cylindrical hole arrived in a field *not* so far, far away.

BARBURY

The following year, on the night of July 16, 1991, a triangular crop glyph appeared in a Wiltshire field of wheat just to the northeast of Avebury. That night local residents had heard anomalous rumblings in the sky, unlike any usual aircraft, and in the morning the sheep grazing in the adjacent field had put as much distance between themselves and the glyph as they could. When discovered on the morning of July 17, it looked like an equilateral triangle, but over time it was felt to be more accurately a 2-D representation of a 3-D tetrahedron—with embellishments.

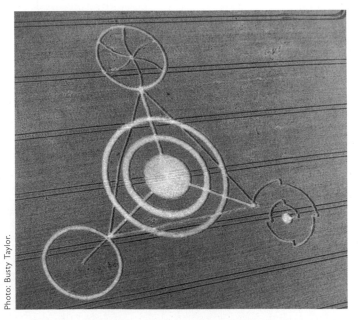

Photo: Busty Taylor.

Figure **11.14.** Barbury Castle crop glyph, Wiltshire, England, July 16/17, 1991.
Located at the foot of the ancient hill fort of Barbury Castle, north of Avebury,
it acquired the name of the Barbury Castle Tetrahedron.

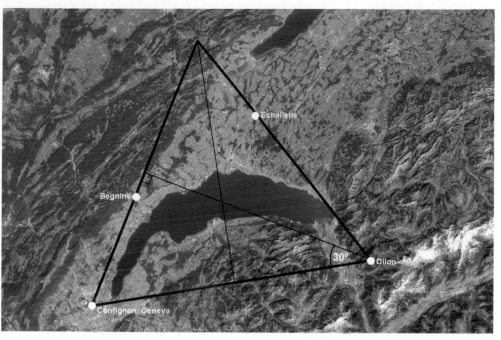

Image: Google GeoBasis-DE/BKG (©2009) Landsat/Copernicus, Google Earth, 2016.

Figure 11.15. Hindsight on four holes and a glyph. The location of all four cylindrical holes of missing mass in Switzerland imply the basic triangular form that will be revealed in the Barbury Castle crop glyph discovered on the 17th day of July, 17 months later. The Cydonian right triangle with Ollon 1981 as the 30° spiral mound angle at the southeast; Begnins 1982, near to the 90° Face angle; Confignon 1990, as the D&M Tor 60° southwest angle; and Echallens 1972, which started the cylinder series, is two-thirds of the way along a line creating the equiangular triangle that connects all four locations.

With hindsight it has been possible to see how the geometry of this glyph picked up on both the arrangement of the cylindrical holes that had been emerging in Switzerland and aspects of the special right triangle found at Cydonia. Which is fascinating, especially remembering that in England at that time the Swiss holes were not on the public radar. Yet matching this glyph to those cylinders brings out remarkable similarities. The Begnin, Ollon, and CERN locations recall the relative positions of the Cydonian Face mesa, the spiral mound, and the D&M Tor, respectively.

Everyone who visited the Barbury glyph was so stunned by its sheer beauty and size that a salient fact went unnoticed: this glyph had arrived overnight on the date that Apollo 11 had been scheduled for launch. Then not being into numerology, most people did not notice that the twenty-two years separating the

Apollo 11 launch from the arrival of the Barbury glyph summed to the second in the series of Glickman's master numbers, hinting if not at the master builder, most certainly at the combination of ten and twelve, the very numbers present in the Confignon CERN cylinder's measurements. Not being aware of the saga of the holes in Switzerland over the last three decades, few of the croppies at that time made the connection. There is a further link between the 1990 Confignon CERN cylinder and another glyph that had arrived in England the year *before* on August 12, 1989.

As already stated, in the early days crop glyphs tended to get names attributed to them according to their location and/or their pattern by those first on the scene. This glyph was the first ever seen to have straight lines incorporated within its circumference. The glyph's principal circle had been divided into four quadrants and in each quadrant the crop was lying at right angles to the next quadrant. At the very center was a small circle. It had arrived in a field very near the ancient site of Stonehenge and because of its form and proximity to that solar temple of Apollo, as the ancients called it, the crop glyph was immediately dubbed the Swastika. The dimensions of this amazing glyph were twice the diameter of the Confignon/CERN cylinder hole but the right angled quadrant lay of the crop immediately brought to mind the notion of latitudes and longitudes—tens and twelves—which of course, were the Confignon cylinder dimensions. When added together these also summed to twenty-two, while those same dimensions multiplied connected the cylinder to 2/3rds, in that 120 is 2/3rds of the degrees in an equiangular triangle making connections back to Switzerland and the form of the Barbury Glyph. The number 120 is also within a smidge of the eleventh iteration of that 5:6 series of numbers and as such completes at a larger scale the eleven-year solar cycle hinted at in the sum of the first iteration. So, to labor the point, we have the last of these cylinders in Switzerland, near to a site dedicated to nuclear physics "held" by two important crop glyphs, neither of which were claimed by human artists and both of which had associations with solar matters, the god Apollo, and the space mission of that name (the Apollo astronauts were issued Omega watches, as noted earlier, and the CEO of Omega had a replica lunar module at his headquarters in Berne, along with an Apollo spacesuit). And all this brings another interpretation of that swastika symbol to the table. The Sanskrit word *swastika* was commonly used as a greeting, with connotations of good luck, success, prosperity, and health, which is encouraging for those holding the hypothesis that all the truly anomalous events, including these early crop glyphs, are manifestations of ETI communication. However, given the Stonehenge connections with ancient man's tracking of the Sun and the Moon across the skies and the modern pre-occupation of attempting crewed space travel to the Moon and Mars, it might not be entirely coincidental that historically, the swastika symbol was abused by the Nazi party and that Nazi scientists formed the core of NASA's early space programs, both for rocket engineering and for space medicine concerning all bio-organisms, including the astronauts. Finally it is of interest that the first three Swiss cylinders appeared in the solstice months of June and December, while the fourth appeared two days after the February Celtic festival of Imbolc.

It's evident that anniversary rituals are observed by the space agencies, so the arrival and timing of those cylindrical holes and these two glyphs should have been noticed and connections made. In which case, were any past misdeeds generating guilty consciences, the joy might not have been quite so unconfined.

Enter cosmonauts Testudo Horsfieldi: in September these small four-toed Russian Tortoises were the first terrestrial organisms to orbit the Moon. They had returned alive, even after a failed skip re-entry of the Zond 5 spacecraft led to their 400 gram bodies experiencing a 16g ballistic re-entry. (Note those four, five numbers again.) They were shortly thereafter killed in the name of science and the forensic examination took place in October 1968. Publicly it was said that they had suffered some mild damage and weight loss due to a lack of nutrition during the trip. The actual results of the forensic examinations were released over a year later at the end of 1969. Post Apollo 11. When it was revealed that the weight loss was not minimal for these Russian Tortoises, it was a dramatic 10 percent loss over a total period of thirty-nine days from pre-mission to return to Earth. Furthermore, these two turtles showed damage to

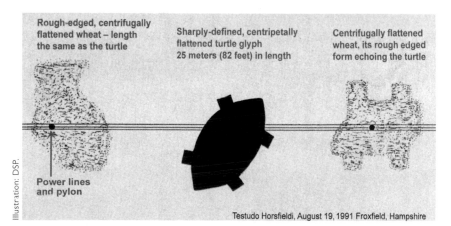

Rough-edged, centrifugally flattened wheat – length the same as the turtle

Sharply-defined, centripetally flattened turtle glyph 25 meters (82 feet) in length

Centrifugally flattened wheat, its rough edged form echoing the turtle

Power lines and pylon

Illustration: DSP.

Testudo Horsfieldi, August 19, 1991 Froxfield, Hampshire

Figure **11.16.** The very fine floor detail, overall forms and positions, strongly inferred a link between the three glyphs and the overlying energy source.

the intestines, spleen, testes, liver, and kidneys. To those who knew the signs, all of the medical details were indicative of radiation poisoning. And this in reptiles reputed to be able to withstand very much higher levels of radiation than the 3.5 rads officially attributed to their lunar jaunt. The reasons for this discrepancy were not provided in the report and none of the physical damage was attributed to radiation; in fact, conditions in space were not mentioned in this report. But it certainly alerted the authorities to the fact that bio-organisms do not do well in space because as author Garvit Rawat notes: "Literature going forth from the turn of the decade on the biological aspects of space travel would be deliberately vague on the issue of the dangers of ionizing radiation."[19] One Russian Tortoise had the identity number 22. Taken together then, the arrival of these Swiss cylinders with missing mass, or weight, along with two glyphs relating to nuclear activities and the Apollo space program of the 1960s, combined with the Meier UFO images and the SETI search for ETI interactions, surely gave the authorities more than a pause for thought at some level. Indeed, the Swastika glyph's name could just as well have been plucked out of the planetary collective consciousness that is another aspect of ETI, but on a more restricted, denser level. Bearing in mind that Switzerland was designated a neutral country during WWII, the message of these cylinders might also be interpreted as a neutral offering, communications designed to inspire us to think differently about many matters, including space travel and how we use our energy resources and how we manage our communications.

Three years after the arrival of the CERN cylinder in 1993, the CERN project finally changed from being exclusively the domain of academe and was made freely available to everyone, which is also apt, because in that year the fields of southern England were also experiencing something of a free-for-all. Since 1989 and the advent of the Stonehenge Swastika glyph and with the absolute shock created by the arrival of the Barbury glyph in 1991, skeptics and supporters of the status quo, some equipped with planks and ropes in order to make their own poor imitations, had been doing their best to take control of a phenomenon that, just like the World Wide Web, was becoming just a bit too "out there" for these flat-earthers. By 1993, when it came to crop glyphs, the creative "believer" was quickly dominated by the linear "realist," ensuring that *hoaxing* would be considered the *apparent* prime candidate responsible for flattened crops. But in their haste to manage this intrusion into the British countryside, via the media or within the fields themselves through simulation, those who wished it would all go away forgot, or rather ignored, the fact that in order to hoax or copy something, there must be an original.

Photo: Lucy Pringle.

Figure 11.17. The Spyder's Web glyph, Avebury, Wiltshire, England, 1994, arrived over a period of two days in a field tangent to the northeast quadrant of the Avebury stone circle. It showed no trace of human fabrication. One interpretation of this glyph was that it represented the ten-based binary system used for computing and the web. Another, that when seen from above it modeled ten meridians and ten parallels, and taken together the glyph inferred the spy and communication satellites orbiting Earth.

And if it's so necessary to copy such events while pretending you're not, then the "real thing" must be of importance and out of your control.

Whether these anomalous Swiss and British events, together with those connected with farms and farmers, are about matching the anomalous event with specific timings created by human conscious and subconscious thought (as we most surely affect the environment through our intentions as well as our actions), or whether the incoming information from "out there" is mirroring events (not necessarily overt) that have occurred or are occurring, either here on Earth or in space, some kind of action-reaction seems to be taking place.

PS AND QS

Berners-Lee had named his putative web project ENQUIRE, after the title of a hugely popular Victorian book on etiquette. Berners-Lee thought its title, *Enquire within upon Everything,* was suggestive of magic, and as the book served as a

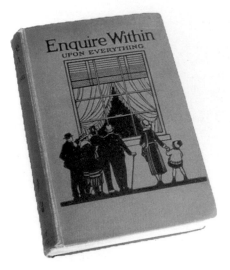

Figure 11.18. *Enquire within upon Everything,* originally published in 1856.

portal to a world of information, for him it was a good analogy of his web project. Not perfect, but good enough for his start-up.

The word *etiquette* refers to the manner of communicating appropriately between human beings in various groups and diverse circumstances, so it is indeed entirely adapted to both Berners-Lee's analogy and to the anomalous events surrounding the gestation and birth of worldwide electronic communications in Switzerland and elsewhere. And for these anomalous events, there were other books in gestation that were also going to help the inquirer along the way.

As for SETI, by the 1990s, everything was on course for continuing the search; well, sort of. NASA historians state:

> Through a methodical process that moved from a small in-house feasibility study, through a clearly-articulated design study, to a series of in-depth science workshops, Billingham and his colleagues built the foundation for a NASA-sponsored [SETI] search that commenced on October 12, 1992, the five-hundredth anniversary of Columbus's arrival in the New World.[20]

For those unfamiliar with the backstory, this again infers that NASA didn't get totally on board with serious SETI money until 1992. In fact, not even counting the expense of its three in-house facilities at Houston, Ames, JPL, and NASA had spent nearly $57,000 desperately seeking ETI during the nineteen years it officially funded SETI. But if ETI was only a part of the SETI agenda, what had been happening in the exobiology domain since Viking?

HUNGERFORD

The NASA newsfeed gave the impression that NASA scientists were weeping into their soup over the negative results of the Viking Labelled Release (LR) experiment concerning, as they put it, the "search for life's biomarkers." Not everyone had felt moved to tears, though; for some it was worse than that. The principal investigator of this experiment designed to check the Martian soil for signs of microbial life, Gilbert Levin, was shocked by the announcement that after being subjected to "control procedures" his experiment had provided "no evidence" that life existed on Mars.

Levin and his colleagues knew that this was untrue.

The difficulties involved in human space travel is yet another reason why NASA was not overly keen on finding signs of ETI or even bio-organisms out on Mars some eight years later.

The post-Viking policy on Mars dictated that the biologists turn their backs on Mars and let it become the fiefdom of the geologists. The question of life on Mars moved from the present to the past and turned into a discussion on the criteria required for life to have existed there in the first place. And it would stay that way for another twenty years, with a small disturbance in 1986 when Levin delivered a lecture at the tenth Viking probe reunion party.

Author Michael Brooks recounts that Levin set up fifteen reasons why his experiment could have produced a false positive and then proceeded to demolish every single one. He drew the obvious conclusion that his experiment had indeed detected life on Mars. The reaction to this talk was more than unfavorable. He was not invited back.

The rocky boat of "Mars science-on-our-own-terms" held steady.

Until 1996, when NASA "discovered" a signature of life inside a thirteen-thousand-year old meteorite from Mars (note that special number: 13). During the rumpus that ensued over that Antarctic find, labeled ALH84001, the press asked if NASA had changed its mind about life on Mars. And what about those earlier Viking samples?

It was then admitted that when it came to Levin's findings, the Viking verifying procedures "hadn't been sensitive enough to rule anything out." Which is a bit different from categorically stating that there had been no evidence. Yet that 1996 acknowledgement changed nothing for most scientists or for Levin. Given this hypocrisy by NASA and the moral abuse of its scientists, let alone the science itself, it is unsurprising that to this day, many scientists are not persuaded of the validity of NASA's claim for ALH84001.

It wasn't until ten years later, and a full thirty years after the categorical "no evidence" for life announcement, that a 2006 statement from NASA recognized that the experimental gear used to verify Levin's findings was calibrated *several orders of magnitude lower than originally thought.*

Writing in detail of this incident in his must-read 2009 book *13 Things That Don't Make Sense: The Most Intriguing Scientific Mysteries of our Times,* Brooks concludes, "There is now no one who wants to stick their neck out like Levin did. And no one has to. If not a scandal, it seems a shame."[21]

Well, let's go for scandal. This entire "life on Mars Levin incident" gives the strong impression that a climate of ostracizing peer pressure ensured the application of NASA policy, even when the science was an inconvenient truth, to coin a phrase.

An officially dead Mars resulted in a loss of interest, not just from biologists but also from the general public, who were well trained since the Apollo scenario that dead planets are boring, desolate places. So from the point of view of public perception management strategies, for NASA, no life on Mars was a good call. With people less inclined to ponder on the possibilities of life ever having been on Mars, the images from Cydonia, including human-looking sculptures, became even more difficult to consider seriously. Nevertheless, the next Mars project, the *Mars Observer Geoscience/Climatology Orbiter,* was considered a high-priority mission designed to expand the work of the Viking probes. Commonly referred to as Mars Observer, this project had little to do with biological research as such, so the word *expand* didn't really apply. Was this going to be the beginning of another round of covert recognition of things Cydonian?

Let's find out.

12

Observers, Surveyors, Dodmen, and Dreamers

Observations and Parallels between the US Pentagon Renovation Works and Orbiter Time Lines around Mars

MODUS OPERANDI FOR *MARS OBSERVER*

Whatever is said in public, it should be clear by now that the powers that be in the US space program and other branches of the US defense industry cannot resist encoding their preferred numbers and ideas into their projects, whether it is through their architecture (the buzzword in space hardware planning), their landscaping (hiding in plain sight, wrapped up as a cultural icon modality), or their timetables. Furthermore, as it's best to not build a program on the back of a failure, *Mars Observer*'s high-priority mission status infers that those Viking

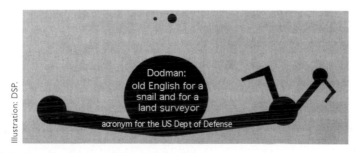

Figure 12.1. The Snail crop glyph, Alton Barnes, Wiltshire, England, July 9, 1992. Dodman and Hodman-dod are old English names for a snail, and a land surveyor also earned the epithet the Dodman.

results were rather more valid than was suggested publicly. Whether intended or not, the exploits of the Mars probes of the 1990s, *Opportunity* and *Surveyor,* bear comparison with the definition of a very good magician's best trick, as described by novelist Christopher Priest in his 1995 novel *The Prestige:*

> An illusion has three stages. First there is the setup in which the nature of what might be attempted is hinted at or suggested, or explained . . . as the trick is being setup the magician will make every possible use of misdirection.[1]

THE SETUP

In the intervening years since Viking, technical issues with conventional blood-and-thunder rocketry (and the 1986 *Challenger* disaster) had created launch problems for NASA across all its programs. So much so that by 1987, the decision was made to delay the *Mars Observer* 1990 launch until September 1992. Destined to reach Martian orbit in August 1993, *Mars Observer* launched just eighteen days before that grandstanding NASA-SETI funding announcement of October 12, 1992. While neatly re-aligning with the SETI funding timetable, this timing also conveniently honored that much-touted seventeen-year gap between Mars trips. Unaware of these events, having learned months before through intuitive thinking and the "internet without wires" that their book should be published by August 20, 1993, authors Myers and Percy were preparing the manuscript of *Two-Thirds* for publication. When *Mars Observer* finally arrived at Mars, three days prior to orbital insertion around the planet, on August 21, 1993, at 01.00 UTC* the spacecraft ceased communicating. It was never heard from again. Two months later, Congress voted out any NASA funding for the search for ETI.

VARIATIONS ON A THEME

Priest defines the second stage of an illusion as the performance, "Where the magician's lifetime of practice, and his innate skills as a performer, conjoin to produce the magical display."

---

*For everyday purposes, Greenwich Mean Time zones based on noon at this location are equivalent to UTC. However, UTC (which means Coordinated Universal Time, called Zulu time by the US military) is a 24-hour time standard, regulated by atomic clocks combined with Earth's rotation. Introduced in the 1960s at the dawn of Apollo, it is used in all digital applications by the military, the aerospace industry, and by astronomers. It still has its critics and decisions as to its viability are going to be reviewed in 2023, at the dawn of the Artemis lunar expeditions.

NASA's habit of slightly skewing published information means that when the agency offered several official reasons as to why *Mars Observer* went silent, it was rather difficult not to reach for the salt pot once more. Especially when the postfailure report stated that *"contrary to all normal practice,* communications were deliberately switched off just prior to this event, so unfortunately there is no record of what happened to the craft" (emphasis added).[2] Even the armchair rocket scientist would hardly credit that, nor the fact that the probe had been fitted with an engine derived from an Earth-orbiting satellite. These types of motors are not designed to remain dormant for months before being fired. If that were not enough, there were a collection of bizarre incidents and an unexplained glitch in communications just twenty-four minutes after launch, and finally, the front runner: faulty engineering had resulted in fuel leaks irretrievably damaging the craft.

Now you're looking for the secret . . . but you won't find it, because of course you're not really looking. You don't really want to know. You want to be fooled. But you wouldn't clap yet. (Christopher Nolan's 2006 screenplay for *The Prestige,* referred to in Priest's novel.)[3]

And indeed, the Cydonia enthusiasts were not clapping. They had been locked in a war of attrition with NASA concerning the re-imaging of the Face mesa since 1976, and many of them were not entirely persuaded that poor engineering and dubious flight control decisions had done the trick.

MARS GLOBAL SURVEYOR

Enter the magician's assistant: three years after the *Mars Observer* transportation blip, *Mars Global Surveyor* was launched in November 1996. Arriving at Mars, it successfully went into a highly elliptical capture orbit around the planet, much to the jubilation of the ground crew at JPL, because, technically, that was the point at which the *Mars Observer* probe was declared "lost" in 1993.

Coincidence No. 1: *Surveyor* (as it soon became known) had left Earth on a trajectory that had been designed to position the craft into insertion orbit at Mars for 9.11.1997. Whether this was intended as a cry for help or a message from the war office, September 11, 1997 was a remarkably prescient date when taken together with the next coincidence.

Coincidence No. 2: *Surveyor* reached Martian capture orbit fifty-six years to the day from when the initial Pentagon build commenced. Was NASA hoping to send an assertive message to anyone who might be observing the agency's progress, merely by the act of inserting this probe into orbit on this particular day? Given that astronomers consider it highly unlikely that there is any ETI life present in the solar system and that SETI-type endeavors should be aimed at the stars, this thought is apparently absurd. But then again, NASA astronomers and life scientists had by now established mathematically, to their own satisfaction, that civilizations capable of communicating must exist in the galaxy and have been present for the last five billion years.

Having arrived at Mars on time, *Surveyor* was supposed to spiral slowly down over successive orbits to its required mapping altitude, allowing drag forces to slow its speed.*

> The spacecraft must be flown at altitudes, or more correctly dynamic pressure levels, sufficient to produce the desired orbital decay rates, but not so low as to damage or destroy key spacecraft components in the presence of the anticipated atmospheric variation. To successfully establish the desired mapping orbit, the MGS spacecraft must aerobrake from its initial capture orbit apoapsis altitude of 54,000 km down to an apoapsis altitude of 450 km; or equivalently, an initial capture orbit period of 45 hours down to an orbit period of 1.9 hours.
>
> Once the spacecraft reaches an apoapsis altitude of 450 km, the spacecraft [*sic*] propulsive capability is sufficient to establish the final mapping orbit. . . . Additionally, the duration of aerobraking is constrained by the time it takes the local mean solar time of the descending node upon arrival at Mars, which is near 5:45 pm, to transition to the desired 2:00 pm condition—a four and a half month time period.[4]

Navigating space probes at a distance requires accurate and foolproof programming and clear communications, not only between the probe and its base controllers but also between all the flight controllers on the ground. The twenty-four-hour clock had been adopted by the armed forces specifically in order to avoid the confusion created by having the same number representing a period of time separated by twelve hours. So it is very odd to see these documents refer to 2:00 p.m. and not 14:00 hours; 5:45 p.m. and not 17:45. Then we ask what *about*

---

*The depth and quality of the Martian atmosphere requires a specific procedure for descending from orbit down to the Martian surface. The descent is organized so that the friction produced by brushing against the top of the atmosphere slows the spacecraft, and that results in lowering its altitude. Maximum drag and a measure of control as to the drop rate is obtained through the additional adjustment of the probe's solar panels. The aerobraking occurs in three phases as it encounters the specific layers of the Martian atmosphere and is controlled from Earth by NASA's Deep Space Network.

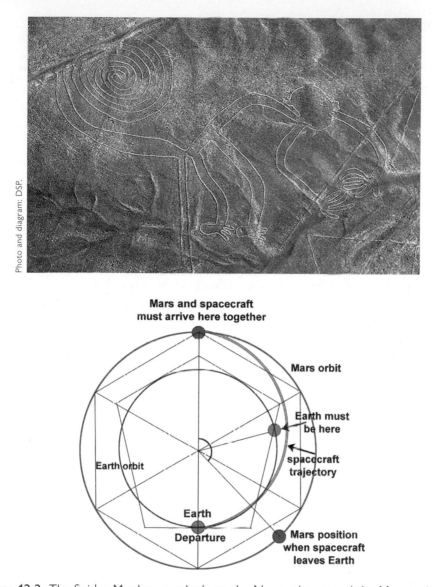

Mars and spacecraft
must arrive here together

Mars orbit

Earth must
be here

Earth orbit

spacecraft
trajectory

Earth
Departure

Mars position
when spacecraft
leaves Earth

Figure 12.2. The Spider Monkey geoglyph on the Nazca plateau and the Mars trajectory diagram. The Mars trajectory diagram strongly infers the five-sided Tor on Mars and the Pentagon five-sider here on Earth. Also when the Nazca plateau geometry is examined in detail (see Aulis Online) the location of the Monkey geoglyph at Nazca finds the analog position of the five-sider D&M Tor on Mars. "Foreign" is the Pentagon inhabitants' term for incoming craft of any unknown provenance, including sightings of extraterrestrial spacecraft. When human beings travel to Mars, they too will be extraterrestrials, and upon arriving at the red planet, they will be traveling in an incoming "foreign" spacecraft. The most intelligent of the New World monkeys, the spider monkey is totally foreign to Peru and to the desert climate. It rarely comes down to the ground, living in the high canopies of Brazil, Central America, and Southern Mexico. It does not have an opposable thumb, using its four long fingers as a hook, and when meeting other spider monkey troops, hugs are exchanged in order to defer aggression and convey greetings.

the proliferation of the numbers 5 and 4 in this description. Even in the last sentence where, suddenly coy, 4.5 months is written out in words.

Hold those thoughts because those significant numbers 4 and 5 are going to feature again and again in this chapter and later in this book.

DRAMA QUEENS

Martian dust storms have a seasonal period of roughly October to April, and scientists at JPL stated that the dust storm threat to *Surveyor* was of particular concern. However, this probe was intended to gather data about the red planet's atmosphere, gravity, and magnetic fields, and *Surveyor*'s trajectory was specifically designed to be aerobraking through the Martian atmosphere at the height of the dust storm season. So one can legitimately conclude that the use of the phrase "of particular concern" was merely titillating phrasing used for dramatic effect. Any organization that wittingly exaggerates the reporting of totally expected conditions or skews its reports to the public, even by a narrowest of margins, cannot be surprised when such statements are held up to scrutiny.

THE EFFECT

The aerobraking procedures for *Surveyor* would normally have taken from September 1997 to spring 1998 to accomplish. Unfortunately, the master plan didn't go the way the agency intended, or so we are informed. Damage to one of *Surveyor*'s solar panels forced a major reevaluation of the probe's mission and timetable. JPL redesigned the orbital descent procedure to accommodate the delicate solar panel, which resulted in less aggressive aerobraking and thereby added a year onto the probe's actual arrival into its required mapping orbit. Remaining in a polar orbit and still operating as Sun synchronous, the daytime equatorial crossing was now established for 2 a.m., instead of 2 p.m. (14:00). The descent procedures therefore became a mirror image of *Observer*'s original plan, thus ensuring "a constancy of shadows and radiation for the various scientific instruments and cameras" in the words of JPL.[5]

Another pinch of salt is needed here. Had it been decided at some level that prior to entry into Martian orbit, *Observer* would "go dark," thereby allowing it to acquire data off the record, while thoroughly examining the anomalous region of Cydonia and that Face mesa and without being accountable to NASA's vociferous and formidable public critics? If so, a solar panel "problem" would be the perfect moment for discreetly adjusting the *Surveyor*'s orbital plan to mirror the *Observer*'s track. As the original orbital plan with its 2 p.m. equatorial crossing had been that of *Mars Observer*, then it would make sense to use the same plan—and the same numbers. Conversations between controllers using the civilian clock time of two o'clock would not necessarily reveal the presence of the hidden *Observer*. And only the magician and his assistants would be in on the trick. Granted, NASA could not actually release another image of the Face until a following mission had successfully imaged the region, but that's where *Surveyor* came in.

> The third stage is sometimes called the effect or the prestige, and this is the product of magic. If a rabbit is pulled from a hat, the rabbit, which apparently did not exist before the trick was performed, can be said to be the prestige of that trick . . . audiences and critics were most intrigued by the setup and the performance, for the performer the prestige is the main preoccupation.[6]

PHOTO-FITS

During *Surveyor*'s revised descent into the designed orbit, the engineers suspended aerobraking twice: for a period of three weeks from October 12 to November 7, 1997, and then for a period of some six months for the science phase orbits scheduled from the end of March/early April 1998 to September 1998. Throughout these two periods, the instruments and cameras were switched on. Here it should be remembered that during the three years prior to the departure of *Surveyor*, there had been further intense discussions between NASA and space enthusiasts over the acquisition and publication of images from Cydonia in general and the Face in particular. One NASA representative, debating the matter on a prelaunch US TV show, met his match when the program's host, during a discussion that was getting nowhere, tried to resolve the standoff. Having understood that the official line was that the Face was just a lump of Martian mesa, he asked, "Then why don't you just take the pictures and let everyone see them for what they are?" The NASA spokesperson, dumbfounded, offered no response! His reaction conclusively proves that the authorities are perfectly well aware that this mesa is not simply a lump of degraded rock. Yet, while everyone endlessly debates the merits of taking these images, no one is focused on the real point of the discussion, in this case the potential revelation of intelligent design in the images returned from Cydonia. And so the argument becomes that of the "believer" versus the "realist," and we are back to the *Brookings Report* model of responses to ETI interaction. At this point, the *Brookings Report* is looking increasingly like a management plan rather than a prognostication.

It was a comment made by Michael Malin when looking at the Billy Meier UFO images in 1981 that made it clear that all this fuss might be shadowboxing on the part of NASA. Back in 2015, when I wrote this chapter, Jim Dilettoso's account of Meier's UFO images had included this comment made by Malin:

> Photographs are pretty, fun, and impressive. But photographs of anything are poor evidence. In a hierarchy of probative value, photographs and sound lie at the bottom, because they both are a recording of an ephemeral event, *subject to manipulation after the fact* (emphasis added).[7]

FACE OFF

By the time *Surveyor* arrived at Mars, the attitude from NASA and Malin's camera team had become not so much unbridled enthusiasm as something more in line with "it's not a priority but if we have time we might try and get something." (As the principal investigator for the Mars Orbiter Camera on *Surveyor,* Malin kept his team at a distance from NASA headquarters. His Malin Space Science Systems (MSSS) was located in San Diego, California.) Yet despite the professed lack of interest from NASA, the Cydonia Colles region had been scheduled as one of *Surveyor's primary photographic targets* during that second aerobraking pause. So that just after midnight on April 5, 1998 (here are those numbers again: the 4th month, the 5th day) the Face turned out to be among the earliest of the images returned to Malin's team. *Surveyor* photographed the Cydonian Face from a distance of just over 443 kilometers (275 miles) and demonstrated to the entire satisfaction of NASA that this was nothing more than an eroded knob. This geologically correct description of the mesa utterly discounted the right-angled contour still very evident at top left of the image (and strongly recalling the Manos geoglyph at Nazca) as well as all the data acquired by the US military via its RV PSI-Ops programs. It also contradicted digital image processing expert Mark Carlotto's conclusions about the Cydonia complex.

However, nothing phased Malin: conveniently forgetting his own criteria for probative value, he rather rudely said that anyone who disagreed with his definition of the new image was themselves a "deficient knob head." While Michael Carr of the USGS, after studying the *Surveyor* image, said that the geological features were the same as those elsewhere on Mars, which was technically correct as far as the geological composition went. Then ignoring the merits of keeping the American taxpayer interested in Mars exploration, he added hopefully, "I don't see a face. Do you?" But then, Carr also hoped that this *Surveyor* image would "scotch this thing for good," which turned out to be a vain hope. And neither did all this fuss remove the public interest in the other anomalous forms at Cydonia, including that pesky five-sided D&M Tor.[8]

Surveyor had also captured the D&M Tor in greater detail than had been possible with the lower resolution Mariner and Viking cameras. The resultant images revealed an extension to its northwestern slope that was previously little appreciated. As space imaging anomalies specialist Keith Laney has pointed out, this Martian extension bears a remarkable similarity of form to the Mall Avenue component at the Pentagon.

Image (left): Mars Orbiter Camera 2-484, NASA/JPL/Malin Space Science Systems, 2003. Image (right): Google Earth, 2018 38°52'15.38"N 77°03'21.44"W.

Figure 12.3. The D&M Tor, Cydonia shown here alongside the Pentagon with the renovated Mall sector brought into prominence.

DOWN TO EARTH

As it turns out, both the *Observer* and *Surveyor* probes are linked to the Dodmen's five-sided house via the repetition of those numbers 4 and 5 so prevalent in the Martian aerobraking procedures. These are laid out in the landscaping of the Mall sector, creating the Pentagon's analog to the D&M Tor extension. The special attention paid to this particular section, along with the timing of all aspects of the Pentagon Renovation Program, reveals more of the hidden rituals we keep stumbling across in US government programs. What looks like numerology, superstition, or one coincidence too many to the outsider is really a form of practical magic that gets the job done while reinforcing the history of the building and of the US armed forces in the minds of those who currently live or work within its confines.

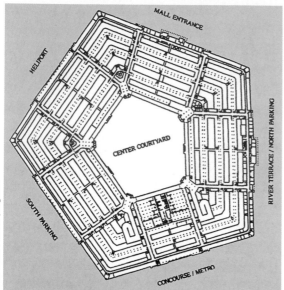

Illustration: PenRen program.

Figure 12.4. The Pentagon with the sides named. Note that illustrations of the Pentagon as with this one, do not necessarily reflect its actual orientation relative to the planet's axis. The angle between the river side and the mall side is 30° east of due north, while the mall center line is 7° west of due north.

Image: PenRen program.

Figure 12.5. PenRen program, Pentagon, with Mall sector.

The much-needed upgrading of the Pentagon's infrastructure was originally scheduled to last sixteen years—one for every month of the original build.
Furthermore, the earlier *Observer*'s September 25, 1992, launch had coincided
appropriately enough with the scheduling of the preliminary works: tackling the
heating and refrigerating environment. Then when *Surveyor* arrived at its capture
orbit in accord with the fifty-sixth anniversary of Pentagon, those environmental
preliminaries were supposed to be completed. These implied in-jokes and timings
lead us back to Earth with a thump as we return to the Pentagon to examine its
renovation schedule more closely.

From the building's creation in 1941, the principal landscaping around the
Pentagon had been concentrated around the river side. In status terms, the river
side was home to the "top brass"; its entryway and gardens fixed the sightline across
the Potomac toward Washington's principal buildings and linked that side of the
Pentagon with the artificial lagoon. Then going counterclockwise came the Mall side,
the Arlington (heliport) side, the south parking side, and the Concourse (metro) side.
(Interestingly, London's new MI6 building, open for business in 1994 and looking
exactly like a Sumerian ziggurat, is known colloquially as the River House).

ALL THE WORLD IS A STAGE

Originally planned for 1990 to 2007, this total renovation of the Pentagon program (immediately called the PenRen by all concerned) was going to take place
in exactly the same circumstances as had much of the original build. Slab-by-slab
from bottom to top, with most employees in situ, the noise of jackhammers and
construction work all around them.

The PenRen's seven build stages were split into three acts: design, procurement,
and completion.

Pentagon Renovation Program Schedule

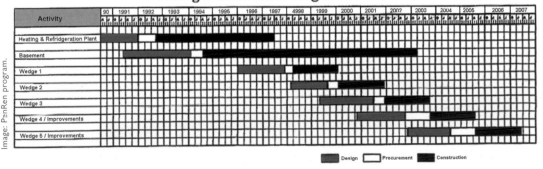

Figure 12.6. Pentagon renovation schedule. This schedule has the year 1998 printed as 4998.

The first of these seven stages was for the preliminary environment (the already mentioned heating and refrigeration infrastructure). Then came the basement, followed by five corner wedges (not the five sides as one might suppose). For those who noticed such things, these groupings of three actions for each of the seven build stages and five corner wedges were poetically appropriate for a building schedule, as they incorporated the Freemason's favorite numbers of 3, 5, and 7. And the seven build stages also recall the ancient practice of dividing the Northern Hemisphere into seven latitudes, which also has a correspondence with the equally ancient notion of seven energy centers in the human body from the base of the spine to the crown of the head.

Looking at the finer detail of this schedule reveals other known Masonic building practices. The basement works would start in the northeastern corner, traditionally the location for laying the *cornerstone* of a building, while the renovation of the aboveground wedges would start in its opposite corner, at the southwestern wedge. Food for thought: in engineering terms, this arrangement aptly recalls Nazca's Spider Monkey glyph, situated to the southwest of the Nazca plateau, and Manos, located to the northeast of Mono, as the Spider Monkey is called in Peru. And both these geoglyphs have 5-4 finger arrangements on their left-right hands, respectively.

Then comes a Martian aerobraking connection: wedge numbers 4 and 5 would incorporate site improvements. This term covered all exterior modifications, repairs, storm drainage, revised traffic circulation, landscaping, and redesigned parking areas. Site work, except for the terrace entrance under wedge one, would be constructed during phases six and seven.

Phase six was scheduled from late 2000 to mid-2005, and phase seven from late 2002 to mid-2007. These details are of particular interest because photos taken of the Mall Avenue section show landscaping replete with those four and five motifs starting *well ahead* of this prescribed scheduling plan.

TALKING ACROSS THE VOID

Author Steve Vogel records that politico-economical discussions stopped the PenRen preliminary stage from getting underway until 1993, with the basement area under the northeastern corner between the Mall and the river side only starting in 1994.[9] Yet the original schedule indicates that the prelims were only a few months behind schedule and right on schedule for the 1994 start to the basement.

During the PenRen program, the river side landscaping and its access were altered, turning this former ceremonial entryway into something less accessible and even more private. And the Mall section now got a major landscape upgrade.

October 2012

Four grass squares
become five and
the helipad H
is rotated 90°

March 1949

Four grass squares
'Battle of the Pentagon'
the anti-Vietnam
protest of
October 21, 1967

April 1988

Three grass
squares and
the parking lot

April 1999

Remote Delivery
Facility Project underway
with completion of
underground docks
at the end of 2000

August 2004

The French château
-style garden is laid
out in fours and fives

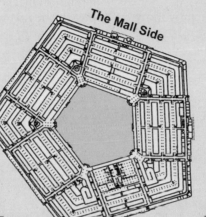

The Mall Side

April 2000

The RDF soon
became known as
Doc's Dock

December 2003

A pentagonal concrete
helipad covers up all
preceding pentagons
including five created in 2002

September 2001

Two nested
pentagons
now visible

July 2001

Four grass squares
lead to three
nested pentagons

Doc's Dock

Doc's Doc
The entryways of the
underground docks
– as seen from
the northwest

Figure 12.7. The Mall section renovation, with aerial photos showing layout and
construction changes for key dates from March 1949 through October 2012.

Looking at figure 12.7, one can see how the evolution of the Pentagon's Mall section works. Defined by the busy roads that run below the boundaries of this section of the Pentagon's grounds, in March 1949 it appears neglected. All that can be seen of note are paths forming four squares below the Mall entrance plaza.

On October 21, 1967, this area was occupied by antiwar protestors who were tacitly allowed into this section and who subsequently scaled the side of the Plaza to access the Pentagon itself before being forcibly removed. Fully described by Vogel in his "biography" of the Pentagon, ultimately this Battle of the Pentagon does not reflect well on anyone, either military or protestors. Ironically, the military were trying to present a stoic attitude toward the protestors, who were happy to use civic violence (spitting and verbal harassment) in order to express themselves. The military in some cases responded to such provocations with their instinctive training. There were no positive outcomes. Until the PenRen got underway, the Mall section had remained a forlorn memorial to the events in Vietnam and the differing aims of the military and the civilians back home. The black-and-white image in 1988 (figure 12.7) shows this desolate section: parking lots abound, and one of the four squares has succumbed to concrete.

The PenRen utterly transformed the Mall Avenue sector, and it took virtually the entire duration of the PenRen program to do so. Once completed, the Mall Avenue section had acquired a ceremonial area some five hundred feet in length. A length that recalls the five-hundred-year life cycle of the phoenix from ashes to rebirth. Unsurprisingly, the rebuild of the southwest section, which had been damaged during the 9/11 attack, was called the Phoenix Project. The refocus of attention toward the Mall Avenue landscaping effectively turned the Earth-based Pentagon into what could be described as a visual replica of the Mars-based D&M Tor, as imaged by *Surveyor*. For those with an interest in space exploration, this was seen as a remarkable coincidence, internal rituals, or a deliberate nod to this extraterrestrial landform. Remembering the storyline accompanying *Surveyor*'s trajectory to Mars, when taken together with the landscaping that would occur at the Pentagon, one has to wonder if the PenRen schedule and the Mars work schedule were discretely reinforcing and reflecting the alliance of military might and space exploration. Because the renovation of the Dodmen's purpose-built five-sided HQ and the construction of NASA's purpose-built four-sided Washington headquarters were taking place over the same decade.

This is not to say that Masonic practices are still the rule at the Pentagon or anywhere else in Washington, DC, yet Masonic ideals had influenced the choice of the location of the government's capital as well as the design of its principal buildings. They had also been overtly manifest during the Apollo space program. So it is perfectly in order to wonder if the quasi-ritualistic underpinning of intent and willpower, in which the actual process of building is as important as is the aesthetic result of these same endeavors, has been here made manifest through architecture and landscape on Earth and in the timing and actions of spacecraft around some geometric landforms and landscapes out on Mars.

Those images of the transformation from the old Mall Avenue to the new Ceremonial Mall certainly require a closer look.

Moving forward a decade and looking at the first color image (figure 12.7) of April 1999, things have started to move: work was underway all over the far end of the Mall Avenue triangle, and all that concrete was being ripped up. A year later (April 2000), the works had scalped the terrain as far back as the edge of the Plaza. And by the end of December, the upper surface was marked out for its geometrical designs.

DOC'S DOCK

While Vogel's history of the Pentagon does not enlighten us as to any esoteric encoding concerning the Pentagon building, thanks to his hard work it is possible to ascertain why the Mall Avenue landscape had been worked on so much earlier than was provided for in the original PenRen schedule. Politics would seem to play its part, although once again, that salt pot might come in handy.

The story goes that the bombing of the World Trade Center in 1993, the Oklahoma City bombing of 1995, and the bombing of Khobar Towers in Dhahran, Saudi Arabia in 1996 had already alerted the Pentagon's director of administration and management for the secretary of defense, David O. Cooke (known to all as "Doc," or the "Mayor of the Pentagon") to the dangers of delivery trucks approaching close to the Pentagon. After each of these bomb attacks, Doc had taken his concerns to Congress. In his opinion, the delivery bays around the South Entrance were far too close for comfort, and he wanted more stringent controls on the 250 trucks arriving *daily* on the south side. What he really wanted was a receiving bay or annex built at a safe distance from the main building. But all the Mall annex plans already prepared for the PenRen had been withdrawn at the end of 1993 for politico-economic reasons, and no one on the Hill was listening hard enough.

It was not until dual attacks on US facilities in Nairobi and Tanzania on August 7, 1998, that Doc finally got their full attention. That autumn, "signals from the Hill," as Vogel puts it, suggested that he was "likely to get the money for his annex," and that was enough for Doc to reactivate his slumbering Remote Delivery Facility Project. The name is entirely descriptive of Doc's purposes but equally applicable to Martian adventures.

The new annex quickly acquired the name Doc's Dock of course, and in figure 12.7, it is viewed from the northwest and the Mall side of the Pentagon is at the top of the frame. It was built between the end of 1998 and 2000 underneath a portion of this Mall segment. If that explains how the aerial views of the Mall are out of sequence compared with the original schedule, it doesn't explain the continuing symbolic geometric landscaping on the surface of the Mall gardens. The overhead views from July 2001 still show significant ground works underway, but now there are three nested pentagons at the far end of the layout. The four grass squares have been redefined and relocated farther from the Plaza, and again that specific numbering comes into play, as we now see four cul-de-sacs with five sides. The first rows of shrubs have been planted in a 3-6-3 arrangement. From the air, the outline of the works look rather organic, although some were reminded of a mummy coffer casing from Egypt. Interestingly, this July 2001 image no longer features in the historical imagery of the Pentagon available on Google Earth Pro. Although for that same year, exceptionally, the historical overheads show *three* images of the Pentagon all taken in September. The first was taken on the ninth, two days before 9/11. The next two were taken afterward, on the twelfth and thirteenth.

As for the Mall works, by September only two of those nested pentagons were still visible (figure 12.7). During 2002, these were again transformed and redefined using concrete and earth to create five smaller pentagons. In February 2003, the central pentagon seems to have undergone a change, and then by December 2003, the snowfall in the photo clearly shows that the whole lot had been covered in a concrete pentagonal shape defining the helipad and inscribed with an *H* facing the building.

Designing nested pentagons and then going to the trouble of laying them out on a structure *in the process of construction,* only later to cover them with concrete seems to be a discreet memorial to something else. What? Receding snows also revealed that the four grassy squares had now acquired a fifth member installed in the gardens below the Plaza. By August 2004, a formal garden in the French chateau style had been laid parallel to the Plaza in another arrangement of fours and fives. And two more rows of shrubs, this time numbering of 2-5-2 and 6-12-6, had joined the earlier 3-6-3 arrangements, completing the landscaping of the Mall's main ceremonial area through the use of numerical sequences of shrubs laid out alongside pentagonal and square forms, the fives arranged in fours and the fours arranged in fives.

The whole pattern working as mirror pairings.

Then all seemed quiet with the landscaping, until eight years later, in 2012, when an aerial photograph revealed that the helipad's *H* had been rotated 90°. Yet the former *H* remained, thus forming another set of four squares. One cannot accuse the landscapers of sloppiness, so obviously it was intended to leave the former *H* visible. The October 2012 photo in Fig 12.7 was taken on October 13, that day and month memorable for the decimation of the Knights Templar in France. Remembering the Pentagon's link to those five-sided bastions of France, the Pentagon's French-style landscaping is yet another nod to their confreres across the pond and the Axe Majeur park layout: the Pentagon's endpoint helipad matching the park's *H*-shaped laser light endpoint also infers a connection to that west–east Cydonia sightline from the City to the Wall. Which brings us to the shrubbery layout at the Pentagon and to exactly what the Mall Avenue landscaping signifies. And for clues I looked to the former ceremonial side, that of the river frontage. That garden had also been redesigned—planted with flowers set out as gold pentagonal stars against a red background, overtly—that seemed to be picking up on the insignia of the high-ranking officers of the armed forces, the Pentagon's five sided form, and even its five-floors-above-ground architecture. The colors used might have alchemical or administrative references for those in the know. And from that any planting in the Mall area would also convey meanings to those in the know.

At first glance, the Mall's grouped shrub numbers look to be a construct of Freemasonry rituals concerning master, senior, and junior wardens. And they result in a hidden reference to the first ETI "contact number" referred to by Glickman: $5 + 6 = 11$, the intuitive and the apprentice for numerologists and Masons respectively.[10] As the first geometry of the D&M Tor revealed it to be a 5:6 pentagonal-hexagonal structure, this is apposite. As is the fact that the number 12 is found throughout the Mall shrubbery, and that takes us to the rotated helipad and to the Hermes/Mercury laser receiver, the twelfth component of the French Cergy l'Axe Majeur layout—again.

None of this would necessarily be of significance here, except for that original Pentagon anniversary link to *Surveyor's* tour of duty and the fact that, as stated, changing the aerial imagery of a piece of land throughout these renovations—imagery that would eventually be completely covered up—is obviously not a structural necessity. And structure and measure—especially height—were the next key to this arcane landscaping.

IN MEMORIAM

While preparing these photos for this section, David Percy noticed that the layout of the gardens is somewhat similar the profile of the now defunct space shuttle at launch. It is possible to see that this area is indeed a memorial garden to those who have gone before, in that the shuttle originally designed for the ISS matches its outline, the two outer sections of green planting indicating the solid fuel boosters.

Furthermore, this area is also reminiscent of a version of the Ares Space Launch System that NASA is trying to build in order to transport astronauts to Mars. Whether Ares actually gets built as planned is another question, but now that the greater detail from *Surveyor* has been incorporated into the Mall extension on the US Pentagon here on Earth, if ever intention was laid bare for "no one" to see, it is there.

The military and the space agency want to go to Mars, and in the Pentagon Mall gardens, the aspirations of the space agency relative to Mars, the politics, and

Photo: Google Earth, 2018 at 38°52'26.41" N, 77°03'23.30" W. Overlay: DSP.

Figure 12.8. The shape of the Pentagon Mall gardens suggesting the outline of a space shuttle at launch.

Photo: Google Earth, 2018 at 38°52'26.41" N., 77°03'23.30" W. Overlay: DSP.

Figure 12.9. The shape of the Pentagon gardens also suggests the outline of the Space Launch System's first stage with boosters.

the military might involved in getting there are hidden in plain sight. Furthermore, the building itself shares common ground with the geometric model of the orbital timings for a successful Earth-Mars trajectory. In this model, the circumscribed Pentagon represents the Earth orbital data. The circumscribed hexagon represents the Mars orbital data. The model also reproduces the pentagonal-hexagonal combination found by researchers when they looked at the first images of the Martian five-sider back in the 1970s (see figure 12.2, page 275).

This is only a brief analysis of the hidden messages encoded through the Pentagon renovation schedule. It is, however, instructive in that the Dodmen and the Fargonauts of the twentieth and twenty-first century are fully aware of the secret ways of encoding hopes and aspirations as practiced by the Founding Fathers, whose ideas had already influenced the choice of the location of the government's capital and the design of its principle buildings. Yet these eighteenth-century men were only continuing practices established by even earlier cultures across the world. The ceremonial founding of a building through laying the cornerstone on a certain day at a specific time in accordance with the astronomical configurations which prime at that moment has always been deemed to transfer and harness the power of the land into the building, in order that its users can benefit from the environment created within the building to achieve their aims. And that is at the root of the Arthurian myths, the land and the king as one. On a bigger scale landscaping is intended to achieve the same result, the power being translated through the design and the interface between nature and man is achieved by inhabiting the entire location. Here at the five-sided Washington building and at the D&M Tor on Mars, as in the Arthurian cycle, we can hear the faint echo of once and future kings.

The timing of the arrival of *Surveyor* at Mars on September 9 does have links to the *human* space program. Had *Surveyor* been able to enter into orbit as originally planned, it would have been taking the Cydonian images in year seven of the original PenRen plan. As we have seen, the Mall's end point of the helipad's rotated *H* produces both an *I* and an *H,* while year seven of the PenRen not only recalls the crown chakra in esoteric terms but also the fact that the start of crewed space flight began with the selection of the original Mercury Seven astronauts. Which further reinforces the links between the space program and that hermetic Cydonian layout over at Cergy in France. The astronauts recruited for the Gemini and Apollo phases of the space program were called the Next Nine, and 9 is a number that breaks into 4 and 5. The Martian journey is currently expressed as sending at least four astronauts to Mars, but if the crewed space projects are slow in manifesting, NASA was not going to be backward in coming forward when it comes to design layouts incorporating hidden wishes and messages. The imitation of the Cydonian landscape by these architects of time and space, whether in America or in France, means that it has been quietly acknowledged that the Cydonian landscape was indeed laid out by a cognitive intelligence. And with that thought it's time to cross the river and check out the space adminstration's four-sided new build.

INDY

In 1987, NASA finally received the go-ahead to do exactly what the DOD had done forty-six years earlier: consolidate all its Washington offices into a single, new building. Construction began in the summer of 1990, and the first employees moved in during 1992. The move took place in stages and was completed in 1995. The new NASA building has nine floors above ground level, and its four sides unite with the five-sided Pentagon as nine. While the 555-foot length of NASA's new headquarters is just six inches shorter than the height of the Washington Monument, it is also the average of the two lengths given for the Pentagon's renovated Mall section. And 555.56 feet also happens to be the diameter of the British chalk mound, Silbury Hill, over in Wiltshire, which itself is an analog of the anomalous cone or spiral mound on Mars. All of which fits right in with the hypothesis that nothing about these programs is coincidental.

For those who appreciated the esoteric significance of the European Union's administration offices being located on the thirteenth floor of their large building, and for further esoteric study, the NASA administrator has his personal office suite in the northwest corner of the building's ninth floor. The Columbia Café (named after the shuttle) is also situated on this floor, while at the Pentagon, the Café Ground Zero is situated in the center of the courtyard.

So we find parallel timings for the Dodmen's repairs and the Fargonauts's new build, and we find the aptly named *Surveyor*'s areobraking numbers, along with a lot of heavy symbolism in NASA's new address. Listed as Two Independence Square, when it comes to the actual policy governing ETI events, the Hollywood attitude to ET as expressed in the movie *Independence Day*, might be too near the knuckle. At any rate NASA generally prefers to cite its Washington location as 300 East Street SW.

MORE FRENCH CONNECTIONS

The NASA document aptly titled "Hidden Headquarters" is a history of its various Washington office buildings prior to the construction of its Two Independence Square new build.[11]

If one cares to notice, it also features information as to the connections between NASA, the Pentagon, the ESA, the military, and France ancient and modern. This document, said to be the result of a brown bag session (an informal discussion over lunchtime sandwiches) has a section titled "Fun Facts—A Tree and a Bridge." With no explanation for this choice, the tallest tree in the world, a giant redwood, is used to illustrate the information concerning the topping-out ceremony for this building.

Why have a redwood and not the actual evergreen tree used during the ceremony? Was the color red important? It references immortality in alchemical terms, but it is also the color of that footbridge over at Cergy in France. Yet this redwood photograph is placed next to a picture of the Millau viaduct also in France and the tallest bridge in the world. So height might have something to do with this odd pairing, but rather more likely, time is the key to this juxta-position. A redwood tree can be some 2,000 years old and the Millau viaduct started construction in 2001. Could it be that the reference to a space odyssey is the actual hidden meaning here? The NASA document tells us nothing but infers that this juxtaposition is illustrating the moment in the topping out ceremony when a building girder is painted white and everyone, from government officials down to the construction crew via the architect, signs the girder. The NASA document practically begs the reader to seek out the reasons why these two images are being shown together. So one might speculate that the progress of these two builds also infers that all the officials on both sides of the Atlantic were involved in some project together. It may be a stretch too far for some, but the fact is that NASA launched the remainder of the Surveyor project on April 7, 2001, and called the orbiter 2001: Mars Odyssey for good measure. Over in France the ESA's *Mars Express* would launch in 2003 during the viaduct's construction, while the nominal end of the 2001: Mars Odyssey mission coincided with the completion of the Millau viaduct in 2004. Another Fun Fact: Norman Foster's architectural firm was behind the Millau Bridge project and would be behind Spaceport America, a project conceived within California's Stanford University Engineering department in the 1990s but which would not get under construction until 2006.

Coming back to NASA's Hidden Headquarter's document, one might ask: Why not simply show a photograph of the actual girder signing ceremony? Well, NASA's signed girder became part of the auditorium, and all those signatures were forever hidden from view. Apparently the great and the good wished to remain incognito.

However, substituting this high-flying French bridge for this ceremonial moment makes sense from other historic and Masonic perspectives. The Millau viaduct links to Washington, DC, through its cornerstone and inauguration ceremonies. Both were conducted on December 14—the date that Founding Father and first president of the United States, George Washington, died at Mount Vernon.

This bridge not only acts as a metaphor for the Great Architect in the sky, as per the Masonic ideal, it is also a witty pun on the bridge between the powers that be in the United States and in France. On the very next page of the Fun Facts section, one learns that the NASA headquarters has been designed with its very own "bridge," created because the West Lobby and the Auditorium are both two stories tall. Their footbridge is the part of the corridor that connects the west end of the second floor with the second floor west elevator lobby. Again, we are back to France and that other footbridge, red in color and the eighth incomplete component in the Axe Majeur park. Even more solutions to this odd juxtaposition of tree and bridge can be found by consulting the history books. The Millau viaduct is constructed near to five famous Knights Templar villages, and from there we come full circle back to that redwood tree, native of the coastal region of central California and southern Oregon. Why? Because the American power players of the modern world hold annual meetings with their international contemporaries among the redwoods of their very own hidden headquarters, on the Bohemian Grove estate, in Monte Rio, California.

RUE FUSS TWO FUSS

Returning to the notion of a message to ETI laid out in the Pentagon/D&M Tor analog, the Mall garden layout might not be so much the shopping center of future space hardware but more of a scrap yard memorial laid out as a visual acknowledgment of the technological limitations of rocketry. The layout at Cergy indicates the need for alliance, at the very least, with the ESA. So it is also possible that NASA is gradually overcoming its "scientist's trauma" and cautiously seeking the help of ETI in getting to the next stage. *Surveyor* perhaps was sent out to Mars as the representative of this emergency signal. However, the secrecy of this silent messaging service is commensurate with not wishing to acknowledge publicly that the agencies have been watching out for contact with ETI since at least 1947.

In the light of the Meier photographs and other information concerning spaceflight, including some acquired through less than conventional means, it would be nice to think that the down-to-earth Pentagon and the dreamer of the night, NASA, have finally come to accept that not all extraterrestrials are out to get them. As already noted earlier, scientists for the most part consider that civilizations capable of developing advanced technologies for space travel, having survived long enough to do so, are by definition mature and are therefore likely to use such power wisely—for exploration and not for domination. In support of technical advances going hand in hand with societal maturity, these same scientists have stated that any unexplainable UFO phenomena have not posed a threat to Earth and that there is no necessity to monitor UFO activity since it has also been established that most sightings are in the imagination of the public. Most. Yet these opinions have not prevented the PR machines of the space industries from holding the public's attention by deliberately exploiting the UFO mythology and the media when promoting space projects while ostensibly ignoring any signs of information input from ETI. Whether generated by ETI per se or merely our collective consciousness, the cumulative effects of decades of anomalous events have apparently stimulated the creative instigators of these governmental landscaping projects into resorting to a method recommended in the movie *Field of Dreams:* "If you build it, they will come."

The Name Game
and the A-Word

Defunct Space Shuttles, a Spaceport, Star Trek,
Black Ops, and Remote Viewing for Answers

FIELDS OF DREAMS

If the crop glyph phenomenon were to be recognized as containing valid transmitted data and information, it might become clear that there are details contained within these designs relating to advanced technology for spacecraft propulsion and the essential health requirements for astronauts traveling into deep space. Scientists and engineers could actually use such information to construct a totally new generation of spacecraft capable of human transportation through interplanetary (and interstellar) space in a safe and timely fashion. To date, this has not been the case. Symbolic gestures (often fatuous, such as Elon Musk's space launched Tesla stunt) do not change the fact that when it comes to plans for human travel to the Moon and Mars, the development and upscaling of chemical rocket designs continues. When research projects are given billions of dollars to study advanced technologies, these are more applicable to space warfare than human transportation.

Space exploration by human beings is a rather two-faced effort: the showbiz front office and the back office nuts and bolts. In the back office, while research and development of rocket-based launch vehicles to lunar and Martian orbits by the so-called private sector is not discouraged, the development of launch vehicles primarily limited to accessing LEO and thereby ensuring a viable link to the ISS is the priority. That's the serious side.

In the front office, in an attempt to keep space glamorous in the public's eye, private companies are being encouraged to invest in commercial space projects. Billionaire Sir Richard Branson's Virgin Galactic is trying to emulate the single-seater flights that USAF X-15 pilots were performing decades ago. In the 1950s, three hypersonic rocket-powered X-15s were built by North American Aviation. Based on a design by Wernher von Braun's colleague Walter Dornberger, this experimental aircraft had its first flight on June 8, 1959. Neil Armstrong flew the X-15, and it was retired from service in December 1968. Scaling up this flight tech to develop civilian space tourism by climbing just fifty-six miles up to the Kármán line at the edge of space in order to experience weightlessness for a very brief and expensive moment is proving rather hard to achieve.

Research and development into hyperfast planetary flight for passengers and cargo—that Australia in two hours ticket—is also part of the front office program. Passing all these projects across to the private sector is a clever ploy on the part of NASA. For however much these organizations might receive by way of direct or indirect funding from the government, it is the faces of the billionaire and his cohorts that hit the press when things go wrong, and inevitably, things do

Photo: Nigel Young, Foster + Partners.

Figure 13.1. Spaceport America hub in New Mexico. Launch site for Sir Richard Branson's company Virgin Galactic; designed by Foster + Partners.

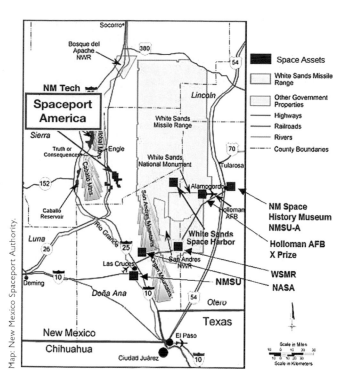

Map: New Mexico Spaceport Authority.

Figure 13.2. Spaceport America location, set between the town of Truth or Consequences and the White Sands Missile Range.

go wrong. As it turns out, scaling up is one of the big problems for space engineering. Virtually all these private start-ups have run into difficulties and have found that chucking vast amounts of money at a project doesn't remove the difficulties inherent in getting off the ground and into space. As Branson put it, "Space is harder than expected." Branson has had spectacular and tragic difficulties with building even a small passenger spaceship using chemical rockets, although it's fair to say that his competition hasn't been particularly successful either. Here we focus on Virgin Galactic because it was this company that was ostensibly responsible for the commissioning of a visually spectacular spaceport from architect Foster + Partners.[1]

Branson called it Spaceport America, and again, in line with the principle of "If you build it, they will come," the set was built, even though there was no cast, no crew, and very little sign of a decent production team. Having signed a twenty-year lease as the anchor tenant early in 2009, by early 2020, Virgin Galactic still hadn't managed to launch any passengers from its hub in the New Mexican desert. And by late 2020 the financial management of the spaceport was coming under intense scrutiny. But never mind those practical issues; for all those interested in the interaction of Earth and ETI it would appear that the instigators

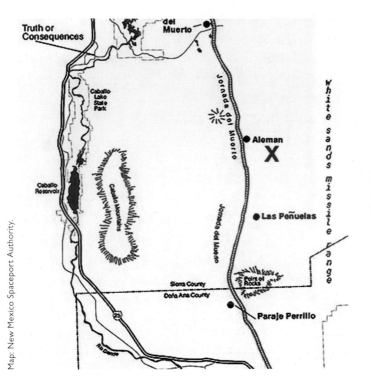

Figure 13.3. Spaceport America location detail; X marks the spot.

of the Spaceport America Project at Stanford University had consulted with the Stanford Reasearch Institute and Hollywood, because Spaceport America had chosen a thrilling location in New Mexico, a state forever associated with space, rockets, and the Roswell UFO saga of June 1947.

From the point of view of the US authorities, 1947 had been a bit of a problem[2] as the year had also included the famous Kenneth Arnold sighting of nine UFOs around Mount Rainier in Washington State. At the time these events created a media storm but were then swept under the carpet by the authorities. Fast forward to the twenty-first century and at Spaceport America, things have changed albeit somewhat confusingly.

While the general public is still slapped on the wrist for believing that flying saucers and ETI might have visited Earth in the past, down at Spaceport America, the public is encouraged to think in exactly these mythic terms as concerns the future. Norman Foster and his team have designed a spaceport that looks not unlike a hulking great UFO or, in the vernacular of the USAF, a not entirely friendly "bird." American Air Force slang for an aircraft, "bird" has confusingly been adopted to describe a satellite. Even more confusingly, in combat it might refer to an air-to-air missile or an ICBM. The "bird barn" is slang for an aircraft carrier.

The image Virgin Galactic has chosen for its spaceship line is that of a giant blue eye, an ancient symbol designed to ward off evil that is often seen painted on a boat's prow in the eastern Mediterranean to this day. When thought of in flight terms, the link to NASA's Egyptian mythos emerges: this eye becomes the Eye of Horus the Hawk.

However, this spaceport also has New World links. The runway has been laid out at an offset of 15.49° *west of due north*, while Teotihuacán, the ancient City of the Gods, north of Mexico City, is offset by 15.49° *east of due north*. The principal features of this new spaceport are also aligned to the two-thirds ratio observed by the three major monuments at Teotihuacán. Measured along the Avenue of the Dead, from the Quetzalcoatl compound centerline to the Pyramid of the Sun is two-thirds of the distance from Quetzalcoatl to the Pyramid of the Moon. In musical terms, the ratio 3:2 is ascribed to the perfect fifth. In the diatonic C major scale, this is the tone G. Hugh Harleston Jr., the surveyor of Teotihuacán, thought the tone G to be associated with the city.

Did the architects take this analogy of a solar system into account when designing this new spaceport? It is known that Lord Norman Foster is sensitive to local traditions, having consulted Feng Shui masters when designing the Hong Kong and Shanghai Bank. He was also very much aware of crop glyphs; as well as knowing architect and croppie Michael Glickman, when flying over the Wiltshire countryside on his helicopter commute to London he had experienced for himself their visual impact. So these UFO/Mythic associations of mine are not entirely unfounded. And when the three principal spaceport components are overlaid on a map of Teotihuacán, they match (from left to right) the Plaza of the Moon, the Pyramid of the Sun, and the asteroid belt (as represented by the blue river), which separates the analog inner solar system grouped on, and within, the Quetzalcoatl compound. And the word *Quetzal* is of course the name of a bird—and half of the name of the mysterious culture hero who is reputed to be the ancestor of all the Mesoamerican peoples. Quetzalcoatl is interpreted literally as the feathered serpent, but *coatl* also refers to the Pleiades star group, and the juxtaposing of bird and serpent should remind everyone of the Egyptian crowns with the bird-serpent motifs. We are in the domain of consciousness transferal via symbols.

Teotihuacán's layout encodes, according to some researchers, the distance between the planets of the solar system, measured in Hunabs of 3.4759 feet (1.0594 meters). This measure is also the width of the sarsen ring's lintel stones at Stonehenge, and when expressed in meters, the number 1.0594 turns out to be the first digits of the twelfth root of 2, and that is the basis of the Western equal temperament musical scale.

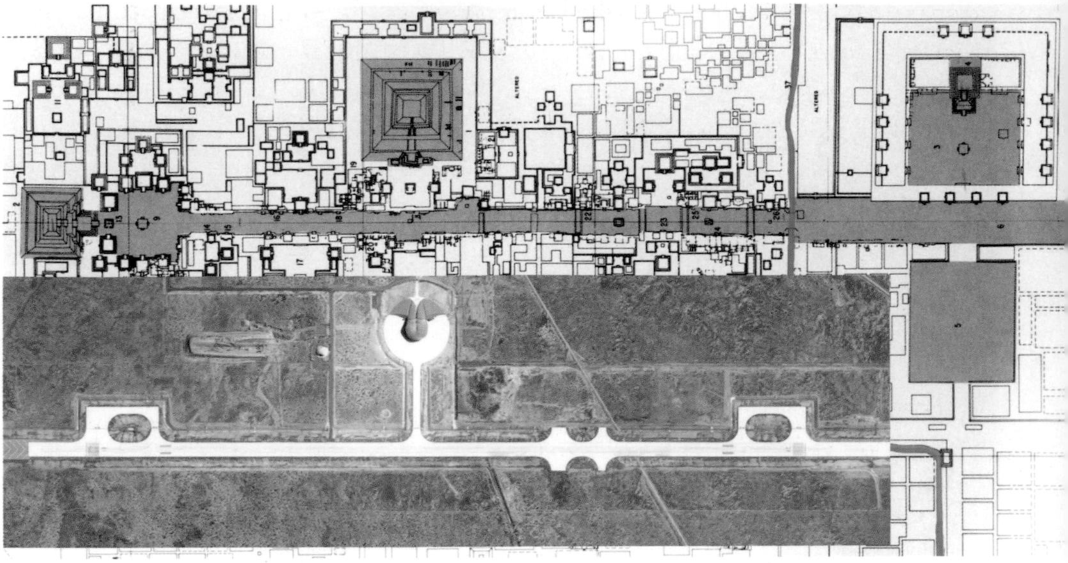

Illustration: DSP. Map: René Millon, *Scientific American*, June 1967. CC BY-SA 4.0.

Figure 13.4. Spaceport America (brown overlay) and layout of Teotihuacán, showing how the three principal spaceport buildings match with the Plaza of the Moon, the Pyramid of the Sun, and the asteroid belt at the ancient site.

SO LET IT BE SAID, SO LET IT BE WRITTEN, SO LET IT BE DONE

Spaceport America is a slightly less poetic name than is usual for NASA and its cohorts to select. Closer scrutiny reveals that it fits right in with the secret traditions of an American philosophy that had decorated its capital city with zodiacs. In the Western tradition of considering the twelve zodiacal constellations as triads representing the elements of fire, earth, air, and water, then Washington ostensibly selected the earth triad of Taurus, Virgo, and Capricorn, if the constellation of Virgo is as predominant in Washington as author David Ovason asserts.

Branson's professed ambition is to open up the boundary between air and space to the general public, all of which is perfectly expressed in the design of this spaceport. Here a spaceport represented by a bird connects to the air triad of Gemini, Libra, and Aquarius. This is not so straightforward; tradition has the twins of Gemini with one twin in space, mirroring Branson's aims. The constellation of Libra was once the claws of its neighboring constellation, the watery Scorpio, and Aquarius was the water bearer of the gods. The Romans called the chief supervisor of an aqueduct the *Aquarius.* This interesting conflation of air and water is resonant even with spaceflight when we remember that nautical terms are also used in astronautics. Space itself is described in terms of depth. While spacecraft are said to navigate through space on voyages measured in nautical miles.

If the Teotihuacán design is representative of a solar system, given the dual nature of the Gemini twins (wherein the earthbound twin was war-like, while the harmonious heavenly twin played the lyre), Spaceport America's layout, along with its location and its bird-like motif, might be referring to a hawk and the words of President Lyndon B. Johnson, Master of the Universe,* rather than the quetzal bird.

I can hear those less interested in symbolism stirring.

"Really? Are you not reading too much into this? Perhaps the Teotihuacán analogy is merely homage to the First Peoples, and the Eagle bird is also sacred to the Native Americans, don't you know?"

"Yes, I do. The history of the colonization of the American continents is not lost on me, and in that regard, it is unlikely to have been memorialized at Spaceport America—unless to warn off ETI. But with regard to using names to encode their own mythos, NASA has form."

"OK. Show me."

NAMING THE SHIP

The US space program has adopted the mythic gods, planets, constellations, and the locations visited by the ancestor heroes for its own projects and programs. The names of space rockets such as Juno, Jupiter, and Titan, entire astronaut projects such as the one-man Mercury, two-man Gemini, three-man Apollo, and the future four-man Ares programs, along with a space shuttle named *Atlantis,* all testify as much. When specific names are used for important missions, such as the use of *Odyssey* for the Apollo 13 command module and *Aquarius* for its lunar module, this heroic theme is reinforced and for those in the know, forward planning is also revealed.[3] As for the whole saga of making Hollywood movies associated with the space program, the links between Hollywood, government projects, and the transfer

---

*In January 7, 1958, Johnson gave a speech at the Democratic Caucus that contained these words: "Control of space means control of the world . . . from space the masters of infinity would have the power to control the Earth's weather, to cause drought and flood, to change the tides and raise the level of the sea, to divert the gulf stream and change the climates to frigid. There is something more important than the ultimate weapon. That is the ultimate position—the position of total control over the Earth that lies somewhere in outer space. . . . And if there is an ultimate position then our national goal and the goal of all free men must be to win and hold that position."

of information/propaganda to the public via sound and light are well understood by the agencies if not by the public. The seminal movie *2001: A Space Odyssey* was made during the crewed space programs, and screenwriter Arthur C. Clarke and film director Stanley Kubrick were fully aware they were participating in a parallel exercise of creation: "saying, then writing, then making" a space age version of the old hero's myth. But that's a subject requiring further study.*

When it came to naming its spacecraft, NASA was simply refining a practice that had started much earlier, and like Clarke, it originated in England. John Montague, the fourth Earl of Sandwich and First Lord of the Admiralty at the time of the American War of Independence, named his ships by opening his trusty copy of *Lempriere's Classical Dictionary* and picking the first names that he came upon: thus he named the HMS *Bellepheron, Orion,* and *Leander,* among others. Adopting classical references has been maintained by the United States, but the randomness of this procedure has ceased. Nowadays the naming of spacecraft appears to be very well thought through.

How, then, did it come about that the name of *Atlantis* was chosen for a space shuttle? After all, Plato's account of an ancient island civilization vanishing overnight from the face of the Earth under a tidal wave is a concept for which most scientists have no regard at all. Indeed, Atlantis as a topic can seriously damage the professional reputation of those who express a belief in the historical reality of Plato's lost civilization. So to find the world's most powerful space agency adopting this particular name for a spacecraft (even one limited to visiting the ISS, at the shoreline of space) is somewhat intriguing.

NASA's PR people state that the ships of the Shuttle Transport System(STS) were given a serial code of two letters and three digits together with a name, the choice of which related to various other sailing ships that had previously been used for seafaring explorations. So far, so orthodox.

In chronological order of delivery, the six shuttles that replaced the Saturn V rockets of the Apollo project were listed as:

OV101 *Enterprise.* The test orbiter *never* flown in space and named after the fictional spacecraft from the Gene Roddenberry TV series *Star Trek.* Ever popular, this name would be adopted by NASA friend and sometime

---

*With regard to space travel and the 1960s, the following articles on AulisOnline cite researchers currently examining Kubrick's work: "Stanley Kubrick and Apollo: Links to the Films of Stanley Kubrick?" "After the Ball: Kubrick and Georges Méliès: Links to the Works of the French Filmmaker?" and Jarrah White's "Kubrick Appendixes."

alleged foe Richard Hoagland for his website, and Richard Branson named Spaceship Two, his first Virgin Galactic test vehicle, the *VSS Enterprise*.

OV102 *Columbia*. Launched April 12, 1981, on the twentieth anniversary of Yuri Gagarin's space flight. Named after the Apollo 11 command module and the Boston-based privately owned ship that circumnavigated the world. Although only the Gagarin flight, the American sailing ship, and this shuttle were circumnavigational firsts.

OV099 *Challenger*. Delivered in July 1982; named after two ships: the British HMS *Challenger*, which explored the Atlantic and Pacific from 1872 to 1876, and the last of the lunar modules Apollo 17. Both were scientific expeditions, and interestingly, the 1872 *Challenger* had fifteen of its seventeen guns removed in order to make more space for scientific gear.[4] The discrepancy in the serial numbers rather indicates that *Challenger* was not initially a part of this shuttle build sequence.

OV103 *Discovery*. Delivered in November 1983; took its name from two ships. The British HMS *Discovery* was Henry Hudson's ship of 1610. During the search for the Northwest Passage, the crew mutinied, leaving Hudson to his fate in the bay that would take his name. And James Cook's HMS *Discovery*, the ship that discovered the Hawaiian Islands. All very fitting when thinking about returns to Earth from space and that difficult passage through the upper atmosphere.

OV104 *Atlantis*. Delivered in April 1985 and named after the first research yacht built for the Woods Hole Oceanographic Institution (WHOI) in Massachusetts. This sailing ship was very specifically named *Atlantis*, and after her sale in 1964 (to the Argentinians), subsequent WHOI research vessels have also been named *Atlantis*.

OV105 *Endeavour*. Built to replace the defunct *Challenger* and named after HMS *Endeavour*, used by Cook on his first expedition to record the Venus transit of the Sun out near Tahiti and to look for unknown lands to the south of New Zealand.

These craft all have plausible and earthly histories for their names, and note that three of them were named after seafaring ships belonging to Her Majesty's Service—the British Navy. Delving into the history of these ships produces interesting connections with their space-age equivalents, some of which are mentioned above, and that surely would be sufficient justification for these sometimes rather odd names. Given that these shuttles were released into service over a period of years, the idea that the chronological list of the shuttle names might collectively enlarge on the connections already found seems perhaps far too fanciful a notion to be taken seriously. Having got this far, the reader will begin to recognize a pattern.

Taking the ever-present Masonic ritual magic model, it could be construed that the shuttles' alter egos were indeed "wish-upon-a-star ships." Their order of arrival on the scene reiterated the wider aims of the space program. Taken literally, "star trekking" is the script, for both the fiction and the fact. The **enterprise** of getting humans first off the planet, then getting the first American into low-Earth orbit (that Apollo 11 **Columbia** reference). Then comes the **challenge** of getting a geologist to the Moon. Followed by the final **endeavour**: the **discovery** of the lost continent: **Atlantis.** That would be a mighty endeavor if ever there was one, since it is not clear whether that might be found on Earth or elsewhere, perhaps even on Mars—the ultimate destination for astronauts for the next hundred years or so. When taken together, the shuttle names certainly provoke thoughts of exploring impossible locations for that LEO-bound spacecraft.

From the logic of using myth for the naming of things, it wasn't thought necessary to justify the hardly sensible naming of a shuttle *Atlantis* after a denigrated lost continent, but interestingly, someone *did* think it necessary to defend the idea of calling a shuttle after a fictional spacecraft from a famous American TV series. It is on the historical record that *Enterprise* was originally to be called *Constitution,* after the US Navy frigate (which famously went to war against the British in 1812, defeating five British warships and earning the nickname "Old Ironsides"). Despite the esteem with which this frigate is held to this day, apparently real events did not prevail over the *Star Trek* TV series and science fiction. It is said that the public clamor for *Star Trek's Enterprise* won over the White House, and thus it came about that the name *Constitution* was dropped.

That PR story about the change of name for the first test shuttle from *Constitution* to *Enterprise* requires a very large dose of salt, it having more to do with old irony than Old Ironsides. Especially since the person said to be whipping up public enthusiasm was that seemingly ubiquitous Cydonia researcher, erstwhile NASA consultant and media event organizer for the Goddard Spaceflight Center, Richard Hoagland. Furthermore, although NASA had held a public vote, it didn't mean the agency would necessarily abide by the results. Clause four of the voting rules stated:

> NASA will take into consideration the results of the voting. However, the results are not binding on NASA and NASA reserves the right to ultimately select a name in accordance with the best interests of the agency, its needs, and

other considerations. Such name may not necessarily be one which is on the list of voted-on candidate names. NASA's decision shall be deemed final.[5]

So the "public vote" might have taken the pulse of public opinion while creating product awareness, as they say in marketing, but ultimately the names were decided on by NASA. To find the name *Enterprise* replacing that of a frigate honored for beating an enemy—the British—one might wonder if this very first test shuttle's name reiterated that United States intention to dominate space, leading me to musings about the actual functions of these transporter shuttles. With hindsight, we now know that the NASA shuttles were transporting satellites for the DOD and that all the crewed space stations presented as "scientific research stations" were intended to function as spy stations in LEO by their respective governments. So that back in the present day, the international designation of the ISS does not exclude it from operating as a "lookout." The naming of the *Enterprise* then becomes more significant, if it was facilitating the installation of technology designed to signal anyone or anything intent on disruptive visits to the shores of Earth. The in-joke being that any unwelcome visitors would be treated in much the same way that the British had been received when they crossed the Atlantic in 1812.

DESTINY

Thinking in less aggressive terms, although the shuttle could boldly go only as far as the ISS, the facts that the density of the atmosphere decreases as the altitude increases and that the ISS docking pod for the shuttle is called *Destiny* are two other little niceties of the space program's naming game. Upon its return to Earth, the shuttle transferred from its rocket-thruster mode at around seventy-six miles from Earth, and it then used atmospheric drag at the fringes of the atmosphere to reduce its speed.

Destiny is an anagram of *density*.

It's easy to see how confusing things can get when every bit of data is open to such paradoxical encoding/decoding. But confusion doesn't automatically mean an inaccurate or incorrect interpretation of that same data; when a paradox emerges, it's best to pay extra attention. In fact, looked at from the perspective of ETI rather than any Cold War scenarios, the whole of the shuttle and space station project seems to be far closer to the storylines generated by Roddenberry's original TV series than has ever been recognized publicly.

Star Trek storylines were generally concerned with exploring other planets, meeting other civilizations (good and bad), and policing space. So another version

Figure 13.5. Gene Roddenberry, creator of the original *Star Trek* TV series.

of this shuttle saga might read: *"Enterprise* and *Columbia* are the birds used to *Challenge* any *Discovery* other than those that we authorize concerning the location and putative inhabitants of *Atlantis."* The naming of these shuttles perhaps acting as a philosophical driver or another inside joke for those in the know, and Hoagland's now defunct Enterprise website would fit right in with this interpretation.

Gene Roddenberry is on record as having had several sessions with the experienced American deep trance medium Phyllis Schlemmer. (She preferred the term *transceiver* since she did not communicate with spirits but received information directly from an ETI source.) Roddenberry spent much of his time with Schlemmer asking questions about ETI and related matters, and *Star Trek: Deep Space Nine*— to which Roddenberry gave his blessing shortly before his death in 1991—especially comes to mind in this regard. As does the fact that Arthur C. Clarke was also a friend and helped Roddenberry by introducing him to his own agent.[6]

OCEANOGRAPHIC INSTITUTION

All of which still leaves the question of *Atlantis*. Why had the NASA decision makers been inspired by the WHOI's ship, and why was *that* ship thus named?

Designed especially for marine biology, marine geology, and physical oceanographic research, *Atlantis I* served until 1964 or 1966 (depending on your source). *Atlantis II* came online in 1967, built by the US Navy, and *Atlantis III* replaced her in 1997. These last two ships also worked with the US National Deep Submarine Facilities submersible *Alvin*. Interestingly, when compared with the space race, officially, the dream of building an underwater piloted deep ocean research submarine dates from 1956, but the *Alvin* was not built until 1964. The latest incarnation of the research vessel *Atlantis,* although operated by the WHOI, is actually owned by the US Navy.

What's going on? Cynics have suggested that maybe they just liked the name, to

Photo: USGS Ships Photo Gallery.

Figure 13.6. The original Woods Hole Oceanographic Institution's research vessel *Atlantis*. This 142-foot steel-hulled, ketch-rigged sailing ship took eight months to build in a Copenhagen shipyard at a cost of $175,000.

which I repeat the fact that, outside of a classical literature or philosophy classroom, the reality of Atlantis is treated with considerable derision. Anthropologist Paul Jordan is fairly representative of this stance. Author of the 2001 book *The Atlantis Syndrome,* he takes a very strong position against the views held by the pro-Atlantis authors, although often preferring sarcasm rather than coherent argument.

This emphasis on Navy research suggests that Woods Hole is a government organization—or does it? Complementing its 1903 West Coast cousin, the Scripps Institution of Oceanography near San Diego, California, the Woods Hole Oceanographic Institution, based in Woods Hole, Massachusetts, was set up in 1930 as an independent not-for-profit organization with initial funding of $2.5 million from the Rockefeller Foundation. Its mission was to represent the United States' East Coast's share in oceanographic research, and the WHOI was to report back to the National Academy of Sciences, which *is* a government institution—and the host to that Einstein zodiac memorial.

To this day, Woods Hole describes itself as "the world's largest private, non-profit ocean research, engineering and education organization." The foundation of WHOI may have been inaugurated with independent money and it is still run by a trust, but it has been contracted to government projects since 1940 and has received government funding since 1950, which rather takes away notions of total independence in either project choice or finance. The reasons for taking on government sponsorship are given as: "After WWII oceanographic research became too expensive to undertake without government money." Since government projects were already being undertaken during WWII that statement is not entirely accurate. Currently WHOI is financed through 85 percent federal funding and 15 percent private money, and that 85 percent government funding means that it does not *have* to turn a profit. Clearly, it may look like a philanthropic educational institution from the outside, but internally, WHOI has been under the US government's overlook virtually from the outset.

In this regard, it is not unlike the so-called civilian agency NASA, which presents a casual shirtsleeved front office to the public, although its CEO is appointed by the president and most of NASA's projects are undertaken for the boys in the back office—the US government and defense departments.

Space research is also too expensive to run without the help of government. As already observed, even in the now burgeoning private sector (well, it might burgeon if they could actually get the large-rocket systems to work safely), many space companies and a number of the start-ups, including Virgin Atlantic, were vaunted as being independent of government agencies. Until they realized that at least in that start-up phase, most of their commissions could only be sourced from NASA and other government agencies. So when it comes to both the naming and the execution of these highly expensive explorative ventures, it's a good idea to remember that underneath all the apparent civilian philanthropy, as always, he who pays the piper . . .

IMMEDIATE RELEASE   July 29, 1958

    James C. Hagerty, Press Secretary to the President

- -

THE WHITE HOUSE

STATEMENT BY THE PRESIDENT

    I have today signed H. R. 12575, the National Aeronautics and Space Act of 1958.

Figure 13.7. First paragraph of the National Aeronautics and Space Act, signed into law on July 29, 1958, by President Dwight D. Eisenhower.

This is not to diminish all the valuable work WHOI and its associates have done and still do today; it is to make the point that the naming of things can generally be an indication of their purpose, that one occupation does not preclude another occurring alongside or hidden within the prime project, and that the overall purpose of a long-term program can be reflected in the collective naming process—as with those NASA shuttles.

ATLANTIS RISING

As for exploration by the seagoing research vessel *Atlantis,* from 1931, she spent busy summers working up near Massachusetts and quiet winters doing "a wide variety of work in the Gulf of Mexico and the Caribbean." However, this same historical document states that "WHOI was a summer institution open only three months of the year," which infers that those prewar quiet winters were taken out of office hours, as it were.

These winter locations have links to the infamous Bermuda triangle, and research vessel *Atlantis* was in action during the decades when the American psychic Edgar Cayce was delivering channeled or trance predictions. Cayce had government contacts, and there can be little doubt that he was monitored, especially in the light of the remarkable degrees of accuracy he achieved during his trances. Among the subjects on the table was the lost continent belonging to the civilization of Atlantis. According to this sleeping prophet (as Cayce was dubbed), the Atlantean Halls of Records was held within the Yucatán, the Gulf of Mexico, and the Bimini islands. The very areas visited by Woods Hole's *Atlantis* during its "quiet winters."

The choice of the name *Atlantis* by Woods Hole and NASA indicates that both scientists and government were, and still are, using this mythological moniker for a very good reason. If that reason was the importance of Cayce's Atlantean insights, then it makes sense of the storyline from WHOI, which tells us that one of its trustees, Alexander Forbes, having bought himself a schooner named *Atlantis,* was even willing to rename it so that the name *Atlantis* could be transferred to the first WHOI research vessel. Years later, collating diverse comments from other sources, reveals that the 98-foot schooner *Atlantis* had been built for Columbus Iselin in 1927 and he sold it to Forbes before the founding of the WHOI. In 1930 when the Rockefellers funded the building of the WHOI laboratory, wharf, and its first research vessel, Columbus Iselin was sent to Denmark to sail the ship back to the States. Launched in December 1930, it took an exceptionally long time to cross the Atlantic, arriving in Woods Hole in August 1931.

Iselin would be the onboard captain for the next two years. Seamen think that changing a ship's name is to bring bad luck to the ship, so Forbes must have had a very good reason for relinquishing the name of his own schooner to this WHOI vessel. And Iselin a very good reason for asking, over and above that of affection for a former sailboat. As Forbes would become a trustee of the institution and Iselin was famously involved with the institute for all his life, taking over from Henry Bigelow to become its second director from 1940–1950, and just at the time the institute was undertaking defense work, it begs the question as to why the WHOI history section is reluctant to tell the story as it really happened, stating only that the first research vessel was eventually named *Atlantis*. Despite all the good work done in the name of science, the lack of clarity around this naming of a ship that passes its name on to successive vessels and a shuttle infers a continuation of that very good reason. Perhaps a search for underwater traces of that lost continent's civilization was part of the sales pitch that got this institute its initial funding and took up a lot of that eight months travel time across the Atlantic. A search then taken to the shoreline of space by the shuttle *Atlantis*.

Apart from designating locations for the Halls of Records, Cayce had also predicted that Atlantis was due to re-emerge in 1968 or 1969. The remarkably apposite timing and location of NASA's Florida launch pad for Apollo just across the water from Bimini leads to speculation: Exactly how much of that speech by President John F. Kennedy about "getting into space before this decade is out" was tied to another more discreet space program, which was designed to comply with the Cayce prediction, albeit not in quite the way that most Cayce devotees would have expected? This would not be the emergence from the brine of a lost continent, the salty waters streaming down its cliff faces as it rose slowly and inexorably from the depths. This would be the emergence of a fiery beast from the marshy lands lying adjacent to the sea and west of Bimini, the gases of chemical rocket fuels bubbling from its base as it ascended to the skies with (hopefully) astronauts strapped inside its nose. Symbolizing the re-emergence of the travelers from the lost lands of Earth on the first stage of the heroic journey that would eventually lead back home—to Mars.

Were the Mercury, Gemini, and Apollo space programs actually intended to show anyone observing this planet from afar that we had taken heed of a trance prediction? And was it the continuation of this theme of off-planet communications (via the intermediary of human beings such as Cayce and Schlemmer) that led to the incorporation of this name *Atlantis* into the shuttle fleet? After all, *Enterprise,* associated via Roddenberry with Schlemmer, was the first shuttle. *Atlantis,* associated with Cayce, was intended to be the last shuttle. This is not to impute either of these transceivers with the political decisions, actions, or intentions of their government. It is merely to point out that governments took communications from people of their caliber far more seriously than is divulged to the public.

Perhaps the apparent fascination that Atlantis seems to have for the US authorities has less to do with the supposed geographical proximity of this fabled continent to the United States and more to do with the associated legend of a mysterious and powerful crystal technology. Knowing how important crystal is becoming in the computer industry (crystal can store vast amounts of digital information), it could be that the authorities in the States really believe these tales or think them worth verifying. Perhaps they assume that this technology is also associated with PSI abilities and the future technologies needed for human space travel into the solar system. One of which would be something that enables humans to communicate easily and in a timely fashion across vast distances between space and Earth.

As already discussed, as communication signals travel at the speed of light, a future astronaut asking a question from Mars can take at worst forty-eight minutes for the conveyance of a single item of information back and forth. From this point of view, even the highly skeptical have understood that PSI abilities have the potential to become a useful if not an essential tool of space exploration. Getting that understanding into a practical reality is more problematic.

INCOMING/OUTGOING

It is a notable aspect of channeling that such messages are received telepathically and then the individual amplifies the message, either by audibly relaying it to those gathered around or by writing it down. Since apparently we do not know how to use telepathy to *send* long complicated messages, it is claimed by the logical-minded that a human receiver is merely transmitting messages from his or her own brain and as such, the acquired information is worthless.

The exploration of outer space requires a full awareness of the capabilities of our own inner space. In other words, what we really think and how we actually feel when faced with the idea (let alone the actual occurrence) of interaction with ETI will govern our reactions, whether aggressive or peaceful. Even if the official consensus is to scoff at the notion of extraterrestrials being capable of interacting with us and our environment, behind the scenes it seems to be a different matter. And the same is true of PSI activities. In the same decade that Mariner and Viking imaged Mars, the US government was intensifying its research into the applications of focused consciousness.

THE PROJECT

In the 1970s, long before the internet and Google Earth, the United States set up a program to train intelligence and military personnel to spy from afar, using meditation and advanced psychic techniques or, more officially, using "cognitive processes to establish the human capacity for accessing events across space and time." In order to mitigate the general distrust of most military personnel to the subject of good old-fashioned clairvoyance, the process was referred to as remote viewing, the personnel as remote viewers. Soon referred to as RV programs, and primarily designed for military spying targets, the protocols formulated at the beginning of their project are also relevant to this book, because as it turns out, ancient sites and the pyramids here on Earth, together with locations on Mars, were subjected to the same scrutiny.

One of the most talented psychics associated with the beginnings of the RV programs was Ingo Swann. Although usually described as an artist, in fact, Swann was also working in the PSI labs in the early 1970s. Along with Hal Puthoff of the SRI, Swann developed the basic method for vectoring into a target site without giving anything away to the remote viewer. In the RV jargon, anything that might give a clue to the target is described as front-loading the remote viewer. In order to avoid this Swann took the map coordinates of a target site to encode the location. This basic method of focusing on the target was given the project code name SCANATE. This code word was derived from SCANning by coordinATE.

```
                    Project SCANATE*

            Exploratory Research in Remote Viewing


        As a result of the experimentation carried out on what might be
    termed micro-abilities, Swann expressed the opinion that the insights
    obtained had strengthened a macro-ability which had been researched
    prior to his joining the SRI program; namely, the ability to view remote
    locations.¹   In order to test the above assertion, SRI researchers set
    up a series of experimental protocols on a gradient scale of increasing
    difficulty.
```

Figure 13.8. First paragraph of the Project SCANATE document released by the CIA on November 13, 2002. Note that RV experiments were done prior to Swann's arrival in 1970.

The in-joke no doubt appealed to the military mappers as the name implies the human process of scanning the supplied map coordinates while the compression of two words into one also implies data compression. Which is exactly what the protocol was designed around. Puthoff's report to the CIA on this method provides examples of this methodology[7] but it is likely not the full explanation of the *actual* method used, because obviously, just presenting the coordinates as provided in an atlas would be too easy; the remote viewer could decode a lot about the location just from those six number groups. However, there is another method. Latitude and longitude consist of three components each. Multiplying the latitude's three components (degrees, minutes, and seconds) into a single figure and then doing the same for the longitude results in two number groups.

This might be sufficient in some cases, but anyone with a little knowledge of the Earth's mapping coordinates and some skill at math could easily back engineer these numbers. It is therefore advisable to add another layer of complexity, and two further operations can be adopted. Taking these two numbers and then dividing the greater figure by the lesser, a single number string can then be given to the remote viewer, who uses it to vector in on the target. In principle, this method effectively masks the target and prevents the remote viewer from *intellectually* deducing the relative location. Thereby ensuring that any deductions made by the trainee viewers were being acquired psychically.

It is also possible, before starting the whole exercise, to manipulate the longitude by varying the prime meridian. For example, and taking Earth as the planet in question, should one wish to keep security tight, while using a prime meridian with relevance to particular project, instead of using the Greenwich prime meridian, one would simply designate another meridian as the "RV prime meridian"—say through Giza, Egypt. A full analysis of the subject is not possible here, but a brief look at the RV method is relevant because all number combinations and specifically the constants we use, such as pi (3.141592654 . . .), start to take on another look.

MIND OVER MATTER

This highly secure method of designating a location meant that any RV trainees with insufficient security clearances could still be run against ultrasecret targets and "foreign" sites such as Cydonia on Mars. The success of SCANATE infers that the final string of numbers had compressed the target data in such a way that, when the numbers are contemplated in focused meditation, the remote viewer's brain acted as the decompressor, unzipping the file, and expanding the triggered information into visual or sensate form.

By 1972, the protocols established by these US researchers were sufficiently robust and the results must have been pretty good, because the CIA, in collaboration with other unnamed agencies, officially funded it for the five years from 1972 to 1976. After which the Army funded the next eight years. And so it went, with funding from varying US intelligence agencies and the project changing its top-secret name at each changeover. (Note: in order to distinguish between the funding organizations and these projects, the top-secret project codenames are in italics.)

1972–1976, *SCANATE* was financed by the CIA and other agencies.

1977 *GONDOLA* was the code name for the recruiting period for the next project. Financed by the Army Intelligence Unit (INSCOM).

1978–1982 *GRILL FLAME,* also financed by INSCOM.

1983–1985 *CENTER LANE,* again, funded by INSCOM.

1985–1991 *SUN STREAK* funded by the military and the Defense Intelligence Agency.

1991–1995 *STAR GATE* funded by the SAIC.* (The last phase of this program has been referenced by military Remote Viewers as two words *STAR GATE.* This code name's meaning alters slightly when written as Stargate and adopted by Hollywood. Although it could be that the birth of Reagan's 1983 Star Wars project also prompted the change of coding from two words to one).

1995, The RV program is returned home to the CIA.

The CIA then ordered a project report, asserting that those making the evaluation would be given access to all the documentation relating to the now-declassified program. It is reported that three large boxes of documentation were under consideration by the reviewers, not a lot for a program officially running for twenty-three years and very much longer unofficially. On the basis of the report's findings, the RV program was shut down. Officially, anyway. Apparently, "after five years of operating the program the CIA abandoned it because it was not helpful." Not helpful to whom? It would not have been

---

*SAIC is the Science Applications International Corporation. Created in 1969, it is headquartered in Reston, Virginia. It provides government services, information technology, and engineering support to both civilian and defense departments. In 2013, it was split into two companies: SAIC and LEIDOS. Derived from the word *kaleidoscope,* this choice of name was intended to reflect the company's effort "to unite solutions from different angles." However, when it came to their actual business, observers have noted that both SAIC and LEIDOS are more opaque than reflecting, as of course befits companies straddling the civilian and defense worlds.

passed along to all these other agencies if there had not been some value in the research.

As far as it goes, this account is truthful in that the CIA did operate the program for five years, from 1972 to 1976 (hello, Cydonia, that's the Mariner through Viking period), and in passing it over to the Army, it was indeed shut down by the CIA. Perhaps those three boxes of documents were only relative to their part of the whole program. The documentation for the years 1977 to 1995 might have something to do with the 99 percent of data, the actual files, the mission details, and methods of RV still classified as of 1997, according to Joseph McMoneagle, the experienced remote viewer and military intelligence officer. This method of publicly denouncing a subject as useless when the circumstances and/or experiments have demonstrated exactly the opposite has been applied to every difficult subject. Officially condemned, the project then continues under the radar, and as it has officially gone dark and is funded by black money, no more questions will be asked on the Hill. McMoneagle was also associated with the Monroe Institute, which specialized in training the mind and took government RV trainees into their programs.

On September 4, 1993, I was visiting a ten-petaled pentagonal crop glyph at Bythorn Cambridgeshire, England, with friends. While we were inspecting it, two strangers approached and asked if I was Mary Bennett, of *The Only Planet of Choice*. I said that I was, whereupon they handed me a large brown envelope, told me "You should go there," turned on their heels and left. Somewhat used to spooky behavior within the crop glyphs I let them go. The envelope contained a brochure and "there" turned out to be the Monroe Institute. I didn't go, but this odd incident prompted me to dig much deeper into governmental research into consciousness along with those RV programs.

Blowing apart the RV program—and by implication psychic subjects in general—means that most scientists would be risking their reputation or career in pursuing subjects that can never be proven according to the criteria applied to them. For the public, choosing to first reveal the project's existence, then its failure to deliver, and finally its demise ensures that the majority of people will join in the scientific consensus that matters PSI are not helpful—as the CIA would have it.

As a result of the "closure" of the RV project, many of the remote viewers went on to write about their experiences. Taken overall, these books reinforce the required consensus view. Their accounts underline the difficulties of living with PSI abilities and contain enough discrepancies in working methods as to ensure that the public get the message: PSI attributes are difficult to manage and unhelpful for everyday living.[8]

The same tactic was employed on a smaller scale when the scientist, physician, and author Dr. Andrija Puharich, who along with Edgar Mitchell had been responsible for bringing Uri Geller's talents to the attention of the US intel agencies, wrote up his experiences with Geller—whose talents, as everyone who has had dealings with him well knows, are the real McCoy.[9] Although even such enlightened beings as Russell Targ persist in dissing Geller's spoon bending, Geller is not a cheat. At the end of a private meeting David Percy and I had with Geller, as an afterthought, Geller asked if we wanted him to bend a spoon. I had brought one along just in case but had completely forgotten about it. I hastily fished it out of my bag and handed it to him. I was sitting next to him. He took the spoon and stood up, so my face was virtually next to his hand when he rubbed the spoon with one finger so very lightly—and it instantly bent. I have the spoon to this day.

However, in writing of his association with Geller, Puharich succeeded in making Geller's talents and the events that occurred around him appear so outlandish that they were off the scale as far as the scientific community went. Given the label of a showman, and a possible victim of Puharich's psychic research methods, the book perfectly fulfilled its task—not only did it make Geller seem ineffective (and therefore safe and able to continue any avenue of PSI work he agreed to), but this treatment also relegated the actual ETI interactions around Geller to by-products of the Puharich-Geller association and therefore *nothing to do with ETI*.

So everyone could go back to sleep.

THE PROGRAM

What was never mentioned when the media reports on the RV program emerged was the fact that the remote viewers saw UFOs. Author Jim Marrs writes:

> Initially ordered to locate high-flying high performance aircraft in order to view new Soviet technology, the psi spies were amazed to discover craft that did not originate on Earth.[10]

Given that some of these guys were looking for "new technology," seeing a UFO fits the bill rather better than a Soviet aircraft, but it was not only those particular remote viewers who saw UFOs; everyone in the program did at one time or another.

Readers of *Two-Thirds* will remember that there is a correlation between an advanced spacecraft's drive function and the spin rates required for consciousness transmission, which begs the question as to whether the vision of a UFO is the by-product of a certain mindset associated with intention.

Does the intention to use clairvoyance create a hyperfunction of the brain-wave being used to scan the target and explain the experiences of Billy Meier? Before Meier set out to meet and photograph a UFO, he got a strong internal mental thrust. If his brain was operating at the same rate or frequency as that of a remote viewer, then he too was visualizing like a remote viewer, and being equally unlimited by time and space, he would have experienced the space-time travels that he describes. That he considers it all to be three-dimensional reality might be because the experience is very strongly imprinted, perhaps in order that he could also function in the real world. Because unlike the military remote viewer, comfortably installed in a darkened room, Meier had to ride out to the UFO meeting place on his moped (a motorized bicycle), and then manage his camera on the slopes of a Swiss Alp. And Billy Meier only had the use of one arm.

Now whether the UFO imagery comes with the territory, that is, whether it is a reflection of the brain's status when remote viewing or whether the experience is also external to the mind's eye, and whether the various shapes of UFOs (which differ over time and place) carry further information—these are all relevant questions. However, the military reports infer that their remote viewers could not make the distinction between apparent and real, considering the UFOs they saw to be nuts-and-bolts technology. (Interestingly, this is like those who judged my completely luminous Avebury sighting to be a solid craft, so now we know the source of "because it was wanted that way"!) The experiences of the remote viewers, combined with the demonstrated ability to negotiate space (distance no object) and time (past, present, and future), apparently led to a conclusion that should come as no surprise: the exploration of space would be the most important and cost-effective application of RV protocols.

Cynically, one might say that any publicized reports of off-planet surveying might have been manufactured out of whole cloth to support both the RV program and the space program agenda. After all, these PSI spies are run by intelligence agencies. Yet, if one is inclined to dispute the motives, one cannot dispute the method: no government is going to spend decades and valuable public money on something that does not work. The sheer number of ordinary people who have RV experiences bears testimony to the efficacy of the method. As does the absolute insistence of the Intel remote viewers that *no one* can call themselves a remote viewer unless they are able to perform under laboratory conditions. This is rich, because these same Intel viewers openly acknowledge that psychic viewing and RV are two descriptions of the same thing, and it is also well established by their own researchers that repetitious performance under lab conditions leads to boredom and the blunting of psychic abilities. Then again expert remote viewer Joseph McMoneagle, batting for the government, is of the opinion that other than fully accredited government RVs, 98 percent of those who call themselves remote viewers are "kooks."

REMOTE VIEWERS

McMoneagle's book *Mind Trek,* the first of many on the subject of RV, was first published in 1993. Which means he started writing it prior to the RV program windup. So the question here is: If much of the RV program was officially classified up until 1995, then how is it that in 1993 McMoneagle was allowed to publish *Mind Trek?*

Coincidentally, 1993 was the year that NASA's continued refusal to commit to rephotographing the Cydonian Face by *Mars Observer* (the first probe since Viking scheduled for the red planet) resulted in scathing criticism from Stanley V. McDaniel, professor emeritus and former chairman of the Department of Philosophy at Sonoma State University. McDaniel published a paperback book: *The McDaniel Report: On the Failure of Executive, Congressional, and Scientific Responsibility in Investigating Possible Evidence of Artificial Structures on the Surface of Mars and in Setting Mission Priorities for NASA's Mars Exploration Program.* And remember, that same year, the *Mars Observer* space probe went AWOL.

Bearing in mind that McMoneagle was an intelligence officer, it looks as if the 1993 release of *Mind Trek* was another PSI-Ops program designed to manipulate public opinion. The book's cover featured the Devils Tower National Monument, recalling the movie *Close Encounters of the Third Kind.* While its title, recalling the TV series *Star Trek,* invited the reader to make connections both to policing the galaxy and to interplanetary exploration of all sorts, mind and body included, the inside contents would bear the assumption that *Mind Trek* also intended to incor-

Figure 13.9. Remote viewer and military intelligence officer Joseph McMoneagle at the Monroe Institute, 2011.

porate references to orbiting space stations, the shuttle *Enterprise,* and Hoagland's Enterprise website, with its interest in artifacts from Mars and the Moon.

Unsurprisingly, inside the pages of *Mind Trek,* a RV session on Martian landforms is to be found. Here McMoneagle strongly recommends *The Face on Mars,* that Mars study fronted by Randolfo Rafael Pozos. Noting that if his readers might conclude that essentially the RV information featured in his Martian RV session could have come from his own knowledge of the Pozos conference material, in fact, McMoneagle assures us, he had experienced the Mars session as "real." Hmmm. Why does he even need to make this distinction?

In the same way that McMoneagle refers to the potential influence of the Pozos report on his subconscious, so it is with *Mind Trek.* Even before readers start on the text, *Mind Trek* has inserted iconic images and symbols into their minds via its cover. However, in case these were not noticed, Courtney Brown, associate professor in political sciences at Emory University, in Atlanta, Georgia, reviewing the 1993 *Mind Trek* publication on his eponymous website, was ready to help them toward the correct interpretation of this book's cover and contents.* Here is an extract from Brown's review:

> Most of the targets described by Mr. McMoneagle are of a sort known as "verifiable," which means that the targets are of well-known places for which there are detailed photographs. But Mr. McMoneagle does not close the door on more far-flung targets entirely. In one of the most intriguing sessions, this one conducted at The Monroe Institute in Faber, Virginia, Mr. McMoneagle describes what appears to be an ancient (and dying) civilization on Mars. Rather than try to force the reader to accept these perceptions as real and accurate, he leaves it up to the reader to decide how to respond. Since the description of this target occurs near the end of a book in which the reality of the remote viewing phenomenon is so strikingly portrayed, Mr. McMoneagle's description of a long-lost Martian setting acts to keep the reader at the edge of his or her seat, begging for more.[11]

---

*Brown established the Farsight Institute in 1995 and claims to have the best remote viewers working with him on such projects as viewing Cydonia. The Farsight Institute asserts that it is a civilian organization using military methods for its RV sessions. In 1996, Brown's Farsight Institute conducted a series of RV sessions resulting in the claim that there was an intelligently guided artificial object allegedly four times the size of Earth (making it just shy of thirty-two thousand miles in diameter!) following the Hale-Bopp comet. Despite the resulting media storm, the Farsight Institute did not consider itself responsible for the decision taken by members of the Heaven's Gate cult who committed suicide in order to join that spacecraft.

Two years later, in 1995, the rug would be pulled from under the feet of these particularly hungry readers when the CIA report on the RV program emerged. The scientific consensus that RV doesn't contribute any useful information to the intelligence community potentially hit those who had appreciated McMoneagle's book particularly hard. Especially when it was found that the Mars section of *Mind Trek* uses erroneous coordinate data. The interesting scatter of Martian coordinates set up for McMoneagle all turn out to be applied to photographs from the same location—Cydonia. This is not necessarily McMoneagle's entire responsibility, and should NASA disagree with this conclusion and reiterate that these particular coordinates do come from all over Mars, then there is even more reason to acknowledge the presence of anomalous landforms on Mars that potentially indicate intelligent design.

In 1997, McMoneagle's *Mind Trek* was updated and published with a new chapter, "Stargate," devoted to that CIA RV report.[12] Perhaps it was named after the 1991–1995 phase of the RV project; for the general public, this name recalled the 1994 *Stargate* movie. That film was all about fear of very unfriendly aliens disguised as ancient Egyptian gods dominating a lesser terrestrial civilization. The cultural inferences could also be taken elsewhere: the use of alien technology to facilitate the domination of an indigenous population living near pyramids set in red deserts—somewhere in space and time—could be applied to a period of ancient Martian history. Whatever the reason, the chapter titled "Stargate" was in some way fitting, since McMoneagle asserted that FEAR was reason for closing down the RV programs. McMoneagle chose to capitalize this word in his text. Fear can generate self-deception at best, intentional deception of others at worst. And his friendly reviewer Courtney Brown is also a proponent of hidden underground alien bases here on Earth, in Mexico, which takes us back to the motivation behind some of these RV groups and that word: FEAR.

Whether a top-secret acronym or McMoneagle the intelligence officer being shouty, he asserts that he meant the fear of ridicule on the part of the organizers. This would infer that by 1997 dabbling with PSI matters was now considered beyond the pale even under the protocols established by the military. So then, attaching the word *stargate* to the 1997 edition of his book front-loads the reader with the information that using PSI methods to investigate off-world matters can have dire consequences, mentally and socially. Discouraged and not wishing to be treated as "kooks," McMoneagle's once enthusiastic readers were hopefully going to re-evaluate their position on RV and for good measure, relegate Cydonia to the dustbin.

THE ATLANTEAN MYSTERY

Having established an RV standard with McMoneagle's oeuvre, the closing of the project also allowed the flood of RV books that followed to create even more confusion. Competing RV methodologies and workshops preoccupied the public forum, leaving the way clear for a discreet follow up on all that still undisclosed RV research.

Previously, it was conjectured that those semisecret PSI-Ops research projects could have been set up to in order to explore access to the Atlantean crystal technology referred to by Edgar Cayce. His pronouncements had generated such an interest in investigating the locations he mentioned during trance that in 1931, the exact same year that the first research vessel *Atlantis* arrived at Woods Hole, Cayce had established his own not-for-profit organization to research and explore holistic health, ancient mysteries, and personal spirituality. He called it the Association for Research and Enlightenment Inc. (ARE). The ARE website shows that to this day, the various archaeological sites he had stipulated as being sources of Atlantean information are still being scrambled over by serious organizations, archaeologists, and scholars. These sites included Bimini and the serious organizations include NASA.

In 2001 and 2002, the Edgar Cayce Foundation paid serious money (in those days, from $12,000 to $28,000, depending on the detail required) when it commissioned the IKONOS satellite to image 630 square kilometers around the two Bimini islands. John Van Auken of the ARE later stated that the order was for an area 243 square miles and that it was paid for by Don Dickinson's *Law of One Foundation*. He is described as a funding angel for ARE projects. Yet there is no reference on the web to a Don Dickinson associated with any such foundation. However there is a website for the *Law of One Society,* seemingly based in London. The content of the website would make a good fit with the interests of the ARE, but it has no data available concerning its founder members or a Don Dickinson.

Apparently, no one at the USGS thought of informing the ARE or Don Dickinson that they could have saved their money, because the seas east of Bimini had already been imaged by satellite over a quarter of a century earlier. Unless of course that information was secret.

This particular Atlantean mystery would take a more than ordinary code breaker to sort out. The data apparently produced as a result of that "first remote sensing imagery" first came to a very small segment of the public's attention when a map was published in a subscription newsletter on matters cartographic. It was

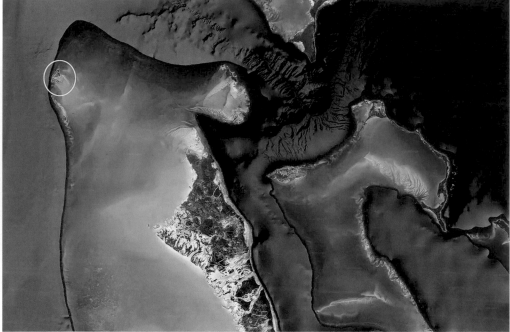

Image and data: SIO, NOAA, U.S. Navy, NGA, GEBCO Landsat/Copernicus, Google Earth 2020 25°44'24"N 79°15'37"W.

Figure 13.10. The white circle around the two Bimini islands defines the area imaged by satellite. The ARE website states that survey was for 640 kilometers and then references 630 kilometers, which would be the 243 square miles mentioned by Van Auken. These discrepancies are seemingly par for the course, as was the lack of agreement over the results of this survey. Andros Island in the Bahamas can be seen to the southeast.

stated that this Bimini map was the result of overflight by the first satellite used to take Earth photographs.

Without more specifics, one was led to surmise that if this was the first of the Landsat program, the data had been acquired around July 23, 1973. That would put it at just over a year after Cydonia had received that *second* visit from *Mariner 9*'s mapping cameras. (The reader will recall that after *Mariner 9* had mapped Mars from 40° N to 60° N at a rate of 12° latitude per day, there had been a two-month hiatus, or period of "recalibration," as NASA put it. after which *Mariner 9* remapped that very same area, 40° N to 60° N, only this time it took *five days* over each degree of latitude.)

Then nineteen years later, in 1992, a hand-drawn map of the seas around Bimini that featured a pentagonal pyramid lying under the Atlantic waters southeast of South Bimini and a flat-topped rectangular form just off the coast of North Bimini is produced in a specialist newsletter. The publisher could not

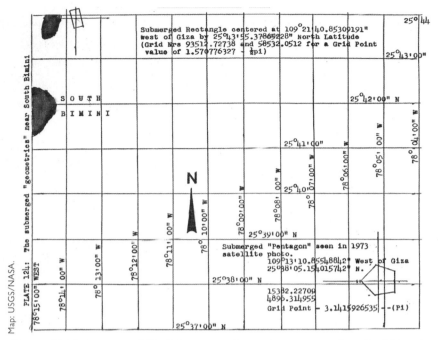

Figure 13.11. Annotated Bimini map as drawn by cartographer Carl Munck.

recall exactly how the map came to be in his files, but found it interesting and published it.

Looking at this map, for Cydonian aficionados the five-sided D&M Tor and the rectangular outline of the Face on Mars sprang to mind. For cynics, this five-four motif reinforced the notion that the emergence of this map onto the scene was a setup.

THE BIMINI BLOOPER

Most people are aware that the Bimini islands are said to form one corner of the so-called Bermuda Triangle, a zone of highly anomalous meteorological effects and as already noted, an area high on the list of candidate locations for the sub-merged lost continent of Atlantis. Yet this map, with its two submerged artifacts, has not made headlines on dedicated Atlantis websites such as the ARE's, nor did it raise ripples among the readership of the cartographic newsletter. Why not? Such a silence raises several issues as to whether these submerged items truly exist or not. A scenario emerges:

They do exist and Woods Hole's RV *Atlantis I* was involved in first identify-ing these submerged Bimini landforms during those quiet winters. As a result of the RV *Atlantis I* findings as soon as it was possible, a Landsat was tasked with imaging this particular area. Which then poses these questions: Did the shuttle *Atlantis* have any specific Bermuda Triangle oversight operations to perform? Does the ISS? Or do we have another scenario altogether; do we have here the manifestation of an obsession over the lost physical remnants of a civilization thought to have been submerged beneath the waves near Bimini? Only later found to be on the ancient shoreline at Cydonia on Mars?

In 1992, having no reason to disbelieve the data in the cartographic newsletter and with no means of digging deeper, there the matter rested—until 2005, when the first version of Google Earth was rolled out to the public. Since then we have come to learn that Google Earth's imagery is compiled from a variety of data sources, depending on the resolution required, so that anything from satellites, aircraft, and helicopters to drones and even cars at ground level all contribute to the end product. However, back in the mid-1900s, the technological landscape was different. And even further back, it was simpler still. As far as those early satellite programs went, NASA websites state that Earth imaging started as early as 1965, using *aircraft* loaded with experimental remote sensing equipment. Then in 1966, the USGS convinced the secretary of the interior that the Department of the Interior should proceed with its very own Earth-observing satellite program. NASA historians made an interesting word choice when they stated that it was a "savvy political stunt" on the part of the USGS that "coerced" NASA into building the Landsat program. As an organization not above savvy political stunts when it suited, "coercion" might better read as "they agreed between themselves."[13]

With the advent of Google Earth in 2005, it seemed sensible to check up on the data supplied by that cartographic newsletter. Bearing in mind that these Bimini features had been picked up by a satellite but being unaware of the resolution capability, it was only possible initially to set Google Earth for the same orbital distance from Earth as the relevant satellite: 559 statute miles (900 kilometers). At that height, the outlines of these two very small islands were barely visible. Zooming in until it *was* possible to see under the surface of the sea around Bimini meant being only ten kilometers off the surface, so either the Landsat image was of the appropriate resolution or the published map was not, after all, entirely produced from that satellite's data.

If the Bimini map *had* originated from Landsat data, the underwater area was recorded via a medium other than conventional photography. Which was also perfectly possible, as these early satellites carried a multispectral scanner in addition to the Return-Beam Vidicon. However, at the beginning of satellite development, this technical information was generally restricted to the US government's domain and was not necessarily available to the public at all. So one wonders why such a map would ever find its way into the pages of a small specialist newsletter ostensibly aimed at the public.

It was time to do some more archive digging.

The next port of call was the USGS website: if this map was from either a Landsat or imaged from the air it should be in the USGS archives, but entering the precise coordinates from the map produced no result. Nor did any other search using all the tools available. Which infers that this 1973 map did not originate in the civilian arm of the USGS but within the Intel agencies. And here's another coincidence: in 1973, the United States government had reorganized its mapping agencies, and some USGS scientists and technicians were given access to film shot over North America by the top secret CORONA satellites. As Cold War historian John Cloud says:

> The initial step was the relocation of the civilian US Geological Survey's National Mapping Division from Washington to the isolated forest suburbs of Reston, Virginia, and the construction of Building E-1, a TALENT-KEYHOLE-level top-secret laboratory. Within E-1, USGS scientists and technicians had access to CORONA film shot over North America.[14]

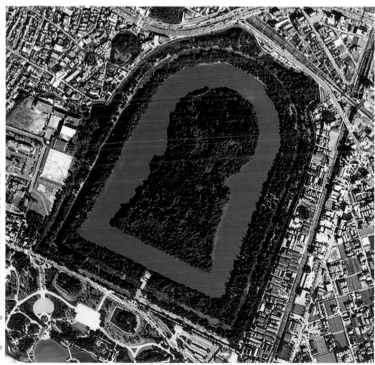

Image: Google Earth 2016 34°33'53"N 135°29'16"E.

Figure 13.12.
The emperor Nintoku's Kofun (tomb). These forested Kofun are full of esoteric meaning. Based on a geometry combining the circle and the triangle and surrounded by water, they are known in English as keyhole mounds.

Using CORONA and its associated mapping application systems, the USGS revised the entire national coverage of the 1:250,000-scale national mapping series not once, but twice. Then CORONA was applied to the national 1:24,000-scale, 7.5-minute topographic map series. Geodetically rectified CORONA photography was projected at the same scale as the maps being revised.[15]

Map revisions were covered by the legend "based on aerial photography and other source data," and as for the other source data, read satellites. Cloud states that "the other source data for all US mapping programs, for the last third of a century, have been CORONA and its successor reconnaissance satellite systems." The successors to CORONA, called SAMOS, GAMBIT, and HEXAGON are, to date, still classified.

Another scenario emerges:

These items do not exist under the waters off Bimini but it was required to place the link between Earth and Cydonia (using the pentagonal D&M Tor and the rectangular Face on Mars) on the public record. The release of this Bimini map occurred three months prior to the *Mars Observer* launch, during that very public argument about acquiring further Cydonian data from the upcoming mission. Perhaps it was intended to make a point about Earth-Mars ETI connections once *Mars Observer* had arrived on-site. Ostensibly then, the apparent loss of the craft demolished this plan.

CAUGHT MAPPING

Thinking along these lines and with the added thought that it would be foolhardy for a specialist cartographic newsletter to issue a map with specific coordinates and for there to be nothing at the location, I returned to the Bimini area via Google Earth. Going over the mapped area once again very slowly and carefully produced nothing, but widening the search area gradually eventually did get a result. Crawling above the ocean and moving away from the coordinates supplied by the publisher finally revealed a flat rectangle at exactly the right latitude off north Bimini and what might pass as a pentagonal form underwater, also at the appropriate latitude. However, both these items were located some ten degrees (six hundred nautical miles) *farther west* than advertised.

The public record turns out to have a problem. The cartography newsletter supplies mathematical calculations relating these two items to a Giza pyramid meridian. And indeed they can be made to correspond with mapping data from the Giza pyramids and from Cydonia, Mars, *but only when those original and incorrect longitude coordinates are applied.* Now this infers either a shift of the planet's skin much beloved by the catastrophe school of thought and those who believe the myth of Atlantis, which of course incudes the Cayce group. Or it implies a very real intention to link Giza, Cydonia, and Bimini together but not necessarily in order to confirm the manifestation of ETI out on Mars. This Bimini mapping "error" looks very much like another "savvy political stunt" to get rid of the Cydonia data. After all, if the basis on which these links have been established by the cartographer are later shown to be demonstrably inaccurate, then the conclusions linking Bimini to Martian landforms and our strangest terrestrial pyramid are also invalid.

This reasoning makes the case for this Bimini map being published as a precursor to a successful *Mars Observer* mission rather strong. Had *Mars Observer* succeeded and returned a post-Viking image of the Face on Mars in 1993 that looked very bad, the newsletter readers would hopefully be inclined to agree with NASA and dismiss Cydonia's landforms as being a natural mesa's trick of light and shade, a figment of the imagination.

However, had the *Observer* probe truly gone out of NASA's control, the result was the same concerning the public release of another Face mesa image. It would look very bad but it would have to wait for the next probe. And so it was that the catbox image was finally taken by *Surveyor* in 1998. But remember, it was only released to the public in 2001 by Malin's team, and while everyone hoped that the excitement over human-looking artifacts on Mars had always been a big fuss about nothing, that was the same year that the ARE Commissioned the Ikonos satellite image of Bimini. Still searching for something.

If this is all too far-fetched, let's stretch it a bit further. It's certainly interesting that the year the top-secret experiment in RV was finally declassified, 1992, was precisely when this Bimini map was finally published. It's a nice touch that the RV initials for the remote viewers are the same as those for the research vessel Atlantis (always written as R/V Atlantis by the WHOI), because it's equally likely that this 1973 Bimini map was generated by the RV form of remote sensing used by McMoneagle for the Martian session. Following up on such dodgy mapping data with the 1993 publication of *Mind Trek*, which contained those allegedly "all over the shop" Martian coordinates, only adds to the confusion for the public and the ultimate dismissal of Cydonia as a location worthy of attention.

Which leads to the conclusion that *the longitudinal error on the Bimini map could have been inserted by a data archivist prior to the map arriving at the newsletter's office, and* those on the receiving end published the map "as seen" either out of ignorance or obligation. Although, even if under obligation to produce an erroneous map in 1992, once the Face on Mars had been buried in the catbox it would have been quite easy to sort out this Bimini blooper.

Yet these erroneous coordinates have never been corrected in any subsequent publication or newsletter that has appeared over the decades. Indeed, this map has been reproduced several times.

It's time therefore to have a closer look at that specialist cartographic newsletter.

Matrix Man

PSI Man Meets the Pi Man

The specialist newsletter that published the Bimini map is called *The Code.* Its contents, copyrighted by Carl P. Munck, now amass to years of newsletters and collected volumes of works. He calls his calculations "Ph.D.-level material." These studies were first mentioned at the Terrestrial Connection event at the United Nations in February 1992 and were made available to the public in June 1992. What comes across from reading the Munck papers is the striking concordance between the RV program's mapping vocabulary and the content of these mapping newsletters. If the published material on RV didn't reveal the entirety of the SCANATE method of RV, Munck's mapping coordinate methodology certainly provided the required levels of security.

David P. Myers and David S. Percy were familiar with Munck's work when they were exploring the Avebury connection to Mars for their book *Two-Thirds,* and Percy had remained in amicable contact with Munck over the years. As the editor of *Two-Thirds,* and while researching this book, I wrote to Munck concerning the remarkable parallels I had noticed between the timing of the PSI-Ops RV

THE SILENT LANGUAGE OF OUR ANCESTORS

VOLUME 1, NR. 1   Copyright © 1992   JUNE 1992

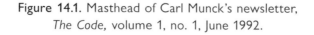

Figure 14.1. Masthead of Carl Munck's newsletter,
The Code, volume 1, no. 1, June 1992.

programs and his own publishing schedule. He declined to reply. Given his previously sociable contact, I took this silence to be significant, but just in case he was very busy, I sent a copy of my findings to his "front office," asking that they forward it to Munck as they advertised. I received no reply from them either. After another two attempts, I decided to go it alone and trawled once again through all of Munck's published oeuvre. He has scattered bits of information about himself across his published works, and I wanted to read it again to see what emerged when viewed as a part of the PSI-Op RV projects. Or more accurately, as part of the education of the public in matters Martian.*

My discoveries were such that the unraveling of this tangled web means some repetition of dates and actions already discussed in chapter 13.

MAPPING THE MATRIX

Carl Munck was in the USAF from 1953 to 1973. The year that he retired, the CIA was one year into the top-secret SCANATE program. Five years later, in 1978, when RV shifted to the Army and GRILL FLAME got underway at Fort Meade, Munck took up his "new hobby" of mapping ancient monuments into mathematical grids.

Figure 14.2. Carl Munck.

---

*Munck only supplied his newsletters and his collected volumes on subscription, for which there is no current contact. He is said to be retired but still active, but his works are no longer easy to obtain. They had previously been available for purchase through LauraLee.com, but this link no longer works. His 148-page 1996 spiral-bound collected publication of *The Code* is listed on Amazon but unavailable. Online searches of his name or "archaeocryptography" will bring up articles concerning his research, while YouTube videos of his presentations are available. And just before going to press I checked again and lo, Munck's work is advertised as a set of four DVDs available from Amazon. These are in fact his videos transferred to DVD, and according to one reviewer, the quality is not brilliant, but the content most certainly is—if you like math.

Munck's professional history with the USAF and his very good relationship with the USGS at Flagstaff, Arizona (used by Munck for data sourcing and verification) hints at the possibility that his new hobby resulted from the mathematical encryption of databases for sensitive locations (such as missile bases or other planets) and the uploading of these codes to satellites or to remote viewers. Given the inserted bloopers into his published output, such as that of Bimini, it would seem that the US PSI projects are multifaceted. Yet seemingly disparate projects might have more in common than is generally assumed, even by those running them. Munck's background gives every indication that he is a part of the story; whether he is in the PSI-Ops loop is another matter, but here it would be advisable to note that the National Reconnaissance Office, a member of the US intelligence community, understood that the personnel in the strategic geographic sciences were members of a "code-word mapping community" and said so out loud in the mid-1960s.[1]

It would be fourteen years from the beginning of Munck's new hobby until the publication of that first *Code* newsletter in June 1992. He often took the opportunity to remonstrate to his readership that he had contacted the authorities with the amazing mathematical relationships he was finding but that his efforts were in vain. Munck's grousing about other authorities is not unlike McMoneagle's grousing about other psychics. And given their acknowledged connections within those same groups, equally implausible. The method Munck had selected for his coordinate mapping project was the same SCANATE method of combining latitude and longitude into a single string used by the US intelligence agencies and military. The prime meridian adopted by Munck—the Great Pyramid—had been used as such by the USAF in the 1940s when it created an azimuthal projection of the world centered near Cairo—on the Great Pyramid. However, in a June/July 1995 newsletter, Munck stated that he first decided that Giza was a prime meridian back in 1957 when the USGS discovered (that is, verified) the 1858 hypothesis that that the geological center of all the land on Earth was near Cairo.

Ostensibly, that Giza map was undertaken as a part of the USAF operation in the Middle East during World War II. Paradoxically, documents relating to the history of mapping state that the USAF was using Lambert conformal conic projections at that time, so it appears that this azimuthal map was created for quite another purpose.[2]

Professor Charles Hapgood has associated the USAF Cairo Azimuthal map with the earlier 1930s American research into the map drawn in 1513 by the Turkish admiral Piri Reis. However, it has been established that the Piri Reis map, which had its genesis among other very ancient maps assembled centuries

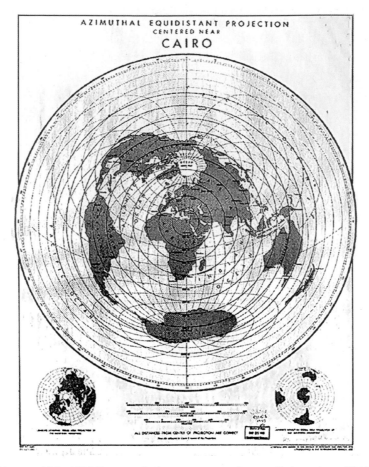

Figure 14.3. USAF azimuthal map centered over Cairo, circa 1940.

earlier, was based on a prime meridian running through the present location of Alexandria in Egypt.*

It is clear that the now ice-covered continent of Antarctica could not have been mapped in such detail for the Piri Reis map without an aerial capability (not possessed by the indigenous population of Earth at the time) or potentially advanced instrumentation and possibly satellites (not possessed by the USAF in

---

*Charles Hapgood worked in the Office of Strategic Services (OSS) in 1942 and was also a liaison officer between the Secretary of War's office and the White House. He anticipated the discussion on plate tectonics by some ten years, writing *The Earth's Shifting Crust: A Key to Some Basic Problems of Earth Science,* published by Pantheon Books in 1958. The Piri Reis map featured in his next book, *Maps of the Ancient Sea Kings: Evidence of Advanced Civilization on the Ice Age,* published first in 1966 and in a revised edition from E. P. Dutton in 1970.

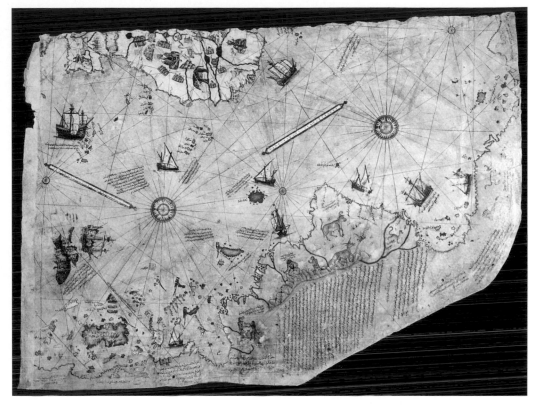

Figure 14.4. Surviving section of the Piri Reis map, completed circa 1513.

the 1940s). From which it is also reasonable to conclude that the USAF Cairo map was also exploring the ETI hypothesis.

Whether Munck had sorted out the development of the basic coordinate scheme for the RV program is not known. That he has sorted out the grid relationships of all the major monuments on this planet and landforms on Mars is a certainty. He has found the mathematical data demonstrating that the Great Pyramid at Giza is the key to the mapping matrix here on Earth and that the huge D&M Tor pyramid is the key to the Martian matrix. Indeed, over time, the Munck papers have linked quite a few of Earth's major ancient sites to the Cydonia region via mapping coordinates, especially those that represent faces or pyramids. Little wonder that those pyramids on Mars were of more than a passing interest to NASA and the military remote viewers.[3] As for this maestro of coordinates, Munck tells his readers that his own awareness of pyramids and their potential had started impinging on his mind much, much earlier than the 1992 date of his first published newsletter. The following is what he conveyed to his readers.

PYRAMIDS

Pyramids just crept up on Munck; his dreams were full of them, and they nudged and nagged him during the daytime as he tried to live his everyday life. He kept asking himself why were they built in this particular four-sided form, with or without stepped levels? Why were they built in such particular locations, generally in clusters and often in numerically significant groups such as triads? The principal pyramids of the Giza complex and the Mexican complex at Teotihuacán bothered him, as did Tikal in Guatemala and the other ancient cities in Mesoamerica featuring groupings of pyramidal forms. And then there were rumors of Chinese pyramids out near Xi'an (see figure 1.7, page 21).

Carl Munck reached the conclusion that precious little progress was being made in understanding the ancient pyramids, monuments, mounds, and effigy mounds by archaeologists in particular, as well as other academics, simply because the right questions were not being asked. He has provided those interested with numerous presentations of ancient sites, demonstrating mathematical connections between the greatest monuments and earthworks. He also confirmed that every single one of the 250 ancient sites he has studied relates mathematically to a prime meridian at Giza, the Great Pyramid. Yet despite making it obvious that this mapping of the planet from the Great Pyramid wasn't possible without the use of satellites and GPS, Munck professes not to be interested in the subject of ETI.

Paradox.

EUMARES—ANAGRAM OF MEASURE

The year that Munck apparently began following up on his pyramid quest 1978, was the same year that the PSI-SPI project GRILL FLAME began in Fort Meade and the same year that Dilettoso and Stevens in the United States started examining Meier's UFO images. Equipped with an ordinary Texas Instruments calculator, lots of paper, and the best maps obtainable (and they really were the best!), Carl Munck sat down at his desk somewhere in North America and began a new hobby. He called it cryptography, and it was the beginning of his mathematical search into the hypothesis that a numerical code system—initially based on map coordinates—could be hidden within the layout of our ancient sites. Here's how he put it:

> I once had a hobby. Cryptography. But it's not a hobby anymore! Back in 1978 I was trying to develop a code system to be based on the language of latitude and longitude (degrees, minutes and seconds), the idea being to mul-

tiply the three numbers in each set to a single number, viz. 15 degrees times 15 minutes times 01.6000 seconds products at 360 (the number of degrees of arc in any circle).

I won't say what I was looking for, but let's just say I was looking for a better mousetrap and let it go at that. It was a great way to pass the time and provided my copious intellect with a lot of fun. That is, until I found my plots getting themselves tangled up with the ancient pyramids.[4]

Upon reading more of Munck's work, it becomes clear that his contacts at the USGS would have been able to supply him with the best maps available, and given the data coming in from Cydonia since 1972 and the fact that Munck was going to find mathematical correspondences between locations on Earth and that "trick of the light" landform out on Mars, his mousetrap and his hobby required closer scrutiny. Not least because data from ETI has informed the PSI seekers that there

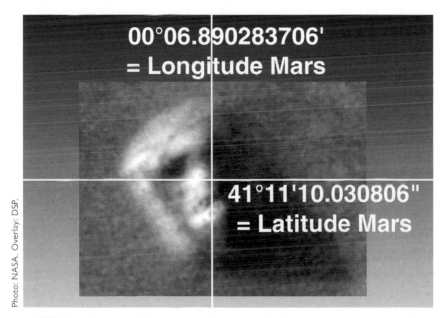

Figure 14.5. The Face on Mars's latitude reveals hidden geometry when subjected to Munck's method. The precise latitude of the Face on Mars is 41°11'10.03080581". Multiplying 41 × 11 × 10.03080581 = 4523.893421. Munck then divided this number by the universal constant pi, namely 3.141592654 (he always used the first nine decimal places) which produced the number 1440. Knowing his Platonic solids, he recognized this number as pertaining to the tetrahedron. The angles of a tetrahedron sum to 720° and the angles of a star tetrahedron (comprising two single tetrahedron interlocked), sum to 720 x 2 = 1440°. (The annotated figure on this image is the precise latitude rounded up at the sixth decimal place.)

are mouse holes in our atmosphere that result from the effects of objects left on other planets in this solar system in the past.

Munck is a number cruncher who states that *everything* is about number, that language is not so relevant to him. That might explain the odd choice of name for his hobby: cryptography. It means developing a code system that does not necessarily conceal the presence of a secret message but does render it unintelligible to outsiders by various transformations of the "plain text."[5] Hmm. That definition recalls those Mall gardens' architectural and landscape transformations we saw in chapter 12 as well as the RV coordinates taken to their third level.

However, Munck was apparently looking for the keys to a code that *already existed*. Surely then, he would be a cryptoanalyst. But as he was sticking with this name of cryptographer, presumably he was indeed trying to develop a secret code system that at first did not necessarily involve all ancient monuments and large landscape features. Or if it did, they were perhaps useful anchor points for a developing technology about which he was saying nothing. Speculation on my part, yes, indeed. But presenting his work the wrong way round from the outset is a red flag and worthy of investigation.

Munck tells us that for some years he compiled his database, cultivated his mind, probed his hypothesis, and developed his already acute sense of humor. Which he was going to need, because ironically, if there was one thing Munck did not appreciate it was working with computers. (With a personal computer now in most homes in many parts of the world and personal digital devices in the hands of tens of millions, there is a tendency to forget that in the late 1970s, this was not the case). While Munck claims to prefer his trusty calculator and/ or his own brain, he then says that he still had to take some notice of the emerging world of computing. Well, emerging for the public, that is. The inception of the CORONA satellite project had taken place over twenty years earlier when the US government under President Jimmy Carter officially acknowledged that it was using satellite reconnaissance, and Munck started with his so-called hobby in 1978. Remember that Carter was the US president who officially claimed a UFO sighting? And remember the Roswell incident, near Corona, New Mexico?[6] And always remember how the US military love their name games.

Munck opined that even if the finer points of his coded system still escaped his grasp, the best way to understand its complexities was to operate *only* with the mathematical evidence and to dump the dogma. He says that this mental focus was to become a considerable asset when his numbers began meshing with the world's greatest pyramids and he encountered the deeply held beliefs of mainstream Egyptologists. Fair enough, lots of dogma there!

All of which inferred that he gave short shrift to the myriad theories, suppositions, and superstitions that composed the accumulated cultural flotsam surrounding the sites he was investigating, and this might be taken to read as "I am a mathematician first and foremost. I am a left-brained person. I only pay lip service to the wilder shores of the imagination."

THE PYRAMID MATRIX

The year that GRILL FLAME came to an end, in 1982 (it would morph into the CENTER LANE project in 1983), Munck's hobby also became a "project" and acquired a new name: cryptography was dumped and replaced with the Pyramid Matrix.

It was at this point that Munck informed his readers that he was aware of the statement made by the psychic Edgar Cayce, who had said that if we wanted to understand the Great Pyramid, we must first understand "the mathematical precisions of the Earth."[7]

Then paradox rears its head again. Munck's presentation of himself as the left-brain rationalist mathematician seems not to have worked in practice. Unless his personal views on the matter of Atlantis, military officers and priest-kings with ancient powers, past lives, and the many other nonmathematical subjects threaded into his published works were inserted merely to attract a certain clientele or *pour encourager les autres.* Such esoteric threads were accompanied by the denial that ETI interventions were the answer to how this matrix was established. Methinks he doth protest too much. Especially because present-day engineers cannot even construct a replica of the missing apex of the Great Pyramid—on the ground—without running into trouble and because all the solutions as to how the Great Pyramid was actually constructed have yet to pass muster.[8]

As if to counteract my criticisms, Munck tells his readers that he thought that Cayce's remark was interesting, for his own numbers were already telling him (as they had other people) that many features of the Great Pyramid, even as we see it today without its original precision covering of limestone, imply that its builders had a considerable knowledge of spherical and planetary mechanics.*

---

*Virtually everyone notes the Great Pyramid is very precisely aligned to the north, east, west, and south cardinal points and that great circles drawn from Giza in those four directions cover more land as opposed to ocean, than from any other place on Earth. The fact that thousands of years ago, due to the Earth's plate tectonic motion, Giza's pyramid was not exactly in the same location as we find it today is totally ignored.

Figure 14.6. Psychic Edgar Cayce, 1910. Cayce said that if humans wanted to understand the Great Pyramid, we must first understand "the mathematical precisions of the Earth."

Munck thought that Cayce's comment required two specific tools: maps and mathematics—not everyone's cup of tea, but Munck's favorite brew. His preceding experiments with the mumbo jumbo of latitude and longitude and his already extensive database, together with his large collection of quality maps and his good connections at the USGS, all gave him an excellent head start. If he came up with the right questions, he thought he might obtain some better answers to the enigma that was the Great Pyramid. He also cherished his hypothesis that there might exist a purposeful relationship between the forms of ancient structures and their positioning. What Munck found would lead him initially to re-examine all the major pyramids and then to take another look at the earthen mounds of North America—and that was just for starters. By 1982, he had come to the following conclusion:

> Someone, long ago, went to extremes in positioning these coded objects around the world and they did so with numbers. Numbers that are familiar to us today in the language of degrees, minutes and seconds of latitude and longitude. Map talk![9]

Munck sounds just like the mapmaker character in *Close Encounters of the Third Kind*. In the film, trying to make himself heard over the hubbub of a room full of experts tracking signals from an unknown source for which they could find no key, he stopped them in their tracks by saying, "Excuse me. Before I got paid to, uh, speak French, I, uh, I used to read maps. This first number is a longitude... Two sets of three numbers. Degrees, minutes, and seconds . . . These have to be Earth coordinates."[10]

PRIME TIME

Munck says that when he set out on his inquiry, he was chiefly interested in whether he could find a *mathematical* reason for the positioning of the major pyramids, any trace of which would automatically infer that there was a coherent global plan to our architectural inheritance across continents and across cultures. One could add: and between planets—Earth and Mars.

Whatever the true origin of Munck's work, the way he explains how he did what he did is nevertheless interesting and useful, as it helps expand our thinking. That this Pyramid Matrix phase change went with the RV program coded CENTER LANE was not a bad match especially as Munck decided to treat Giza's geographical position as if it had been an early planetary prime meridian and then see what that revealed. Or more likely, he followed up on that earlier USAF azimuthal projection of the world based on Cairo (recall figure 14.3, page 332).

While the decision to choose the Giza Pyramid as a prime meridian might seem a little arbitrary to some people, in fact, throughout history, the choice of exactly where to run a notional line extending from north to south through the poles, thereby creating two zones east and west of what we call the 0°/360° longitude (together with its attendant international date line), has been entirely arbitrary, and it entirely depends on which of the political, religious, economic, or cultural interests are dominant at the time of the mapping and how the choice will affect the aims of the interested parties.

Prior to 1884, each nation had its own meridian, and these varied across the country because the Sun was at the zenith (directly overhead at midday) at different times. With the coming of the railways, when the speed of travel required the coordination of timetables, matters changed. In October 1884 the International Meridian Conference was held in Washington, DC, and much discussion ensued, with the Egyptologist and Astronomer Royal of Scotland, Charles Piazzi Smythe proposing that the unique location of the Great Pyramid made it the obvious candidate for the job. However, by this time over two-thirds of the world's shipping was already running on Sir George Biddell Airy's 1851 Greenwich 0°/360° meridian, and everyone else was in favor of Greenwich becoming the world's prime meridian.

Everyone that is, except the French.

FROM AMUN TO THE OGDOAD

The political factors inherent in choosing a prime meridian were well understood by the French back in the nineteenth century when they attended the 1884 meeting in Washington, DC. At an earlier conference, they had stipulated that if Britain converted to metric measurement, they would convert to the Greenwich meridian, but this was strongly resisted by the British. (Today the meter is being introduced in Britain by stealth.)*

As a result, the French were disinclined to concede their national meridian. So it was unsurprising that when the delegates from twenty-four out of the twenty-five nations present voted to retain Greenwich's 1851 marker as 0°/360°, the French abstained from the vote. They sulked for the next three decades, retaining the Paris meridian as their timekeeping longitude until 1911 and as their benchmark for navigation until 1914. The Paris meridian can still be found on some French maps, and there is still a monument to the Paris meridian in Dunkirk. Remarkably, from the beaches of Dunkirk to the foothills of the Pyrenees, each end of the French meridian is associated with the huge turmoil of war. The evacuation of Dunkirk is well remembered; less so, the thirteenth- and sixteenth-century wars of religion that were fought at its southern terminus, around Rennes-Le-Château. Centuries before the US Pentagon was even thought of, according to author David Wood, this hilltop village was part of a huge pentagonal form marked out across the landscape by both natural and constructed landmarks and sightlines.[11] As we have seen at Cergy's Axe Majeur park, the French are right up there with the United States when it comes to understanding the implications of associating the natural power of the land with strategically placed buildings, all of which were designed to protect and enhance the nation via their constructed sightlines and recognized meridians.

This profound understanding gave the French historical and philosophical reasons for their 1884 grump: it was not all down to their dislike of the British Royal Navy and its earlier deeds!

MATH AVOIDANCE

This same understanding led Munck to his idea that the Great Pyramid was the hub of a planetary matrix. His thought that the ancient Egyptians (as we under-

---

*NASA lost the *$125 million Mars Climate Orbiter* because a Lockheed Martin engineering team used British Imperial units while the agency's team used the more conventional metric system for a key spacecraft operation (source: CNN, September 30, 1999). This event has been much touted as justification enough that the entire world should adopt the metric system.

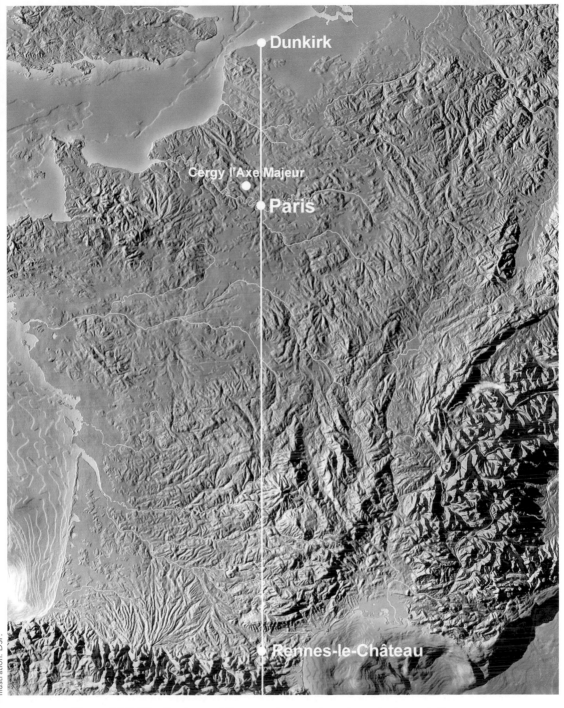

Figure 14.7. The Paris meridian running north and south through France. "The land and the king are one." Note the location of Rennes-le-Château and the Axe Majeur park.

stand them) based themselves along the lifeline of the south-to-north-flowing Nile reinforced his decision to test the Great Pyramid against a 0°/360° longitude. He also knew that when Napoleon's surveyors had surveyed Egypt in the 1700s they had used the cardinal points of the Great Pyramid during their survey of lower Egypt and found that the delta was divided by the center line of the pyramid while lines extending from the northern corners made tangents to the west and east limits of the delta. In order to test the Great Pyramid against his own hypothesis, Carl Munck first had to adjust his Giza meridian from that of our modern Greenwich-based 0°/360° longitude. This involved finding the difference in longitude between Greenwich and Giza. (This is 31 degrees, 08.00 minutes of arc, and 00.80 seconds of arc.) Munck would then add this coordinate to the longitude of any monument sited west of the Great Pyramid. Obviously, if the monument under scrutiny was sited east of Greenwich, this same amount would be subtracted from the modern Greenwich longitude. This exercise produced a "Giza longitude" for each monument on the planet, based on a 0°/360° Great Pyramid prime meridian. Munck shortened this GPPM to just GM, the Giza Meridian. Since the Greenwich prime meridian is also referred to as the GM, this can be quite muddling for his readers, but then we have also seen that official data sources often confuse the public.*

Writing about Munck's work with hindsight, it is easy to forget that for anyone not familiar with math, what he did next would seem to be even more astonishing. Having decided that he would try out the Great Pyramid as the nexus to which all other monuments would "talk" in mathematical terms, Munck selected a site, took its latitude coordinates and its new Giza longitude, and came up with the idea of working them to produce a third number. To do this, he multiplied together the degrees, minutes, and seconds of latitude, which produced his first number. He called this the Grid Latitude. He then did the same thing with his Giza longitude. This second number, he called the Grid Longitude. Then finally, he divided the smaller result into the larger, thus producing an ultimate figure that Munck initially decided to call a Grid Vector (shades of the remote viewer's vocabulary emerging here), and then he later refined the name to Grid Point.

---

*By convention, the world's weather communities use a twenty-four-hour clock, similar to "military" time and based on the 0° longitude meridian, also known as the Greenwich meridian.

Recalling earlier data, but useful here too, prior to 1972, the time zones calculated from Greenwich were referenced as Greenwich Mean Time. With the advent of atomic clocks and the need to keep time for satellites and computers, Coordinated Universal Time (UTC) was introduced. It is a coordinated timescale, maintained by the Bureau International des Poids et Mesures. Also known as "Z time" or "Zulu Time" by the military.

CHALLENGING FINDINGS

Munck asserted that the resulting data were to challenge the basic assumptions of most scientific communities and especially the international archaeological establishment. Having sent out data to organizations such as NASA and the National Oceanic and Atmospheric Administration (NOAA), along with other agencies and various governmental departments, the *National Geographic* magazine, and members of both the scientific and the academic communities, he told his newsletter readers that he had not received any acknowledgement for his efforts. (Much like the lack of response from him to my data presentation.) From their deafening silence, Munck made a deduction (as did I concerning his silence). He concluded that it was not worth bothering these authorities any further. (As indeed had I.) Having deduced that these communities had decided to turn their backs to his findings, he then added this very interesting comment "And pretend that the data didn't exist." (While I looked closer at his own data.)

Munck thereby implied that these organizations were at the very least aware that his methodology had validity. He was, after all, using a minor variation on Cartesian coordinate systems that were well understood by the mathematical, mapping, and satellite communities. And as noted, the spy communities were using the very same principles in their own RV research programs. Cleverly, Munck's remarkable parallels to that secret operation would make his comment that these communities had turned their backs literally true, since these secret programs were going on behind the backs of most of those organizations. But then you might ask: Why contact these organizations and tell them about his work? Well, according to that *Brookings Report,* the notion, even if challenging, that the presence of ETI ought to be taken into account was a given in the estimation of many in the science communities, but not all. The derision of the public over the question of UFOs had tainted the entire subject of ETI for many others. And now we see a reason for Munck's introduction of those esoteric elements mentioned earlier. This was one way of getting information to the more recalcitrant of those scientific communities while not rocking the boat. The "nonsense" elements allowed the truly recalcitrant to dismiss the whole of the ETI premise while absorbing the implications of the mapping data. And from a purely geological viewpoint, the obvious dichotomies would soon emerge.

Consequently, Munck's account of a categorical refusal to take any notice of his work might require another large pinch of salt from that nearly empty pot.

It does not, of course, detract from the discovery that the major archaeological pyramids and monuments on this planet really do relate to one another in a language derived from intersecting latitudinal and longitudinal calculations—*all to an amazing eight decimal places of accuracy.*

Nor does it detract from the revelation that the locations of many ancient sites appear to have been *specifically* selected in other ways: the topological features of these primary sites that were linked to one another shared visual similarities and expressed redundant mathematical constants. Pyramids talk to pyramids, cone mounds whisper to cone mounds, effigy mounds speak to each other, and so on. In fact, Munck found that he was actually rediscovering a code that had been designed and then laid out according to the monument's vector or grid point. So the entirety of these interrelationships must originally have been conceived as a whole and then established piece by piece across the planet. And for some, the fact that *the Cydonian locations also fit into this matrix* is a truly mind-boggling concept.

Of course, there are critics of Munck's work, not least because of all those Atlantean references he makes. Also, much to his disapproval there are also those who have adopted his formula and created their own "matrix data," based on other criteria than Munck's. Yet, if the mathematics and maps published by Munck are, as the Egyptians would say, the openers of the way, and if they are truly the revelation of long-held intentions by those who came before, then it should be possible to turn "if" into a statement of fact through an independent verification of Munck's system. Indeed, one critic's mind was truly boggled to the extent that he claimed Munck's work invalid precisely because it hadn't been subjected to statistical analysis. Evidently, this person was totally unaware of the report by the mathematician and systems analyst Anatoly I. Kandiew.[*]

Kandiew reviewed the latitude/longitude configurations regarding this emerging worldwide matrix and gave independent verification of its validity. He writes:

> The code that Carl Munck discovered was embodied in the grid system developed by the ancients. The "Code" maintains that instead of using conventional locators, in terms of degrees, minutes, seconds and fractional seconds, a grid was used which was the product of the individual terms. While more

---

[*]Anatoly I. Kandiew, Ph.D., was awarded the first ever patent granted for designing computer software. He had been building computers and writing software for over forty years at the time of writing his 1992 paper on Munck. It was first published in the Louisiana Mound Society's newsletter, and David Hatcher Childress included Kandiew's full analysis in the appendix of his book *Anti-Gravity and the Unified Field* (Adventures Unlimited Press, 2nd edition 1998) along with Munck's analysis of the Face on Mars.

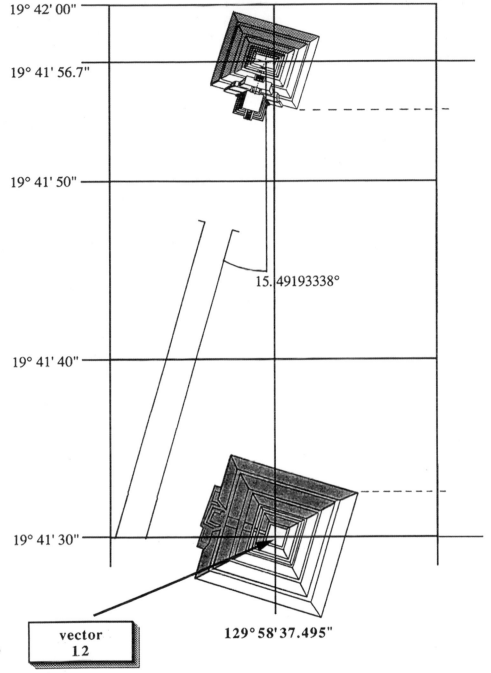

19° 42' 00"

19° 41' 56.7"

19° 41' 50"

15.49193338°

19° 41' 40"

19° 41' 30"

vector
12

129° 58' 37.495"

Diagram: DSP.

Figure 14.8. Pyramid of the Sun in Teotihuacán, Mexico.
According to Carl Munck's mapping formula its longitude and latitude
coordinates west of Giza produce a Grid Vector or Grid Point of 12.

detailed analysis may shed much more insight into the nature of the problem and what the implied messages are, it is beyond any doubt that the codification was as accurate as our present system, yet conveyed additional messages and meanings.[12]

Kandiew's detailed examination of the mathematical requirements involved led him to conclude that without doubt the entire matrix is artificial and certainly not accidental. Kandiew found that the accuracy of siting required is to within twelve inches (one British foot) and that the results of Munck's calculations were beyond chance: the likelihood of properly positioning two structures to conform with this grid matrix relationship would be approximately 1 in 100 *trillion!*

Kandiew adds that this conclusion does not take into account the odds for *more than two structures* maintaining these relationships. Combinations of three, four, and more were found throughout the matrix. He thought that a reason why this matrix has remained undetected for so long was because it incorporates sufficient permutations within its structure to discourage all but the most assiduous code cracker. Kandiew concluded that this matrix cannot have come about by pure chance and that since the math is encoded in every monument it implies that all these interrelated monuments were all built by the same "people."

Kandiew concludes his analysis with this question to the reader:

Now do you think that the Matrix was an accident, or does the work arise from intelligence?[13]

A MAJOR PROJECT

Given the vastly different ages and cultures of these builds and locations, it would be more accurate to state that this grid matrix was initially *devised* by those of the same mindset. And it would be entirely reasonable to conclude that those responsible for establishing such a long-term project must have had major interests in this planet, its cultures, and its future people—us.

The implications are that due to Earth's plate tectonics, these structures were built in meaningless positions that would only take on true mathematical relationships here on Earth *thousands of years later*—at just the same time that we have managed to send adequate imaging probes to Mars, a planet with no plate tectonic activity. The fact that several locations here on Earth that were found by

Munck correspond mathematically with two of the principle Martian Cydonian structures, the five-sided D&M Tor and the Face mesa, suggests thoughtful planning on a truly large scale, both over time and space.

It also suggests that there is far more to this matrix of architectural builds than just the demonstration of intelligent, technical engineering and design by our brilliant ancestors. For all we know, seemingly meaningless positioning of Earth structures morphed through different geometric forms and various relationships as over time this matrix moved slowly into its present mathematical relationships here on Earth and ended up as now, matching immovable surface structures on Mars. It would be astonishing if this highly sophisticated matrix didn't deliver information that so far we have yet to fully perceive, appreciate, or make use of. Physics explores relationships between the mathematical hierarchy and the physical world. And we might be looking at the reflection of just such a construct in these mapping projects on Mars and Earth. From inception to the result we find today, there might be three levels, cumulating (or emerging) from a fourth level, operating very much in the manner of a Combinatorial Hierarchy.*

It also suggests that something like a state-of-the-art global positioning system must also have been deployed by those ancients in establishing the beginnings of this matrix. Surely then, those who were in the top-secret PSI and satellite programs would see that there was much more to Munck's mathematical data than a simple grid map.

Even if Munck reiterates that the scientific and the academic communities contacted by him had ignored the data and the cryptographer, reading his story and running it in parallel with the development of mapping and satellite technology in the United States made the reality of this situation rather less certain.

For a nation whose intention was to control access to space, the Moon, and Mars; who was actively building satellite technology as well as global positioning systems; who had allowed all sorts of nonsense to be stated about ETI presence on the Moon at the time of Apollo; who had found anomalous landforms on Mars that looked like artifacts from Egyptian and Central and South American pyramid cultures to the extent of calling some of them Inca cities while ostensibly ignoring others such as Cydonia—this supposed lack of interest did not entirely ring true. In fact, it sounded more like hogwash.

---

*The Combinatorial Hierarchy was originally discovered by A. F. Parker-Rhodes in the 1960s, and values close to the fine structure constant, and the proton-mass gravitational coupling constant appear in the generation of the Hierarchy.

Munck's work was discovering connections with both Cydonia on Mars and the locations harboring the emerging phenomenon of crop markings in southern England, so *it is far more likely that everyone who was anyone in the world of ETI research, off-planetary exploitation, and PSI exploration was in fact paying very close attention.*

It would be necessary to probe the Munck oeuvre a little deeper, because by 1985 his little hobby became a quasi–full-time research project and his body of work changed its name again. Which is something the military does when they have achieved a basic test phase and proceed to the next stage. Carl Munck shifted from the Pyramid Matrix to the Code, the name he would later use for his newsletter when he started publishing.

He also gave himself the title of archaeocryptographer.

That same year the PSI-Ops program changed its funding source and became project SUN STREAK.

15

What a Way to Run a Railroad

*Mapping the Planet with GPS
and More Top Secret Projects*

WHISPERS

It must be said that the Munck newsletters were produced in a suitably rustic style and then photocopied, and even when he began publishing collections of his material, his works were an extremely challenging read to all but the most dedicated long-number enthusiasts. This was because the rustic style was maintained, making already difficult data even more difficult to absorb. These offerings were also peppered with mathematical bloopers, which were not corrected over time, creating a paradox with Munck's insistence that the numbers had to be accurate to be valid.

Figure 15.1. *Whispers from Time,* Carl Munck, 1997. Periodically Munck published collections of his data, in spiral bound volumes, as illustrated here. The newsletters operated much as a website does today, providing updates from Munck and a small place for reader's comments.

As we saw in previous chapters, this is turning out to be general practice when dealing with subjects outside the norm: creating uncertainty in the data enables the whole structure to be denounced should it be necessary. These numerical bloopers could also serve to see who is paying attention; anyone who wrote in could be noted. An alternative explanation would be that the newsletter may not have been the one-man job it was purported to be, but rather the work of a team of individuals, headed up by Carl Munck.*

THE WHERE AND THE WHEN, THE WHY AND THE WHAT

Having clearly stated his preference for mathematics over language as a tool for understanding the ancients' mindset, Munck then did a complete turnaround when in 1986, having sorted out the where-with-all of a monument through the use of the where and the when SCANATE coordinate system, he decided to ask three other questions. He asked:

- Why are these monuments where they are?
- Why are they designed the way they are?
- What does the design convey?

It was at this point that the Hebrew number/letter substitution system known as gematria was added to his toolbox. Language was important after all!

The Munck papers advise the reader to take notice of gematria as an important subject, but while it might have enlarged his own understandings and might be a helpful tool for many, it is too subjective for others. The interpretation of any such system, whether it is gematria, astrology, numerology, or any other such associative system, is established through the social mores of the culture from which it emerges. Later interpretations can become infused with the social mores of the culture that studies it. If that culture is picking up a universal truth and is merely the mirror through which everyone else can discover that same truth, then assertions such as "144 is always associated with light" should be useful when found in monument measurements. However, the word *light* has many contexts. It could mean the speed of light, the photon field, the light of understanding, day

---

*The data in this chapter is collated from years of subscription newsletters now in private hands that are no longer available to the public. As already stated, at the time of writing the mathematical data they contain is available in the form of DVDs on Amazon.

as opposed to night, or any other meaning that the particular culture has placed on the word.

The same comment applies to another method that the Munck papers applied to the decoding of ancient monuments. The putative archaeocryptographer was told to take into account the number of steps, terraces, corners, or any other obvious architectural or natural features when seeking verification for the findings already achieved through the SCANATE method. This method was also used in the PSI-Ops program, with the remote viewers instructed to draw what they were seeing for verification against the target. Again, while some useful data might emerge, what constitutes the important feature(s) of a monument is in the eye of the beholder and as before, may not necessarily be the feature(s) most relevant to the original builders.

While it would be surprising if there were no course corrections in such pioneering work (corrections that in themselves do not in any way alter the validity of the Giza-based worldwide grid hypothesis), trawling through the available Munck papers from 1992 to 2008 revealed enough contradictions to signal an alert.

If the author comes across as a man who doesn't suffer fools gladly and it is also quite obvious that he has a brilliant mind in the domain of mathematics, his preference to not say exactly what he was doing, "let's just say I was looking for a better mousetrap and let it go at that," sounds like the acknowledgement of secret operations, as well as an open invitation to investigate further.

After all, if it was just a hobby based on mathematics, there was no need for intrigue, and if it was a secret operation, no requirement to disclose the fact. So the better mousetrap might well have been a search for improvements of an extant technology or methodology, but it also involved trapping the attention of his readers in a subtle maze of discrepancies. Notwithstanding any uncorrected mathematics or maps, scattered throughout his work there were snippets of information relating to his own life and professional background, and here the sums definitely do not add up.

Although as a legend to match the PSI-Ops, they can be made to work.

BACK TO THE GRID

We learn that, around the age of ten, Munck became interested in archaeology and the lost skills of the past thanks to his Merchant Navy father, who had brought home small artifacts he had acquired during his shore leaves while serving in the Mediterranean. Two of these items in particular exercised a fascination for his father because they displayed engineering techniques that were not considered to be extant at the time the civilization was in existence.

The young Munck picked up on the idea that advanced technologies were ignored by archaeologists because they don't have the mindset to be able to see them. This is quite sensible, and indeed, this would become a recurring theme in his later work. Munck's working life has been spent in service to his nation. During his twenty years in the USAF, from 1953 to 1973, he was already on the lookout for ancient mounds and mentions those he saw while stationed in Alaska in 1964.

Wendell Stevens, who would go on to investigate the Meier UFO photographs in 1978, was also stationed in Alaska. Stevens worked on the top secret Project Ptarmigan. This research program was ostensibly photographing and mapping every inch of the Arctic land and sea area. It was also registering any external influences caused by UFOs. Anomalous phenomena and any disturbances in the electrical and engine systems of their aircraft were recorded using the equipment designed to capture, record, and analyze all electromagnetic force (EMF) emissions in the Arctic. The data was then couriered nightly to Washington, DC.

It is not known if Carl Munck was involved in this Alaska-based project, but it sounds right up his street.

In his published data, Munck has a representation of the military's technical sergeant insignia. Whether that was his final rank is not made clear. Within the enlisted ranks, that of technical sergeant, TSgt, is the second most difficult rank to achieve. It is one of leadership and carries high status and responsibilities. To be eligible, a minimum of five years service is required, but normally at least ten to twelve years of service will have been achieved before anyone is even considered for this rank.

Interestingly, Munck published this image of his insignia alongside a grainy photocopied image of the Moon from orbit, a shot associated with the *Apollo 8* flyby—visually inferring that the stripes on this insignia matched those of the lunar terrain. Whether he was telling his reader that he gained that rank in 1968 or whether he was obliquely referring to some aspect of that USAF rank relative

Figure 15.2. Technical Sergeant insignia.

to the eighth Apollo mission or the fact that the Apollo program was a cover for another top secret program are moot points. As the Cold War historian John Cloud writes:

> The release in 1995 of previously deeply classified data on the CORONA program makes it clear that the coupling of open and secret science, as in the Apollo program and CORONA, was not unusual.[1]

THE JUNCTION BOX

After his USAF career, Munck spent from 1973 to 1983 working in the rail industry with the Delaware and Hudson Railway. He says that he started at the bottom: cleaning out rail cars. Which usefully provides the context to the analogy he uses to illustrate the fact that the names in the relieving chambers above the King's Chamber of the Great Pyramid are no guarantee that it was built by Khufu as claimed by the Egyptologists (some of whom manage the Giza plateau). Munck said that the fact that some tramp had written his name in a boxcar was no guarantee that he had built the boxcar, only that someone had been in the boxcar and left some graffiti. This could equally well apply to himself. After all, whether a man with the leadership abilities and responsibilities he had acquired in the USAF is going to start in the railroad industry by cleaning boxcars is not so certain. As a useful hook on which to hang an analogy, it chimes with other stories he relates of this time. In conjunction with a scientific discussion on the speed of light, Munck tells of his expeditions to photograph trains and how he filmed and videotaped moving trains from the side of the track.

Munck also tells an incredible yarn that is a little hard to take at face value. Again concerning a scientific fact, in this case magnetism. This tale concerns the huge railway yard electromagnets that are used to lift two tons of scrap metal. We are told that he had taken an hour to collect the tonnage into the required

Figure 15.3. The Delaware and Hudson Railway logo.

heap. He warned everyone not to wear metal belt buckles around these magnets and says that he (and/or his team members) lost *four* watches in one summer, messed up by these huge magnets. Now apart from the change of status from a leader of men to the stacking of metals in a rail yard, we also know that he comes from an educated background where engineering principles were appreciated. He writes that he has a master's degree, but does not say in what subject. Although in another newsletter he writes of leaving college before finishing the course because he didn't like dogmatic teaching. Writing of the archaeocryptography work, he states that it is Ph.D. material. Contradictions all round. Even so, for such an intelligent person to lose one watch to that rail yard magnet might be a possibility (though even then doubtful), but to repeat the same mistake four times in the same summer smacks of stupidity—and that puts Munck out of the game.

However, to experiment (not necessarily in a rail yard) four times with technology in a magnetic field in order to evaluate the results might be a totally different matter—and that puts a different spin of the same yarn.

Following the trail laid down by Munck, we learn that in 1982, at the age of forty-eight and with a family to support, he finds himself chucked out of the railway and "living in Endsville" with no prospect of finding a job. But wait, earlier in his papers he wrote that this railway period went from 1973 to 1983. It was, as he tells it, the result of a management changeover when "last on the Extra Board became first out." The Extra Board is the list of job vacancies advertised by the railways in the United States, and you can find it on the web today. It is a fact that the last chairman of the Delaware and Hudson Railway served in 1982. It is also a fact that Munck really does like railways; he has taken photographs of trains and owns a model railway of some importance. It actually sounds as if *this* is his hobby, whereas the pyramid work is more of a job, for his writing oozes with frustration toward those who do not understand (or will not accept) the importance of his pyramid research, while he says that everything to do with railways is an antidote to the stresses of his everyday life.

DREAMLAND

It is at this juncture, in 1982, while wondering how to make a living, that Munck starts receiving specific dreams—night after night and dream upon dream, all apparently concerning pyramids. However, in another newsletter, Munck writes that in 1980, two years *before* the dreams started and *before* he had come across Cayce's commentary on the Great Pyramid, he had experienced a vivid memory recall concerning Giza at an earlier time in its history, the details of which he remembers to this day. So, as he does not consider that memory to be a dream, what was it? The result of a PSI-Ops RV (his own or that of someone else), an education in consciousness research for the general public, a channeled communication, or a little of all three? It is another moot point.

We'll soon have enough for a meeting in a Moot Hall.

PARTIAL RECALL

Munck's 1980 recall was a vivid memory of being at Giza for a ceremony that celebrated the completion of the Great Pyramid's construction. He remembers being at the head of a procession that marched toward the Giza plateau from the northeast, in the general direction of what is now called Old Cairo. In his estimation, there were thousands of people watching this procession. When they reached a place where they could clearly see this "monument of monuments," as he calls it, the entire procession came to a stop. And at that precise moment, Munck saw a banner unfurl on the northern face of the Great Pyramid. On the banner he saw letters that he read as *JOLDING,* and as he saw this word he fundamentally knew it to be a part of his own name at that time; he also somehow knew that this name was not quite right—either not pronounced the same way as it was written or that there were letters missing or added to the original identity.

CHECKING THE TIME LINE

Two years later, in 1982 and at the ending of his railroad period (give or take a year), Munck decided that one cannot argue with "the head-shed" and made his hobby a fulltime occupation. This date means that the earlier retirement from the USAF of 1973 is thereby discounted on a technicality. Which is just as well, as he hadn't retired his USAF vocabulary. For those not in the US armed forces, the "head-shed" is slang for headquarters, or the command post where the leaders gather. This whole "railroads-to-pyramids" changeover might be another way of implying that he was ordered to do this job by those higher up (terrestrial orders rather than the heavenly orders he otherwise might be inferring).

Ten years into this new job, in June of 1992, the first *Code* newsletter was made available through subscription. With hindsight, this title could not be more apt. Whether good at math or not, humans are constructed in such a way that we resonate with ratios and numbers. Even if the mathematical research into this matrix was initiated for scientific and defense purposes, it has had the unintended consequence (on the part of the military) of waking up the coding deep within our own DNA. Coding that connects us to all that is here on Earth and everywhere else, including ETI.

The timing was perfect; as already noted, it occurred just as the PSI boys turned their project from SUN STREAK into STAR GATE and nicely picked up on the forty-fifth anniversary of the Roswell ETI incident.

We have assumed from his earlier statements that Munck had commenced his hobby in 1978 and that his hobby became a full-time *research* project in 1985. But now he says this happened in 1982 after the dreamtime events. As it turns out, the published works and the private and semiprivate communications made available to his readers provide multiple-choice answers concerning the actual start year of Munck's matrix project. He mentions the specific years of 1973, 1976, 1978, 1981, 1982, 1985, and 1986 along with these confusing general statements: "twenty years ago" as he wrote in 2001 and "over twenty-five years ago," also written in 2001. It becomes clear that Munck's grid network was run in stages. Just like the RV program.

So another pattern emerges: the Munck papers were modified according to the RV programs. Bearing in mind that his Bimini map, which arrived from "elsewhere," dates from around 1973 implies that *The Code* offices or its publisher were already on the radar, as it were. Whatever else he was doing in mapping terms, collating the data from the RV programs would seem to be a part of it. The pyramid grid dreamtime (involving, no doubt, the examination of pyramid forms on both Earth and Mars) belonged to the Army's 1978–1982 Fort Meade project in which Joe McMoneagle was the star remote viewer. When looked at as a whole, the correspondence between the stages of Munck's work, the PSI-Ops RV programs, and the exploration of Mars is striking.

As is the fact that *The Code* was born as a subscription newsletter. In the real world, for such a start-up to be financially viable, it must maintain a healthy base of subscribers interested enough in the subject matter to ensure its continuing production. The Munck papers were always emphasizing the one-man band, financially restricted production schedule. Yet, Munck also states that each newsletter was ready some *months* before its actual publishing date. Starting out in 1992 with a minimum of nineteen years worth of collated data implies a prescribed schedule for the release of information and a healthy pocketbook that doesn't necessarily require subscribers.

TRACKING AND TRUCKING

Returning to the timing problems of this Munck tale thus far: the whole episode of working on the railroad sounds as if it is absolutely true, but possibly out of sequence, those ten years of doing a different job every day around the railroad being back engineered versions of summer jobs taken on while Munck was a student or a glossing of R&D then being undertaken. Then there's that statement: "I once had a hobby. Cryptography. But it's not a hobby anymore! Back in 1978 I was trying to develop a code system" still needs further elucidation. For example, Munck says that in order to pursue his hobby he had to acquire knowledge of several areas of expertise: cartography, ancient metrology (in an early newsletter he titled himself "archaeometrologist" rather than the usual "archaeocryptographer"), archaeology, oceanography, vulcanism, geography, and paleomagnetism. A lot of subjects to learn about, then. No worries, though, as he didn't have to do all of this in the free time dedicated to his hobby because elsewhere in his oeuvre we learn that his job in the USAF concerned or included drafting and mapping. The very skills he would have to acquire for one of his pet hobbies turn out to be the exact same skills he needed for his principle (and responsible) job.

CODE BREAKING

This information reinforces the notion that we should also back engineer that hobby of cryptography into a serious full-time occupation. It might explain a great deal. John Cloud noted:

> The mix of the traditional and the innovative in the practice of military geographic intelligence is reflected in the play between the complex interactions between secret technologies and classified programs on the one hand, and the research laboratories and their publicly accessible institutions on the other.[2]

THE OHIO CONNECTION

But where, back in the days after World War II, would be the best place for a mind such as Munck's or anyone else involved in mapping and math to study such matters?

It is well known by now that at the end of World War II many German rocketry, nuclear, and medical scientists (some of seriously dubious repute) were shared out between Russia and the United States. What is less well known is that in October 1944 a team of some twenty-one Americans led by US Army geodesist Floyd Hough and aided and abetted by the allies was searching Europe for all the geodesic materials it could find. These materials were returned to the

United States, and together with these recuperated geodesic items, known as the "German materials," came German geodesists and mathematicians who were used for the scientific endeavors of the US armed forces. Two years later, in 1946, the Mapping and Charting Research Laboratory was established at Ohio State University, funded almost entirely by the USAF.

Then came 1947.

As we know, 1947 was a highly eventful ETI year for the United States: overall, the number of UFO sightings actually reported to the authorities in 1947 came to around 850. The biggest impact had come from the June 24 sighting of nine disks flying in a five-four formation by Kenneth Arnold around Mount Rainier. On that same day, another eighteen UFO sightings were reported by people totally unaware of Arnold's experience. Some four or five days later, an even bigger impact occurred—literally—when the Roswell/Corona UFO crash was placed on record. All sources give different dates, but here are those four-five number combinations again.

Three months later, on September 18, 1947, the CIA was established, and the US Army and the US Air Force were separated into two entities. The USAF retained its contacts with the Mapping and Charting Research Laboratory while the funding of these Ohio State facilities was taken up by "various nascent research centers of the postwar DOD and was conducted by classified contracts and secret reports," as Cloud writes:

> This laboratory soon acquired the services of the best collection of scientists in geodesy and its allied disciplines in the world. The OSU then opened the Institute of Geodesy, Photogrammetry and Cartography in 1951.[3]

All of this was in place when Munck, back in 1952 or 1953, was ready to consider his USAF career options. Then in 1958, the nominally civilian federal authority NASA, the DOD's advanced research projects agency DARPA, and the aforementioned CORONA spy satellite program all came into being. The very obvious code word CORONA was used for a program officially cancelled by the Air Force, but which had been continued in secret and undercover—a process known as "going black."

The very next year, the first group of special students were drawn from the Air Force Aeronautical Charting and Information Center and enrolled at Ohio State University. In this same year, 1959, the US Army's world geodetic datum was completed. Now, since all of this mapping is principally being underpinned by the defense interests of the United States, it is unsurprising that by 1965, with the wars in Vietnam and Laos going rather badly for the United States, a secret DOD study proposed that the nominally civilian federal authorities should also carry out classified reconnaissance work.

The very next year, 1966, the USGS began building its own block within the newly established National Mapping Division in Reston, Virginia, and five years later, in 1972, as Cloud notes, "Most of the DOD and IC service level mapping and geodesy agencies were consolidated into the DMA, the Defense Mapping Agency." Leaving the USGS to "own" the now-completed World Geodetic System 72 (which would later be replaced by the World Geodetic System 84).[4]

The establishment of a secret and powerful global satellite network seemingly went hand in hand with the rate at which technological programs either "went black" or were handled using the front office/back office strategy. However, the US government's Office of Management and Budget was seemingly not read into this strategy. In 1973, its Federal Mapping Task Force advocated a major consolidation of its federal and geodetic efforts.

As for any public acknowledgement of the role of satellites in all of this, it was only in 1978, during the Carter presidency, that satellite reconnaissance was officially recognized as a tool in the US defense bag. Then, *fourteen years later*, in 1992, the National Reconnaissance Office was officially acknowledged as existing. It took another three years for the declassification of the CORONA "black" program in 1995.

Notwithstanding the cloaking of CORONA installations with Apollo activities, it is more than interesting that the USGS, a nominally civilian federal authority, has charge of off-planet mapping, and that all these dates echo stages of the RV program, the release of information from *The Code* cartographer, Munck, and the arrival of those cylinders in Switzerland from 1972 through to 1990.

Closer Encounters

*Analysis of a Mind-Mapping Exercise
and the Relevance of More
Anomalous Events*

ROLLING GLOBES

Carl Munck's 1978 decision to turn his hobby into a full-time occupation came a year after the US release of Steven Spielberg's *Close Encounters of the Third Kind* and two years after the Viking picture of the Face on Mars was revealed to the public. Spielberg's movie contained much of the latest research on ETI and PSI, including data from SETI. As a tool in the forming of public opinion on the matter of ETI, and especially now that we know of the deep and long-standing connection between Hollywood fiction films and matters off-planet, a closer look at *Close Encounters* will be most informative. We start with that scene mentioned in chapter 14, in which Earth scientists, having been given the clue by a map-maker, used latitudes and longitudes to determine where to meet ETI. Although the filmmaker didn't reduce the coordinates to a single string, *the whole premise of the "rolling globe" scene might just as well have evolved from the SETI program and Munck's work.*

In the film, the inference is that ETI is responding to both the SETI searchers and the PSI-Op RV teams by using sound pulses to code the mapping coordinates of the amazing flat-topped Devils Tower in Wyoming.

However, checking the coordinates given in *Close Encounters* against a globe revealed another astonishment. It was not possible to find Devils Tower by following the film script, which goes like this:

Figure 16.1. Devils Tower, Wyoming.

Photo DSP.

INT. TRACKING STATION—DAY

The Goldstone Radio Telescope (Station 14) top-security missile
tracking complex, a specialist excitedly shares with a colleague
the newest deep space transmissions that have been received.

SPECIALIST

We just received two fifteen minute broadcasts . . .
104 rapid pulses. After a five second interval,
44 pulses. Another five second break and 30 pulses.
Sixty seconds of silence and then an entirely new
set of numbers. 40, break five. 36, break five.
10. A hundred and four rapid pulses. . . Wait sixty
seconds and the whole doggone thing repeats.

SECOND SPECIALIST

Where are these signals coming from?

SPECIALIST

Right in the neighborhood. Light travel time, roughly
seven seconds. It's well within the plane of the
ecliptic.

SECOND SPECIALIST

Are these nonrandom? 40 . . . 36 . . . 10 . . . In
response to that?

SPECIALIST

No. They should be. We've been sending out this
musical combination for weeks. But all we're getting
back are numbers.

SECOND SPECIALIST

This could mean the Indian [from India] sounds
reached a dead end. They don't mean a thing.

LAUGHLIN

Excuse me. Before I got paid to, uh, speak French,
I, uh, I used to read maps. This first number is a
longitude. . . Two sets of three numbers. Degrees,
minutes, and seconds. The first number has three
digits and the last two are below sixty. Obviously, it's
not in the right ascension and declination on the sky.
These have to be Earth coordinates.

THIRD SPECIALIST

Surely, somebody has a map. . . There's a globe in the
county supervisor's office.[1]

Having discovered the key to this code—the coordinates—several people, perhaps far more than necessary, rush to the supervisor's office in a flurry of urgency and excitement. They burst into the deserted office that is dominated by a massive floor-standing globe, roughly tear it from its stand, and roll it to their own workstation. In a shot mimicking the artwork in the Sistine Chapel where Michelangelo's God touches the finger of man, the camera cuts to two forefingers tracking these *movie* coordinates on a globe, from south to north and from east to west, to meet at the Devils Tower location in the state of Wyoming, whereupon, "We're going to need a Geodetic Survey map of Wyoming. I want this down to the square yard."

A line that sounds very much like another Munck moment. Accuracy is everything. Most of the time. The Wyoming Devils Tower finds itself at 104°42'54.52" W longitude and 44°35'24.56" N latitude.[2]

Yet the coordinates Laughlin gives the team are not those. His longitude of 104°44'30" W is 1.3 miles farther west. His scripted latitude of 40°36'10" N is a whopping 275 miles to the south. This false latitude does however, fall into the ballpark of the City at Cydonia on Mars. Once again, fiction matches fact, as it had with the error/actual Bimini map coordinates. For the film actors, given that the north longitude was tracked from east to west, getting those forefingers to meet at the Devil's Tower location implied a start point from that southern

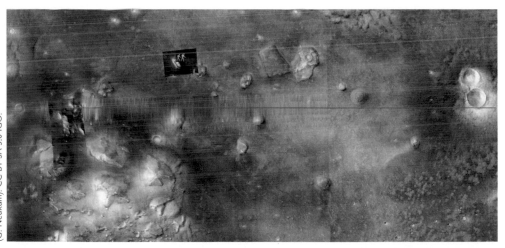

Figure 16.2. The *Close Encounters* latitude starting in the west of the Cydonia complex and ending in the east. With no thoughts of this film's coordinate messaging, *Two-Thirds* authors Myers and Percy (a book with metaphorical constructs) had placed an airport terminal at this western start point, while they had assigned the double craters in the east to farming activities: sheep shearing in the northern crater and cattle feed storage in the southern crater.

latitude and a nod to Cydonia. Checking with Google Mars, the movie's selected latitude references the Cydonia latitude and—incorporating a nod to the movie's longitude by tracking it east—terminates at the twin crater complex.

Whether the profession of that *Close Encounters* mapmaker-turned-translator was referencing both Munck and French scientific and political connections is not known, but the name of Laughlin, with its reference to the eponymous USAF base in Texas and Munck's avowed preference for a Texas Instruments calculator, tempts one to make the link. Unless that too was a calculated in-joke on the part of Carl Munck.

PARALLEL WORLDS

Admittedly speculation on my part, this mix-up of coordinate data led me to think of it as an in-house joke concerning the efficacy of remote viewers, as seen by scientists in the know. It also reminded me of the militaristic inclination to cover up their activities. More seriously, I asked myself, did this coordinate mumbo jumbo mishap actually reveal anything when analyzed?

This query prompted the application of Munck's mousetrap multiplication methods to Spielberg's *Close Encounters* coordinates, and remarkable results emerged, quite apart from all that alliteration. Using the film script coordinates, the latitude for the Devils Tower ($40 \times 36 \times 10$) becomes 14,400. In esoteric systems such as the gematria system promoted by Munck, 144 and all decimal placings of this number (1.44, 14.4, 144, and so on) are said to denote light. So Spielberg's grid latitude of 14,400 is possibly an encoding relating to the idea of light. Or perhaps a reference to Station 14, a.k.a. NASA's deep-space network's dish in Goldstone, California.

Apropos of Goldstone, Mars, and the number 14, this is the *average* number of minutes it takes for a *one-way* radio communication to travel from Mars to Earth. When Mars is at its nearest approach to Earth, it takes four minutes and at its farthest, it takes twenty-four minutes, so the longest time gap is a full forty-eight minutes for a two way message. And, of course, the fourteenth day of the Moon is full moon, at which time, due to its orbital location relative to the Earth and the Sun, its electrical output is considerably greater than usual.[*]

---

[*]Observations by NASA Explorer probes prior to the 1970s have shown that regardless of whether the Moon is in Earth's magnetotail, the Moon becomes slightly electrified by sunlight. The day side of the Moon becomes positively charged as solar radiation knocks electrons from the surface. Electrons build up on the night side of the Moon and give the surface a negative charge. During a full moon, the situation gets worse for anyone or anything on the lunar surface. Data from the Japanese probe

Figure 16.3. Amblin Entertainment logo.

Without going into a myriad of analogies, the important thing is that this erroneous grid latitude conveys meaning relevant to space activities. Taken together, at the most basic level, we have a light Moon/moonlight. There is also the act of moonlighting, to have a second job in addition to one's regular employment, typically secretly and at night—and that is the very best fit for this particular film.

Whether a production in-joke or a subtle but serious point, this example serves the argument concerning the intelligent planning of a worldwide grid based on map coordinates, themselves based on layers of information encoded within these same coordinates.

It also provides a practical demonstration of the hypothesis that such a grid had to be prepared in advance from its cumulative vector, demonstrating fore-thought and deliberate planning: neither the joke nor the message, not even the information concerning the site and its attributes can be decrypted unless the coordinates correspond to, or give rise to, the interpretation of the ideas encoded within the numbers themselves.

Munck's own realization was that the mapping coordinates he had discovered relative to the planet's ancient monuments and sites must have been artificially generated at the outset because *they were worked backward from the grid vector* and

(*cont. from p. 364*) *Kaguya* (published in 2010) found that when the Moon passes through the region of the magnetosphere called the plasma sheet, which runs down the middle of Earth's magnetotail, another relatively intense electric field occurs. The newly seen effect adds to what scientists already know about the Moon's electrical activity. Andrew Fazekas, writing in *National Geographic News* online on November 17, 2010, said, "The big question now is whether a strongly-charged lunar sur-face poses risks to future robotic and human explorers. . . . It is quite possible that electric fields induce a charge-up and subsequent discharge around a space vehicle, which could bring about serious damages to the human missions."

not forward from the latitudes and longitudes. And so it is with *Close Encounters.* Their coordinates are also artificially manipulated, and unless one knows the reasoning *behind* the choice of coordinates, at first glance it would appear that the esoteric meaning of the number combinations was the original requirement, and not the vector itself.

TIME OUT

When Myers and Percy were working on *Two-Thirds* in 1992, neither of them had any idea that the *Close Encounters* coordinates were back engineered, and discussing this movie with both authors in 1994, it had not occurred to any of us to even consider checking the validity of these *Close Encounters* coordinates against a globe. Nor did anyone think of applying Munck's mousetrap method to them. However, if you actually take the trouble to track the film coordinates and set your own fingers ambling down a map of North America, you would find yourself in Weld County, Colorado, and thanks to Google Earth, you would also know that this precise location was in a field next to some farm buildings and surrounded by *circular* irrigation fields just northwest of Ault, which likes to advertise itself as "A Unique Little Town," and whose name, using Munck's gematria system, adds up to the number A1, U21, L12, T20 = 54, or 5 + 4 = 9. We keep coming across that five and four combination we saw at the Pentagon's Mall gardens. Setting ourselves back to the time of making *Close Encounters,* you would have every reason to remonstrate, "Well, I don't know why an empty field in Colorado has any such relevance; sounds to me you're on a fishing expedition." Well, let's see. Quite apart from the Cydonian references already cited, Ault sounds like Altea, and that's a word that was used by an ETI source with associations to the number 9. So if we're playing guessing games or charades, then Ault also sounds like ultimate, and according to previous ETI information published in both *The Only Planet of Choice* and the

Figure 16.4. DreamWorks
Pictures logo.

earlier report from the same group, *Briefing for the Landing on Planet Earth,* the terms Ultima, Altima, and Altean are three aspects of the same idea.[3] The arrival of crop glyphs certainly looks like "a landing on planet Earth,"[4] and this movie is looking more and more as if it contains messaging intended for ETI.*

ASSOCIATIVE THINKING

That gematrian number of 54 associated with Ault and the circular irrigation of fields in that part of the United States produces other connections to matters already discussed. Aficionados of the master numbers will remember that 54 occurs relative to the six series and in the ninth location.

Those familiar with *Two-Thirds* will recognize a number that has to do with the engineering of a key component for space travel. Aficionados of the Munck papers will be reminded of the number of holes relating *directly* to the Aubrey ring at Stonehenge. This Aubrey ring is a circle of holes located between the outer ditch and the sarsen ring. There are fifty-four around the circumference and two more across the avenue entryway into the ring. A ring of artificially engineered holes recalls the disappearance of soil from those massive holes in Switzerland. Despite the fact that the all Aubrey holes are currently capped with concrete, and most of the monument roped off from visitors without special passes, it is still possible to have PSI experiences at Stonehenge. Observers standing on the location of one of these Aubrey holes have experienced shifts in the visual field, producing effects that infer a faster rate of energy emanating from that particular spot.[†]

---

*Some of the characters associated with the adventures recounted by Stuart Holroyd in *Briefing for the Landing on Planet Earth,* according to their later biographies, were not unfamiliar with the agendas of the R&D departments of various Intel agencies worldwide. Also, bearing in mind that the ETI transmitter named TOM in *The Only Planet of Choice* also stated that the word *civilization* refers to dimensions of consciousness, it's of interest that this information comes from a source described as "the Council of Nine," and still playing charades, as "Counsel," that title takes on more meaning.

†Russian scientists visiting Stonehenge experienced the momentary disappearance and then reappearance of a person standing well beyond the monument, according to a personal account given to me in February 2020. And decades ago, on a sunny day with a blue sky and no clouds, I was standing in its vicinity talking to one of the monument's guardians when I saw a wide swathe of very blue light shimmering around the top third of the Heel stone, which is outside the earthworks, in the Avenue. The quality was completely different from the background sky. The guardian also saw it and told me that it was not an unusual occurrence.

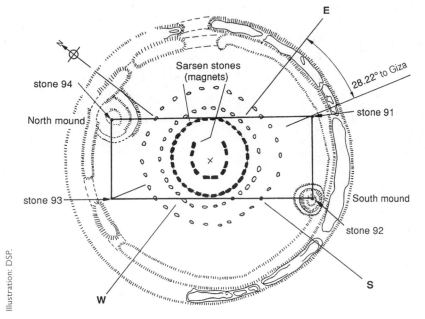

Illustration: DSP.

Figure 16.5. Layout of Stonehenge. The fifty-six Aubrey holes form a ring just inside the inner Earth bank. Stones number 55 and 56 straddle the Avenue (the gap in the ditch at the top of this image). The other fifty-four Aubrey holes (now filled) are intersected by two mounds and four markers (stones 91–94).

Photo: DSP.

Figure 16.6. Stonehenge, Wiltshire, England. The stone just in the frame to the right is the Heel Stone, located just outside the ditch and on the Avenue.

In 1989, twelve years after the release of *Close Encounters,* a grain field in southern England adjacent to Stonehenge received a circular marking divided into quadrants. This was the Swastika glyph already mentioned (in chapter 11) relative to the 1990 Confignon cylinder hole near CERN.

Taking the discussion of this Swastika glyph further, its location cleverly linked past and at the time of its arrival an unknown future, in that Stonehenge

Figure 16.7. The Swastika crop glyph. Found August 12, 1989, just west of Stonehenge at Winterbourne Stoke, a location replete with important Neolithic burial mounds.

Photo: Busty Taylor.

was known to the ancients as a temple of Solar Apollo, and the Moon, always described as a dead planet, is a future Artemis for the future planners at NASA. And the reference to latitude and longitude seen in those quadrants brings to mind the numbers 10 and 12. These numbers reminded me of the ten heavenly stems and twelve earthly branches of ancient Chinese astronomy, and deliberately dividing the heavens by the earth's twelve branches produces the number 0.8333333 . . . , which is one-tenth of the time (in minutes) it takes for light to travel from the Sun to the surface of the Earth.

This glyph could not be explained away as an effect of nature, nor was it claimed to be the work of artistic human beings, as hoaxing had not yet become an issue. So, a *specifically geometric* floor pattern had been impressed into the living crops by methods, technologies, and persons unknown. And importantly, despite this drastic alteration to their vertical status, the crops would continue to live and grow. Biology and physics being what they are, this important aspect of "authentic" crop glyphs has never been replicated by human copycats: hoaxing kills the flattened crops. In the case of strong thick-stemmed crops, such as maize and canola/rapeseed (*Brassica napus*), the damage inflicted on the stems by the heavy-footed humans and their planks can snap them horizontally and often fractures them longitudinally.[5]

Whether those who named this Swastika crop glyph were on nodding terms with the US Intel agencies or whether this was ETI talking back to the RV gang in the United States is debatable. Meanwhile, noting the date of this glyph's discovery, August 12, 1989, led to more information.

August 12 is the start of the grouse shooting season in Scotland, home to the UK nuclear submarine fleet, and back in 1953, it was the date of the first Soviet test of a hydrogen bomb, coming *nine* months after the first American test of an H-bomb. Again, the remonstrance of ETI apropos of our nuclear activities is to the fore.

MICKEY BASIN

Remembering the Chinese astronomers' encoding of latitudes' ten heavenly stems and longitudes' twelve earthly branches, these same numbers would make the August 12, 1989, date even more significant. Six months after the CERN missing-soil event of February 3–4, 1990 (which also incorporated the numbers 10 and 12 in its height/width ratio), on August 10, 1990, another earth-moving event was discovered in the United States. This new geoglyph would reference RV research, ETI connections, and the DOD connections with Hollywood, while its Mickey Basin location brought to mind the Disney-owned Graphics Studios.*

This large geoglyph turned out to be a Hindu symbol known as a Sri Yantra,

Image: Google Earth, 2020. 42°44'16"N 118°18'32"W, Landsat/Copernicus.

Figure 16.8. Mickey Basin, Oregon, location of the Sri Yantra geoglyph.

---

*During World War II, Walt Disney made PR films for every branch of the US military and government. Including the nuclear and space industries. The Disney studio also made training films for the use within the American military and created, free-of-charge, more than a thousand insignia for military units; the designs centered around established Disney characters as well as new ones.[6]

which immediately brought to my mind the SRI initials of the Stanford Research Institute, and its PSI research programs. Appearing in the ancient dry salt lake bed of Mickey Basin in southeast Oregon, the geoglyph was oriented true north and under a well-patrolled Air National Guard training corridor, and it had seemingly appeared during the night of August 9, 1990. Certainly, no pilots had reported any designs in progress during the preceding days. I spoke to the pilots, and they were sincere in their statements and curious as to what it was about. In terms of photography, they were, as one of them informed me, "all over it" for days on end; everyone wanted to get their own pictures. They would surely not have bothered had this geoglyph been made by the five artists who surfaced some forty days after the event, claiming to have made it using a garden culti-vator. Testing the artists' claim and using just such a cultivator produced small mounds of dirt on either side of a plow line. Physicist, meteorologist, and profes-sor emeritus at Oregon State University James (Jim) Deardorff refuted the art-ists' claims publicly because his colleagues Don Newman and Alan Decker had visited the site on September 15, 1990, and noted that the area was noticeably lacking any signs of tire tracks or footprints, even though their own tires had left quarter-inch-deep marks on the crusty surface of the dry lake bed. Furthermore, analysis of the geoglyph's ten-inch-wide by three-inch-deep lines indicated that the soil had been completely removed, just like those Swiss holes.

Figure 16.9. Sri Yantra, Mickey Basin, Oregon, August 10, 1990.
The Sri Yantra meditation is from the Hindu tradition. Note the nine triangles are arranged in a five-four pattern. Using the search term "Sri Yantra, Alvord" will reveal numerous references to this event on the internet.

SHIPS AHOY!

As for more ETI connections, the pilots who had discovered this Oregon geo-glyph were members of the adjoining state's Air National Guard and flew out of Boise, Idaho. Boise was the hometown of Kenneth Arnold and the location of the search-and-rescue flight team he had founded. The 1947 Washington State UFO sighting reported by Arnold talked of nine UFOs grouped in two sets: five in the lead and then four (those numbers again); it took place just before three p.m. near the 14,410-foot Mount Rainer. (That height in feet was encoded in the *Close Encounters* latitude for Devils Tower. It is also reminiscent of the actual 14°14' latitude of another site of geoglyphs, Peru's Nazca plateau.)

When Arnold saw those nine UFOs, he had actually been searching for a crashed C-46 transport plane used by the Marine Corps. Even if coincidental, the association of names is interesting. The 1947 search for the crashed Marine Corps plane is followed by the crashed UFO at Roswell incident. *Mariner 9* photographed Cydonia in 1972, and is followed by the RV start-up project at SRI. Note also that the Smithsonian Institution thought it appropriate to compare the size of the Martian volcanoes against Earth by overlaying the USGS survey map of the Martian volcanic Tharsis Rise onto a map of the United States, with

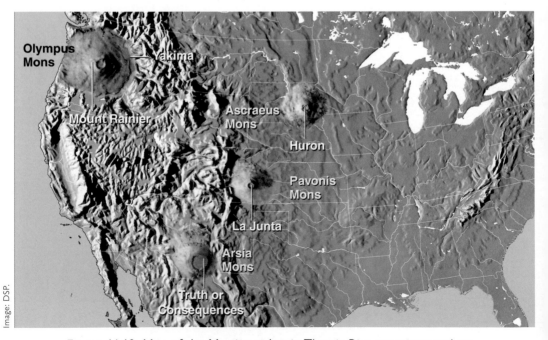

Figure 16.10. Map of the Martian volcanic Tharsis Rise superimposed on a map of the United States, as comparison of relative sizes.

Figure 16.11. Sri Yantra diagram. The four-five symbols manifest
in the putative outgoing messaging of the Pentagon landscape design
are again referred to by the nine principle equiangular triangles
arranged in two interlocked sets: four with the apex upward and
five with the apex downward.

Olympus Mons, the Martian 19.47° volcanic upwelling positioned precisely over
the area where Arnold had seen those nine UFOs flying in a five-four configuration. Those Pentagon landscape projects take on yet more meaning.

Close Encounters was released in 1977, which means it was in production at
the time the Viking images were released. The movie associates a Cydonian latitude with a meeting of ETI at the more northerly Devils Tower. On the same
continent thirteen years later, the Sri Yantra mandala, a symbol associated with
the Hindu chakra system, arrives. The interlocking of the nine principal triangles
produces a total of forty-three triangles, which is also the total time, in years,
from the Roswell UFO crash in 1947 to the arrival of this glyph in 1990. One
cannot help but think again of the cylinders of missing material in Switzerland
and their locations relative to that same basic triangular form.

In terms of potential incoming ETI contact, these facts are relevant: SRI is the acronym of Stanford Research Institute, and home to the initial RV project under Hal Puthoff and Russell Targ. A member of the Intel agencies, Puthoff is heavily involved with everything we are discussing here, including the research of advanced technologies for space travel. And when it came to this Sri Yantra, interestingly, he would have a time warp connection to the geoglyph in that he would go on to co-author a paper on ETI and the potential for extraterrestrial visitations with Deardorff and others in 2005.[7]

The uncomfortable fact that this sign of peaceful meditation had appeared precisely forty-five years after the United States had dropped a second atomic bomb on Japan at Nagasaki likely encouraged the relative lack of media about this event. And if, as was customary, the glyph had arrived the night before its discovery, then we have the exact day of the bombing. The meditative qualities of mandalas in general and the Sri Yantra in particular, when taken together with these other associations, could infer that the application of consciousness is a requirement for advanced deep-space travel for a human crew. The timing could also infer that ETI is not seeking confrontation, but rather it is attracting our attention to be mindful and not pursue nuclear options for travel, either on or off the planet. Four years before this glyph arrived, Deardorff had written a paper for the Royal Astronomical Society on the subject of ETI intentions which is apposite to this chain of events.[8]

Despite the evidence for the Pentagon's undisclosed awareness of the reality of ETI thus far presented, those not fully persuaded that any sort of independent UFO/ETI contact has ever occurred might find another refuge. These seemingly external intrusions onto the land (whether cylinders or markings in crops or on the ground) along with sightings of apparent UFOs could be put down to an energetic "bounce-back," a manifestation of our own collective consciousness. If that suggestion is still too outlandish, then any pragmatists need to have answers ready to explain the following:

- How these anomalous events were actually produced
- What technology was deployed
- Where all the disappeared soil from those Swiss cylinders actually went
- How crops can be braided without any physical contact
- How these anomalous events occur within a time frame of just a few minutes
- Why major biological alterations only affect the crops within the formations relative to control samples taken from the surrounding unaffected crop.

SKEPTICS SCORN

Whether the consciousness aspect of that argument was an essential element in the position taken by scientists influencing the *Close Encounters* scenario is not known, but some of the foregoing was the position of those involved with the production. Apart from the coordinates issue, the film script incorporated elements from physicist and astronomer J. Allen Hynek. In his 1970s account of his long-lasting collaboration on the United States' UFO investigatory committees, we learn that Hynek had mentored the French astrophysicist and computer scientist Jacques Vallée, who was creatively portrayed in the movie by François Truffaut as Lacombe. In real life, Vallée postulated that ETI and UFOs had little to do with the common understanding held by most people. Starting his career as an astronomer at the Paris Observatory, Jacques Vallée had moved to the United States in 1962 and co-developed the first computerized map of Mars for NASA in 1963. At the time of writing his 1988 book *Dimensions: A Casebook of Alien Contact,* Vallée believed that "the UFO phenomenon represented evidence for other dimensions beyond spacetime" [referencing the Einstein definition].

In fact, whatever might be their private and personal inclinations, professionally, Hynek and Vallée were both part of the management systems relative to the perception and understanding of the ETI hypothesis by the public. At the time of his early UFO-busting exploits, Hynek was the director of America's

Figure 16.12. Allen Hynek (left) and Jacques Vallée.

early space satellite tracking program. To the public, this program was dressed up as a part of the International Geophysical Year (IGY). It was during the IGY, which actually lasted for seventeen months, from July 1, 1957, to December 31, 1958, that NASA Explorer probes discovered the first intimations of the Van Allen Belts. As the US government stated:

> Post–World War II developments in rocketry for the first time made the exploration of space a real possibility; working with the new rocket designs, Soviet and American participants sent artificial satellites into Earth orbit. In successfully launching science into space, the IGY may have scored its greatest breakthrough. Overall, the IGY was highly successful in achieving its goals.[9]

And during this time Hynek served on several USAF UFO committees. From 1947 to 1969 Hynek felt that debunking was what was expected of him and he enjoyed the role. "The whole subject seemed ridiculous," he said. This was typical of the scornful attitude within the USAF toward those who reported UFO events (which might explain Munck's dismissal of ETI as not "in the equation"). However, as time went on Hynek's experiences led him to soften his totally skeptical attitude toward the UFO phenomenon. The sheer refusal of the USAF to recognize the possibility that the UFO phenomenon existed and the quality of the data he had assessed over the decades led this wise, open-minded man to change his first opinion on the subject.

Hynek came to the conclusion that there was evidence for both extraterrestrial intelligence (ETI) and extradimensional intelligence (EDI). This related to aspects of UFO interaction that occurred before, during, and after an encounter with a UFO and these included: poltergeist activity often occurring after the encounter; dematerialization and rematerialzation of the UFO; photographs of UFOs not seen by witnesses at the time of taking the photo; telepathic communications; levitation of physical matter or people during a UFO encounter; the sudden stillness that can be experienced when in the presence of a UFO; and the development of PSI abilities after an encounter. He wondered if ETI and EDI were two aspects of a single phenomenon or two different sets of phenomena. When he did start expressing such views in the early 1970s, he was castigated by colleagues for his change of position.

As for Spielberg's Claude Lacombe, a.k.a. Jacques Vallée, astronomer and computer expert were only two of his talents. Vallée had worked for NASA on mapping Mars and was instrumental in developing that precursor to the internet—the Advance Research Projects Agency Network (the Arpanet). In

France as a young man, Vallée had experienced firsthand the censoring by his superiors of anomalous data associated with UFO sightings.[10] This had aroused his lifelong interest in the phenomenon, and over the decades, he evolved his views on the interactions of ETI into a hypothesis that didn't suit Spielberg. Vallée tried to persuade Spielberg that *Close Encounters* would be more interesting if it was about interactions that, although real and seemingly stemming from ETI, were not in fact produced by ETI at all. Something akin perhaps to Hynek's definition of EDI, extradimensional Intelligence.[11] Spielberg, conforming with the nuts-and-bolts merchants (those who would have it that every UFO sighting is of a real and solid spacecraft from "out there"), considered that even if Vallée were right, *Close Encounters* was a product of Hollywood and it should give the audience what they expected.

DETER, DETECT, DEFEND

It certainly gave the former Hollywood actor and sitting US President Ronald Reagan what he expected; he had remarked that if people knew how much of this film was true, they would be astonished. As was I, when I found out that those film-fake latitude coordinates actually led to a field outside Ault, in Weld County, Colorado.

Although the word *weld* goes nicely with nuts-and-bolts merchants, there was little of apparent interest except those circular crop fields. Oh! Crop circles. That's the name by which these extraneous markings in grain fields are generally known. Did these coordinates allude to systems relating to the production of crop glyphs? It's certainly food for thought, and if nothing else, this movie is giving our creative and associative thinking faculties a workout. This Aultian latitude, reflecting that of the *Two-Thirds* storytellers' Cydonia City connection between the "city airport" and the "food storage" crater for the ranch, certainly hinted that we should look closer at this area. Yet, considering that the fake longitude coordinates were so close to the authentic Devils Tower longitude coordinates, there surely had to be something other than a food field that linked Colorado to Wyoming's Devils Tower,* and to my mind, it would have to be *something* to do with the military and/or the Air Force. Tracking farther southward from Ault down this meridian through Colorado ought to produce an interesting result—and it did.

---

*Devils Tower is a laccolithic butte composed of igneous rock in the Bear Lodge Mountains near Hulett and Sundance in Crook County, northeastern Wyoming. It rises 1,267 feet above the Belle Fourche River, standing 867 feet from summit to base. The summit is 5,112 feet above sea level.

It landed on one of the airport runways belonging to Peterson Air Force Base. Good, the fact that we now had an airport and a field, discovered via a longitude echoing a sightline at Cydonia that itself was found via a latitude, showed that the logic was working but also bad, since it didn't mean much more than that. Until one realizes that Peterson AFB is the command headquarters of the North American Aerospace Defense Command (NORAD), whose motto is Deter, Detect, Defend.

To most people, NORAD immediately means the space monitoring facility deep inside Cheyenne mountain that lies to the east of Peterson AFB, and space goes with NASA, doesn't it? Maybe so, but the whole of NORAD is administered by the 721st Mission Support Group of the USAF 21st Space Wing at Peterson AFB. The commander of NORAD is also based at Peterson AFB. As is the US Air Force Space Command

While on the subject of ETI and the military, leaping forward for a moment, on August 29, 2019, the DOD declared space to be a war-fighting arena. Against whom precisely? It was announced that Space Command would be the first new US military service to be formed since the USAF split from the Army in 1947 (and that occurred just after the Roswell UFO incident, as the reader will recall). Hitherto part of the USAF, Space Command would now be the sixth military branch of the DOD. However, the new badge actually states that the Space Command *is* a department of the USAF, so it's not quite the sixth force all that

Figure 16.13. US Space Command 2019 alongside the USAF Space Command insignia. Despite the antecedents, the fact that the USSC shield bears a strong resemblance to the *Star Trek* logo has not been lost on many observers and humorists.

2019 PR suggested. Perhaps they were thinking of the sixth sense and all those ETI-EDI matters.

Back then to *Close Encounters,* the movie. It seems that this little game of false coordinates was yielding rich pickings, more especially because none of this was overtly necessary: the Devils Tower National Monument was already an established tourist attraction and fully capable of receiving the extra visitors it would get as a result of this film's success, so one does have to ask *why it was felt necessary to falsify these coordinates in the first place.*

Was it to convey a particular item, or items, of information? Or was it an exercise in the art of media manipulation? This scene would certainly work as an experiment with the art of coordinate cryptography, testing the delivery medium to see how many people would actually notice. That "rolling globe scene" is a good lesson on how to get information past the observer: create tension (in this case excitement) so that there is no time to rationalize the problems that might otherwise be registered by the target audience. In much the same way as the magician uses distraction in order to create an illusion.

CRITICS CORNER

The reason for making such a fuss about a brief moment in a mere piece of Hollywood entertainment is because we are all well aware that Hollywood movies are exceedingly powerful tools in the management of Western thought and in the promotion of the American way of life. Their output can (and most certainly does) manipulate public opinion via the emotions. It is also true that human beings have been encoding stories with important data to remember and pass on since forever. In Australia, the Aborigine elders taught the tribes the routes to all their important tribal locations using the constellations above their heads as direction guides. This was the dreamtime; the songlines were the routes. Since it was the Aborigines who taught the first settlers how to get from A to B, many of these ancient songlines are now incorporated into Australia's modern highways.

We have this recorded account from Australia, but no doubt our early cultures across the world did exactly the same thing. For example, the layout of the city of Chicago was originally constructed on the same principles, through learning the trails from the native Americans; only later did the city's north-south grid emerge.

In the twenty-first century, it is easy to forget that before the arrival of the printing press and the dissemination of books, few people were literate and everything had to be learned and retained orally. Such practices produced oral histories transmitted from generation to generation, told in a way that facilitated the retention of many facts and, over time, resulting in a collection of myths and legends specific to each culture. People were originally concerned with how to navigate their way, whether by land or sea, to obtain whatever it was that was required for the tribe to flourish at that time. With the passing of vast amounts of time and the development of technologies such as writing and mapmaking, the tradition was no longer kept alive, and these practical memory-storage methods for the location of food, water, and the necessities of life became tales of long ago; over time, they became legends. Then, as they faded even further into the mists of time, these legends became myths. The stars followed by ancient travelers became gods in the sky. Today most people, considering themselves to be very modern, do not really believe these old stories ever had anything to do with any sort of reality as we now perceive it. The living oral histories of the Australian Aborigines (and the mythmaking pursued by NASA in its own navigational sagas) should advise us otherwise.

With the advent of motion pictures that are accessible to all, we have been able to leave the linear storyline behind and can now project ourselves into the future with our what-if scenarios that can be used to both amuse and abuse our minds. This medium of film, using sound as well as sight, can also be used to bring us stories that are seemingly about one thing while really being about another. *Close Encounters* fits this last bill perfectly. While apparently giving us all a gentler and softer look at the idea of ETI contact than the usual sci-fi fare, in fact it was highly ambivalent in its attitude toward the idea of contact with ETI and was definitely threatening in its portrayal of the US authorities and the lengths they would be prepared to go to achieve their own somewhat ambling and ambivalent aims.

With hindsight, and just taking this one scene, it most certainly looks as if the scriptwriters and/or the makers of the movie knew about the secret meaning of numbers as held by cabalists, gematrian scholars, and practitioners of other esoteric disciplines. And they also appear to have been aware of Munck's mousetrap methodology, including the use of the Giza Great Pyramid as the prime meridian. All of which infers that someone, somewhere, seems to have been thinking about such matters as Munck's findings while pretending otherwise much, much earlier than 1978.

It's time to leave North America and head over to Giza and the land of the pharaohs.

Giza and Cydonia

Significant Giza-Cydonia and Egyptian-Cydonian Connections

PRIMARY MARKER

From any point of view, the Giza plateau, with its three primary pyramids, its six secondary pyramids, and the Great Sphinx constitutes one of the most amazing architectural legacies in the world. As such, it has always been a point of focus for human beings inspiring both the artist and the scientist. The play of light across the plateau and the beauty and grandeur of the geometrical configurations that these monuments confer on the landscape have stimulated the observer, resulting in numerous images, books, and TV/film documentaries—some made for love, some for money, and some for propaganda purposes. And when Carl Munck found the planetary matrix, another level of thinking about this plateau began.

Anatoly I. Kandiew's analysis of Munck's work had found that "Munck's Matrix" incorporated enough permutations within its structure *to discourage all but the most assiduous code cracker.* This led Kandiew to conclude that this was exactly what had ensured the survival of the code. Certainly, no one could break this code until plate tectonic motion had done its work, and the relevant monuments were in the right place for the mathematics to function accurately enough to reveal intelligent design. This suggests that there was considerable forethought (other than Munck's) to the selection of both the location and the placing of the Great Pyramid.

When establishing his "CO-ORD" system (a variation of Frenchman René Descartes's use of coordinates and the Pythagorean formula to calculate the distance between vectors), Munck had made an interesting find. None of the intersect coordinates he had derived from the manipulation of the latitude and longitude of all the ancient monuments he studied exceeded that of the Great Pyramid's vector of 248.0502134. He had concluded that the whole matrix had been *worked backward from the vectors, not forward from the latitudes and longitudes.*

While Munck has decided that the Great Pyramid served as an ancient prime meridian, thinking sideways, as Edward de Bono recommends, a 0°/360° marker can mean whatever the designer of the system wants it to mean. From a merely numerical point of view, the Great Pyramid is effectively a hub for the planetary matrix, but there are several indicators that its particular 0°/360° spot has a much wider radius than that of a terrestrial time zone marker.

At Cydonia, there is evidence of a sculpted mesa revealing details indicative of a face, but when the image is analyzed, a humanoid/lion vertical split is revealed; there is a physical sculpture of a lion body/human head at Giza. There are physical landforms creating a pyramid city at Cydonia and artificial buildings creating a pyramid city at Giza.

If the largest of these pyramids, the Great Pyramid, operates as a great attractor to all of humankind, it is all of the plateau that is of special interest to NASA, which is obsessed with the Egyptian Orion/Osiris tradition, while the SRI together with the ARE are equally obsessed with the plateau because of the Cayce physic communications concerning Atlantis and its downfall, with the subsequent emigration eastward of Atlanteans to what is now Egypt. And having written about the ARE in chapter 13, it is only now that it dawns on me: I have suddenly seen that the acronym ARE only requires an *S* at the end and we are back to Ares and the planet Mars. It looks as if these organizations, the spacecraft hardware merchants and the human software merchants, are all looking for something at Giza to help them on their way. So let's look at the Giza plateau and its monuments from these other points of view. What happens when the dots are connected?

GIZA: THE GEODETIC HUB

The nine pyramids at Giza and its Great Sphinx are located on a plateau within a country whose modern boundaries echo the form of that huge flat-topped Cydonian mesa lying to the east of the Face. As stated before, named the Bastion by Myers and Percy, nobody else really cares to mention this particular landform. Which is unsurprising, because seen from the air its outline plunges us into the world of repetition. It is repeated not only by the modern country of Egypt's frontiers but also at a much smaller scale by the side view of the Sphinx's head. The ancient Egyptian hieroglyph for the nose also matches this outline and suggests that this is a right-side view.[1] As this hieroglyph was also used to denote the whole face, inevitably,

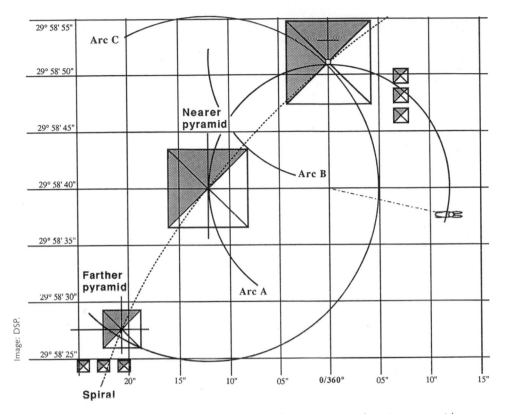

Figure 17.1. Giza plateau and the relationships between the nine pyramids.
A grid-coordinate system is referenced in the two sets of small triplet pyramids
on the plateau. The grid increases by 5 seconds of arc in latitude and longitude.
The 0°/360° prime meridian drives through the Great Pyramid. The arcs relate to
matters discussed in *Two-Thirds* but here can be seen to demonstrate interactions
between the various components of the plateau. Arc A is centered on the prime
meridian south of the Great Pyramid; Arc B is centered on the Great Pyramid and
if completed would take in some of the land north of the plateau. Arc C is centered
on Khafre, the nearer pyramid. Further data can be found at Aulis.com/pathway.

this finding brings comparisons with the photographic right/left matching of the
upturned Face mesa that revealed the hidden hominid/leonine aspects of that landform. On Mars the hominid was on the western side of the Face mesa, the lion on
its eastern side. Matching the Egyptian nose/face hieroglyph to the Sphinx profile,
one must stand on the causeway to the south of the Sphinx to see the matching
hominid. If standing to the north of the Sphinx would reveal the leonine, there is
also the fact that the body of the Sphinx also infers the leonine aspect reflected in
that eastern Martian face. Then, remembering that in this Sphinx sculpture the

Figure 17.2. In ancient Egypt, the hieroglyph for the nose could also signify the face. As the mouth is missing from the hieroglyph, it also looks like the profile of an animal, and given the associations of this Cydonian location with the Giza plateau, this suggests a lion/man or sphinx. For Masons, the flat-topped mesa might suggest an analog to the perfected ashlar.

hominid is really the north–south vertical and the body is the west–east lateral axis of the sculpture is even more food for thought. The fact that the lion is the symbol of courage as well as the top predator in Africa and that humankind is said to have emerged from Africa is interesting, to say the least.

Finding disparate items of the Cydonia complex mirrored at Giza is highly significant, as is the fact that the geometry of that special right triangle is also present. On Mars, the Bastion tangents the midpoint of the special right triangle's adjacent side, which links the Face mesa, the D&M Tor, and that imported cone-shaped spiral mound. Recalling that in *Two-Thirds,* the very large scale model can be found "when Giza is Silbury, Stonehenge is the Face," it's possible to match the position of the Martian Bastion to the Giza plateau. Taking Giza as being halfway along the adjacent side and Stonehenge as the analog of the Face, it is possible to re-create the Cydonia special right triangle and then mirror it to complete the visual of a 2-D tetrahedron overlaid on the globe.

This now produces a very large scale model of Cydonia. *Thus the Giza plateau location matches the Mars Bastion* and earns the same name. The point on the equator in the Indian Ocean becomes the analog of the cone-shaped spiral mound. The Indian Ocean has an area that has less gravity than elsewhere (a point that will become relevant later on). The real surprise of this hypothetical comparison of geometry with cartography comes with the positioning of the analog D&M Tor.

If there were such a construct it would be found on the same latitude as the Great Pyramid but underwater in the eastern Atlantic. This is very much where

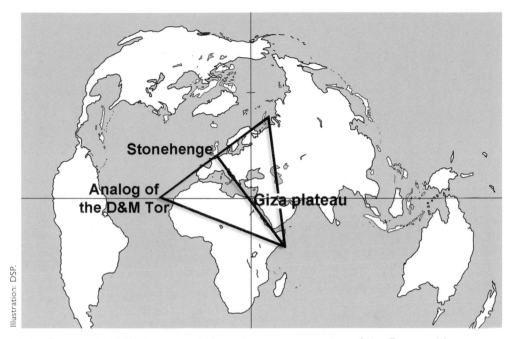

Figure 17.3. USAF map with Stonehenge as an analog of the Face on Mars and Giza as an analog of the flat-topped Bastion on Mars. The corners of this tetrahedron find themselves in three different oceans—the Indian Ocean, the Atlantic Ocean, and the Arctic Ocean—reflecting three different temperatures from warmer to colder, respectively.

Atlantis is supposed to have been, according to legends. It's little wonder that everyone is looking for underground chambers and secret passages on the Giza plateau; the geometry is encoding all of this ancient memory—and is saying so, in that underwater is a metaphor for the subconscious. The Woods Hole group were either looking at the wrong place, in the far western Atlantic, or they were looking at a fractal or something associated with, but not quite the same, as this hub.

Returning to the Giza plateau, what is also of interest here is that while the plateau finds a concordance with the Mars Bastion, it is also self referential: the position on the adjacent side of the special right triangle is found by dividing the square root of 3 by 2, namely 1.732/2 = 0.866. That resultant 0.866 is the ratio of the width to the length of the Giza plateau from its southern end up to the location of the hominid/lion effigy—the Great Sphinx. The large set of the big three pyramids, built over different periods, references the phi spiral generated from southeast of the pyramid group. And reciprocally, taking that ratio of 0.866 and applying it to the nose hieroglyph/Bastion on Mars shows the analog position

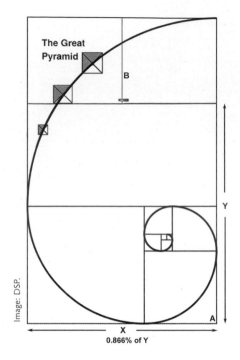

The Great
Pyramid

B

Y

X
0.866% of Y

A

Image: DSP.

Figure 17.4. Giza plateau, Egypt, showing the phi ratio spiral as it relates to the pyramids. A human being's framework is also constructed relative to the phi spiral: from the ground to the navel is 1, and from the navel to the head is 0.618. Combined that is the phi ratio of 1.618. Further matter relating to this subject is available on Aulis.com/pathway.

of the Giza sphinx to be just above the brow—the seat of the pineal gland according to Western esoteric disciplines, the location of the third eye according to Hindu terminology, and the third dantian of the Taoists.

GIZA GAZING:
A CULTURAL PERSPECTIVE

Looking at the Giza architectural complex, it's hard not to come to the conclusion that the key to understanding its past (and our future) might lie in the expansion of consciousness potentially available to everyone as a result of tussling with its very obvious puzzles. Giza's conundrums strip bare philosophical and religious viewpoints, historical time lines, cultural perceptions, and as it turns out, sheer good manners. The ferocity with which opinions about this complex are held is testimony to the difficulties encountered when it comes to encountering different ways of thinking. Changing long-established ideas or even seriously considering another viewpoint when it requires stepping out of the psychological comfort zone of the individual or organization lead to cognitive dissonance. At that point logic flies right out the window. Emotions take over and in order to maintain the status quo, anything goes.

So far, despite any discoveries made in the recent past, publicly at least, the problems raised by the presence of the Great Pyramid and its attendant complex have not changed the thinking of those in charge.

TRADITION

Egyptians are naturally proud of their ancestors' achievements and quite rightly defend the cultural heritage of their nation and its artifacts with considerable gusto. Traditional Egyptologists are focused on the Great Pyramid uniquely as the work of Pharaoh Knum-Khufwy, known as Khufu to modern Egyptologists, as Cheops to the Greeks, and estimated to have been built during his reign of 2589 BCE to 2566 BCE. All the calculations as to how long it took to build it and the means by which it was built are based on this premise. The rest of the plateau has to conform to this dating. While paying lip service to research with updated technologies, these traditionalists have so far refused any other date of origin or any interpretation of the pyramid's purpose other than that of a funerary tomb/vehicle for the safe afterlife journey of the pharaoh.

The hostility of those in charge is perhaps fostered by the very choice of geographical location for this set of pyramids. As already discussed, the siting of the Great Pyramid infers a knowledge of Earth's planetary configuration that fits better with the knowledge portrayed on the Piri Reis map than it does with the days of Khufu. A knowledge acquired from people having a bird's-eye view of the entire planet.

If it should be scientifically and irrefutably demonstrated that the Great Pyramid was not built by the pharaoh Khufu or that any other site on the plateau does not correspond to the traditional time line, not only would all history books have to be rewritten but the authors of books specifically on Giza would have to change some of their tunes, and in the eyes of the Egyptians themselves, it would cast a slur on the abilities of their ancestors.

Should it eventually be discovered that the structures on the Giza plateau, even if built by Egyptians, were primarily designed by unknown others, it will be of no comfort to the Supreme Council of Antiquities (SCA), as in their view it again infers that the Egyptians weren't up to it and were potentially bossed around by outsiders to construct these monuments—and for what purpose? The worst outcome possible would be for parts of the complex to have been built long before the advent of the Egyptian civilization as is understood by our historians.

Yet traditionalists also ignore the fact that, should it turn out to be earlier in date and definitely not associated with Khufu, whomever they were, the builders chose a location *on what was to become Egyptian soil* for the erection of the most amazing pyramid and one of the most intriguing architectural complexes on this planet. In so doing, it would become their inheritance, and the Egyptian people would become the privileged guardians of Giza for the whole planet.

THE BOOK REPORT

When it comes to the primary pyramid on the plateau, and remembering that these comments have been made regarding the seriously degraded pyramid as it stands today, many architects and engineers have already recognized how difficult it must have been to site and build such a perfect structure as the Great Pyramid with the underlying bedrock in situ, how the build was made unnecessarily complicated by the incorporation of the Queen's and King's Chambers' shafts into the structure, and how even parts of the structure that would supposedly be hidden from view were subjected to as great attention to detail as those areas that would eventually be seen. In fact, reading across the literature, it appears that even those on the most ardent of the Khufu-hypothesis teams are not entirely satisfied with the proposal that Khufu's people built this pyramid from the ground up. Especially since every attempt to copy the builder's methods has failed, while revealing engineering problems unsolved to this day.

This remains the case for all such forays into ancient architectural builds. Even when these experiments have added modern tools such as the odd digger, a metal tool, or a hawser to the period's available toolbox. The justification for this lapse into modernism is always "We used a digger due the short period of time allocated to the experiment." In fact, the timing of these experiments is built in from the start: it enables the experiment to look as if it works when in reality it does not. No amount of that other old chestnut, "We used ancient methods just enough to see that they *could* work," makes these situations any better. Other than Giza, experiments that have relied on this formula have all been about megalithic sites. Relative to Machu Picchu and the South American sites, methods of shifting large blocks of stone up steep mountain slopes ended up with the stone crashing into the ravine; for Stonehenge, transporting a smaller bluestone from Wales over water failed, as the stone sank. As for the huge sarsen stones, an experiment involving the rope and roller method of dragging a replica sarsen stone was tried out—for a very short distance along specially flattened ground (nothing like the eighteen miles of undulating terrain lying between the stone's source and the Stonehenge site). Considered a successful proof of concept, there is now an exhibit of a roped stone on rollers at the Stonehenge visitors center. In truth, none of these experiments succeeded, which should prompt us to really think about the data and information contained in these monuments and locations.

When it comes to back engineering at Giza, the experiment to build a replica of the missing apex of the Great Pyramid and then place the pyramidion, its top stone, utterly failed to convince even Mark Lehner, the American archaeologist who is not only a friend of ARE but also considered to be one of the world's pyramid experts.

Lehner wrote up this experiment in his 1997 book *The Complete Pyramids*. At 29.53 feet (9 meters) per base side, this experimental pyramidion was built to the proportions of the Great Pyramid's flat top. Lehner states:

> We had just three weeks and 44 workmen to build our pyramid consisting of 186 stones and measuring 6 m/20 ft high. It would have fitted neatly on to [*sic*] the top of the Great Pyramid, in whose shadow we built it.[2]

Lehner therefore estimates the flat top to be 461.60856 feet (140.69828 meters) from the ground. Which is accurate as far as it goes. The wording of this text glosses over an important fact. When attempting to fit the blocks of this small pyramid, the photographs and the video of this experiment (which was filmed for the PBS-TV program *NOVA*) show that it was necessary for the builders to stand well back from the base of their experiment in order to get the leverage their long wooden poles required; at the top of the pyramid there would have been no foothold to do that.

Lehner placed the report of this *NOVA* experiment in two separate sections of his seminal book on Egypt's pyramids. Having discounted levering as the means by which these upper rows of the pyramid could have been constructed with any safety, he still, wittingly or not, managed to gloss over this salient fact. Furthermore, despite having discounted the use of ramps so high up a pyramid, these experimenters used a spiral ramp for the placing of their own little pyramidion. Overall, this experiment revealed that the actual build would have been impossible using these building techniques. Which surely went some way to agreeing with all those who consider that there never was a pyramidion in the first place: the geometric build's *invisible apex* inferring "otherness" or rather aspects of physics invisible to a human being. Under certain conditions electrical charge effects have been registered by people standing on the top of the Great Pyramid. Remember those cylindrical holes of missing material? This is another version of those same principles. What you cannot see or touch can nevertheless interact with or affect that which you can see and touch.

Back at Giza, if no progress was made in understanding how this pyramid was actually built, there is evidence that this principle of an invisible apex was well understood. We are back to mumbo-jumbo ritual ceremonies: Lehner's little pyramid was made of 186 stones. Given the association of the Great Pyramid with the reflection of light, the myth that the apex was made of gold, that the Sun is over the top of the Great Pyramid at the spring equinox, and the fact that the solar system's light speed is 186,242,000 miles per second, makes this choice of 186 stones more significant.

Lehner's nine-meter base side and six-meter height measurements are related to each other by the ratio of 2:3. Nine is the sum of all the pyramids on the plateau, and six is the sum of the two separate groups of three small pyramids located along the x and y axes. In meters, the base squared is eighty-one, the missing stone would take that down to eighty, and among other things, 80:81 is the ratio of the mass of Moon:Earth.

His nine meters when expressed in feet are 29.5272 inches. The Great Pyramid currently sits at 29.9727° north, so this number recalls its location to within 27.12 minutes of arc. These numbers expressed in days are also within a whisker of the 29.531 synodic and 27.3 sidereal orbit of the Moon.

In all of this measuring of the Great Pyramid, it's easy to forget that despite our modern-day understandings of those figures as being related to values such as pi, phi, radii, royal and sacred cubits, feet, meters, or whatever else we take as our point of reference, it was the original engineers who left the space for a pyramidion that turns out to reflect in its base and height a decimal measuring system and reference to future builds on the plateau as well as matters pertaining to the Sun and the Moon's orbit around the Earth.

That being said, the *NOVA* team did have a choice as to how many stones they would use to make their pyramidon, and clearly, that choice of 186 stones *was* essential to their scheme because even these downsized blocks were unmanageable: getting them from the quarry required a truck with a winch! Traditional methodology: nil; mythmaking: 10/10. If this experiment contained hidden messages or memorials, it would be unsurprising, as the Giza plateau was another of Cayce's Atlantis markers and Lehner is very familiar with the ARE. Whether Lehner has noticed the other correspondences is not known. However, when Lehner's height of six meters is expressed in feet, it comes to 19.6848 feet, and thinking of the pyramid as the ancient meridian brings to mind the timing of things. Interestingly, given that 1968 was the year that Cayce predicted the rising of Atlantis, on December 21, 1968, the solstice of the twelfth month (and from the points of view of northern astronomers, the lowest point of the Sun's path),

NASA scheduled the launch of *Apollo 8,* the first circumlunar flight with a crew.[3] So what is it about this plateau that makes it so special?

PROFILES

The Mokattam formation on which all this architecture is sitting consists of three distinct geological layers. These members, as they are called, are numbered in ascending order from the lowest to the highest. The diagram in figure 17.5 shows a side cut of the geological strata in which it can be seen that the body of the Great Sphinx, considered to be the guardian of the plateau, surprisingly, shares member II with the northeast corner of the Great Pyramid. While the head of the Sphinx is made up of member III material, looking at the scale on the right of the diagram, which is of meters above sea level, it can be seen that the top of the Sphinx's head is level with the forty meter measure (131.232 feet), and that is two-thirds of the way from sea level to the plateau.

The Great Pyramid sits at 196.8 feet (60 meters) above sea level—ten times the height of Lehner's copy of the pyramidion. The Valley and Sphinx temples, the Sphinx enclosure, and the *body* of the Sphinx are one-third of the way up from sea level at 65.616 feet (20 meters), and note that 656.5612701 is the Munck grid vector of the Sphinx-like Face on Mars. Further, referencing the nine principal pyramids on the Plateau, Munck also states that raising the longitude of the Face on Mars to the ninth power produces the distance between Mars and Earth expressed in statute miles.

These are very interesting facts. At Giza before the building of the Aswan dam, the Nile floods raised the water level by some 19.6848 feet (6 meters). That is yet another reference to Lehner's reconstruction and that supposedly missing pyramidion. At the interface between systems, there is a transfer of energy from

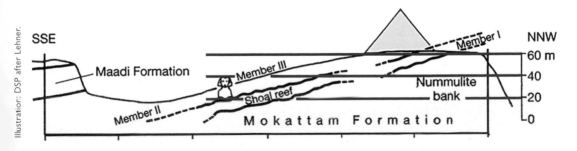

Illustration: DSP after Lehner.

Figure 17.5. Side view of the Giza plateau showing the three Mokattam formation layers relative to the Great Pyramid and the Sphinx.

the one to the other. The Sphinx not only acts as observer of the rising Sun but also sits at the interface of sky, land, and water.

It is quite possible that features found in ancient monuments that we consider overcomplicated and unnecessary offer us the opportunity to expand our mindset beyond the comfort zone of established thought and explore ways of thinking in multiple dimensions. In Egypt, we are reminded of that continuously, as their art depicts people from multiple perspectives: side on from feet to waist, front on for the chest, side on for the face, but front on for the eye. This is not about ineptitude but attitude! The goodness of the heart (the Egyptians made associations of the heart with the mind) and the eye as the mirror of the soul are both expressions of inner intention and communication. Side profiles also demonstrate the practical means by which to achieve said intentions. Perhaps then, side cuts of geology or buildings offer similar opportunities for different thinking. The ancient Egyptians also used animal symbols in hieroglyphic writing to indicate in which direction the glyphs should be read. One starts reading *from the opposite direction in which the animal is facing.* Just knowing this important fact provides further information about the Giza plateau.

There is a Sphinx facing east on the complex: irrespective of when it was built or by whom, we could use this ancient custom to acquire some understanding of the Giza layout. Surely, we should consider one reading of this whole complex as starting from the farthest of the three small pyramids situated south of Menkaure (the smallest of the three main pyramids). As already noted, these three pyramids are arranged along the latitude and thus can be taken to represent meridian markers along an X-axis, whereas the three small pyramids beside the Great Pyramid are arranged along a meridian, representing latitude markers along a Y-axis.

Or in the Roman understanding, the west–east X-axis would be the Decumanus (that of the tenth), and the north–south Y-axis would be the Cardo (that of the heart). The crossing point of these two streets at right angles was the axis of the city. When Munck looked at the plateau, he found a vector for the crossing point of these two small pyramid groups. The reduction of coordinates created the number 5577.09018. Did the Romans merely pick up on matters that had preceded their empire building? And did it have anything to do with those master numbers referred to by Glickman? The first four figures of that vector recall the sum of the fifth and seventh positions in the Master number series.

Extended, these two groups of three meet and form a right angle at a location not far from the Maadi formation (seen to the left of the Sphinx on the geological side cut, figure 17.5).

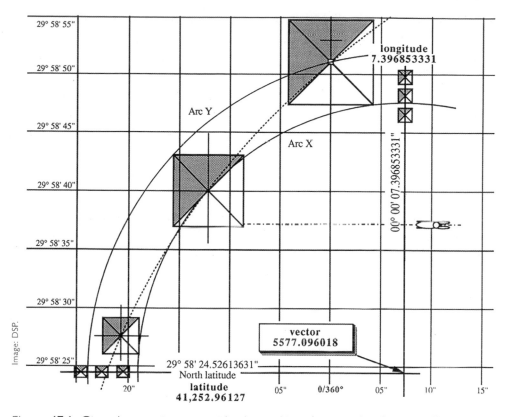

Figure 17.6. Giza plateau nine-pyramid relationships showing the three small pyramids, south of Menkaure (itself the most southern of the big three), and the arcs formed from the geometrical relationships between these three and the three small pyramids to the east of the Great Pyramid. Extending the horizontal and vertical lines drawn through the centers of these two groups produces the vector found by Carl Munck.

RETHINK

Interestingly, author Robert Temple, in his geometric assessment of the Giza plateau, excludes the Menkaure triple pyramids. He also dismisses the man/lion notion of the Sphinx preferring it to represent Anubis, the dog-like guardian to the underworld.[4] He might be ignoring time, but his Sphinx theory does hold some points of great interest when thinking across the board. The key point that emerges, when considering the numerous books and other media about this unique location, is that everyone has a different perspective and everyone is probably right in part when it comes to this great meeting place of humanity.

Traditionalists totally ignore a phi relationship connecting the pyramids, with Mark Lehner preferring the strike of the plateau dictating the positions of the big three.[5] The similarity of the Giza strike line to the *natural* volcanoes arranged in the same triple configuration on Mars relative to its equator will have surely not have gone unnoticed.

When the authors of *Two-Thirds* originally conducted their research and developed their understanding of Giza, they concluded that it was not the strike line but the phi spiral that governed the plateau's geometrical placing of the principal monuments. In fact, it is both. If the phi spiral underlies these pyramids, with all its connotations of golden ratios and light, something else amazing happens when the strike running tangent to these pyramids is drawn: a line from the Y-axis pyramids to the X-axis pyramids tangents the big three (figure 17.8). This strike line of the plateau follows the direction of the African tectonic plate, which is moving from the southwest to the northeast at a rate of about 2.15 centimeters (0.85 inches) per year.

To ascertain if anything truly was encoded in the data revealed by Lehner when building his pyramidion, I also checked Munck's work and found that he had determined a Great Pyramid latitude sited on the northern slope with a difference of 27.15 arc minutes from that of the apex. As his latitude was not on the apex itself, it has to do with something other than the strike line or the phi spiral that I have yet to ascertain, but I did know that Munck hadn't had the opportunity to crawl over the rough top of the Great Pyramid in person, as had Lehner, so he was not aware of the actual dimensions of the pyramidion. That being said, these two people had come up with numbers with miniscule differences. And that led me to believe that I was onto something.

Given that African plate motion is from the southwest to the northeast, I thought that to go back into the past I should extend my strike line to the southwest and according to that 27.12 minutes of arc difference between the apex now and Lehner's pyramidion height. I found myself in the sandy golden desert. Remembering that Myers and Percy had linked Stonehenge to the Face on Mars,[6] and like many other Earth energy researchers also thought the Giza–Stonehenge connection to be of importance,[7] I followed a hunch, and using Google Earth Pro I tracked toward Stonehenge from this desert location. I found that I had followed the geometry discussed earlier in this chapter and that I was picking up on the adjacent side of the special right triangle (see figure 17.9, page 396).

I found that I had also created an angle of 50.72° relative to the strike line. This was an angle that related to sightlines at Cydonia.[8] And returning to Africa,

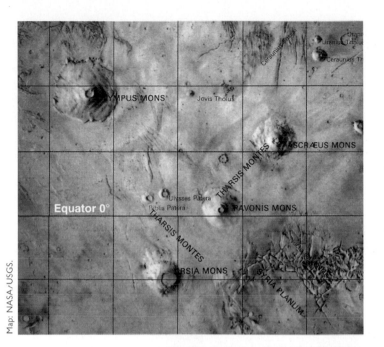

Map: NASA/USGS.

Figure 17.7. The Martian volcanoes form a straight diagonal line.

Photo: Science Photo Library. Overlay line: DSP.

Figure 17.8. The Giza Plateau strike line, which touches the southeast corner of the three main pyramids, forms the western diagonal of a triangle.

I discovered that the Giza plateau pyramids, the Sphinx, and the strike line were part of what Myers and Percy had called a "golden triangle," which repeated other angles out at Cydonia, while picking up on energy lines and geometry referenced by many Earth observers.

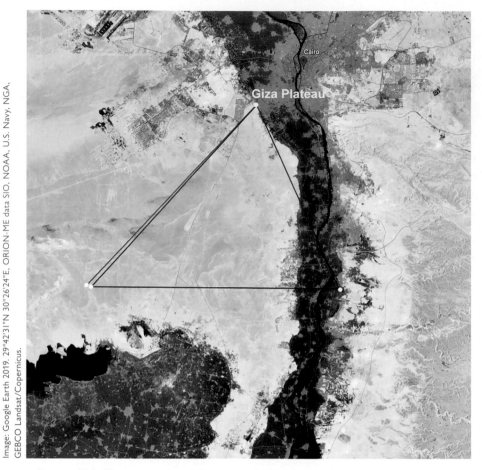

Figure 17.9. Continuation of the strike line in figure 17.8. The dark lines in this image create a golden triangle from Cydonia onto the landscape of Egypt. The western side of this triangle terminates on the Giza plateau at the eastern side of the Great Pyramid; the red line in this image indicates the southwest-to-northeast direction of the African tectonic plate motion.

INTERPRETING AND RESPECTING THE DATA

Coincidentally or not, the intellectual scrapping that has taken place at Giza over the centuries has increased exponentially since the advent of space exploration, and the problems came to the fore when the Face on Mars was compared with the Sphinx in the 1970s. The end result of Lehner's *NOVA* experiment avoided any references to ETI, and if any of the underlying geometries discussed and illustrated here were ever noticed, again, nothing was said publicly. Then the plateau and its attendant monuments were once again disturbed when the age of the Great Sphinx was put into question by geologist Robert Schoch in the 1990s.[9] More recently, another scan of the Great Pyramid has been underway, and speculation is rife as to the reasons for an apparent void said to be located above the Grand Gallery.[10]

The bottom line is that Giza throws back at everyone the same issue: whether you believe that this is the gateway to the next world as seen by a pharaoh or whether your inclinations take you elsewhere, this architectural complex offers everyone great opportunities to rethink and reformulate their own notions about humankind's origins, the ability to communicate with each other, and how to be tolerant of others' opinions while perhaps disagreeing with a given premise and even the data supplied by others. In this regard, reading across the Giza literature, it is remarkable how many different heights one finds for the Great Pyramid and the attendant monuments on the plateau. Sometimes these can even vary within the same book. From such discrepancies it is of course possible to interpret the data according to one's pet theory, and from that arguments erupt among the experts. Although when it comes to measuring, it might be that the very layout of the plateau affects the ability to measure accurately. It was a noted aspect of early crop circle research that what was simple to do outside the glyph was found to be very difficult within its perimeters. Measurements often had to be taken several times for accuracy.[11] For those interested in the overall culture of Egypt, the power games over "the Giza Great Truth" are confusing, and as with other contentious and potentially weird and ETI-related subjects, one is tempted to walk away, if it were not for the fact that reading across these accounts leaves two strong impressions:

1. *Everyone, with no exception,* is looking for something at Giza. Mostly it's the validation of a pet theory—but not always.
2. The relevant Egyptian authorities, namely the Ministry of Culture together with the former Egyptian Antiquities Organization (now called the Supreme Council of Antiquities [SCA]), which is under the Egyptian Ministry of Culture, might be under some duress. Sticking with the Khufu time line come hell or high water (both appropriate similes under the circumstances) might be one way of coping, but it seemingly leads to paradoxical behavior.

COPING STONES

The SCA is besieged by the political and academic interests of those nations and organizations focusing on Giza. It also has to be responsive to the various cultural issues within its own nation, which is by no means unified when it comes to matters of the pharaohs and pyramids. It therefore has a heavier burden than is publicly recognized by its critics. Naturally, it is only right that in terms of research permits, the SCA decides who has access to the Giza plateau and to the rest of the archaeological inheritance of Egypt. It is also normal that before publication, the Egyptian authorities wish to have oversight of any discoveries made by any team as a result of such research.

The maintenance of the academic and ministerial infrastructure surrounding the Egyptian monuments therefore manages the storyline. Many who have been granted permits have found that it isn't unusual to find that their papers are held up or that further research permits are no longer available should the findings not contribute to the traditional storyline. Any researcher who falls foul of the rules set out by the SCA can be banned from the plateau—for years or even decades, depending on the severity of the perceived misdemeanor. In the case of the Great Pyramid, finding unexpected evidence that leads further away from the Khufu thesis can lead to suppression or delays in publishing, together with ad hominem attacks on the researchers, even when the perpetrators of such a crime were officially solicited to come to the plateau to work.

This is precisely the same behavior pattern meted out by NASA to Vincent DiPietro and Gregory Molenaar, the two US government employees who were suddenly "no good at their job" when they examined the Cydonia data. It is exactly the same thing that happens in the UFO world and within the crop glyph community, and it happens whenever there is a strongly vested interest in maintaining the status quo.

In the case of Giza's Great Pyramid, new research and the use of improved technologies is making the Khufu thesis increasingly untenable. The added complication of those seeking that Atlantean Hall of Records, many of whom have strong links to NASA and the quasi-secret PSI-Ops projects, indicates that the authorities are doing the usual thing of using the traditional storyline to keep the lid on Giza while secretly attempting to find out exactly who did build the complex. With specific reference to the Great Pyramid, the engineer Rudolf Gantenbrink's experiences will make the point, as his saga demonstrates the level of discomfort that finding something unexpected can create within the establishment.

Once again, it's show time.

18

Giza Gazing

Exploring Details of the Great Pyramid;
Making Connections with Essential Requirements
for Safe Travel in Deep Space

ROVING

Rudolf Gantenbrink was involved in the 1992 King's Chamber engineering project intended to improve the airflow within the pyramid. Following this, and at the instigation of the authorities, he created a robot to investigate the Queen's Chamber shafts, and work started on March 7, 1993. The robot was called Upuaut (meaning "Opener of the Way"), and everyone was delighted with its work, until March 22, when some 216.52 feet (66 meters) from the chamber, it found the southern shaft completely blocked by a polished stone slab.[1]

At that time, Zahi Hawass was the general director of both the Giza Pyramids and the Saqqara and Bahria Oasis to the south of the plateau. However, over an

Figure 18.1. Zahi Hawass, Egyptian archaeologist and former head of the SCA and Minister of Antiquities.

issue not of his making but for which as director he was held responsible, he had been suspended two days previously. Therefore, Gantenbrink reported his new finding to Rainer Stadelmann of the German Archaeological Institute of Cairo (GAI) and then returned home to Munich.

So far so good.

By April 7, after weeks of debate and at one point stating that the finding was a hoax, the Egyptian authorities had still not produced a press release concerning this exciting new find. Gantenbrink was also surprised that the authorities had never asked for the data he had acquired during his investigation of the Queen's Chamber southern shaft. As author Robert Bauval relates, this inactivity had led Gantenbrink to reflect on his last meeting with Hawass, during which he had felt something was not quite right. He had found it odd that Hawass, although officially suspended, had turned up at Gantenbrink's hotel with Stadelmann immediately after the discovery.[2]

What was also very peculiar was a documented telephone conversation that took place between Bauval and Ulli Krupp, the GAI representative overseeing the three-week project. Bauval recounts that he was seeking to verify precisely who from the authorities and the GAI was in the Queen's Chamber at the time of the blockage discovery. During the course of the conversation, Krupp stated that he wasn't there, having been sent away on another job for the last week of the Gantenbrink project. Then he said he *was* there when the blocking stone was found, then he wasn't sure. And he ended up saying to Bauval, "It's whatever Gantenbrink says." Back to square one; Krupp wasn't there. Which again is not normal procedure. An archaeologist from the GAI was expected to be on-site during the whole project.

Eventually, Gantenbrink decided to go to the press himself, which he did on April 16, 1993, and as a result he was promptly banned from the plateau. However, Bauval later acknowledged that it was he who broke the story of the blocking stone (now called "the door") to the press and thus earned Gantenbrink his ban.

Whether seen from the position of Gantenbrink, who has since taken great pains to show that he doesn't share in Bauval's theories on Giza and its star-like associations to Orion/Osiris; the GAI, which needs the SCA in order to function; Bauval, who wants everyone to know "the truth"; or the Egyptian authorities, who were trying to keep their sanity in the midst of this intellectual mayhem, this entire saga reveals that the authorities are totally unprepared for finding structural anomalies that don't fit the time line. Not only was the blocking stone a major problem, it also had two fixings on it, apparently made of metal. Chronologically, that was another problem.

Hawass was later reinstated, and by 2001 he was the undersecretary of the state for the Giza Monuments, having spent the intervening years procrastinating about any follow-up to the 1993 shaft investigation and fending off inquiries about the Gantenbrink blocking stone. He even asserted at one point that that he did not believe the block was a door or that there was anything behind it. Then "open sesame"! A deal was announced between the SCA and the US National Geographic Society (NGS). The latter's TV and film division was going to transmit live the penetration of the Gantenbrink blocking stone as a big media event. The NGS also affiliated Hawass to the NGS by making him one of its Explorers in Residence, a position he would occupy from July 2001 to 2011. This appointment could be considered at the very least a conflict of interest with his official position, but then again if both groups had the same aim—to control the Khufu legend while profiting from new discoveries in terms of benefits to Egyptian tourism—then there was no conflict at all.[3]

BEWARE THE BOTS

In 2002, a new $250,000 robot was commissioned, this time from the Boston-based robot manufacturer iRobot. It took six months to build and was named the Pyramid Rover. Much was made of the fact that iRobot had supplied the robots used to search for victims after the September 11 World Trade Center attack. Less was made of the fact that the date selected for the broadcast of the show, September 17, 2002, was significant for those who attributed some sort of world calendar to the architecture of the Great Pyramid. In the nineteenth century, in order to establish the Great Pyramid as a prophetic tool, the lengths of various passageways were measured using the so-called pyramid inch (1.00106 British inches), and biblical events were attributed to various points along each of them. Establishing such a calendar for the timing of world events past and future only functions if one has (a) decided on a start date and (b) is working with hindsight, which is not prophecy but "I told you so." The whole concept is flawed and especially so, given that the start point at the entryway to the Great Pyramid was based on the notion that creation of the world occurred in 3999 BCE. The end date, again based on fundamentalist Christian ideas of the rapture (or Armageddon as eschatologists would have it) was determined to be September 17, 2001.[4]

Whether one buys into such a flimsy concept or not, the PR for this United States/Egyptian collaboration at Giza *chose* to trade heavily on the association between a disastrous event that occurred on Tuesday, September 11, 2001, and the exploration of a blocked passageway taking place a year later, on a day and a month linked to the idea of end times, namely Tuesday, September 17. Tuesday, a day attributed to the planet Mars, has associations with NASA, Cydonian research, and Giza researchers. Many of these people consider that Mars "died" by *losing its atmosphere* as a result of a catastrophic nuclear disaster. Whether linked to the past, present, or future of Earth and/or Mars, this layer of covert ritual speaks of an absolute awareness of the significance of the timing and the event.

As does what happened next with that blocking stone.

Despite the proclaimed desire to learn more about the pyramid and do so publicly, when it came down to it, actually finding something unexplainable behind that very challenging block of stone would be very difficult to deal with live on air. It must have been understood that unless the historical development of metallurgy across the planet was misunderstood or incomplete, this stone, with metal attachments considered to be beyond the engineering skills of the indigenous civilization, strongly inferred ETI involvement with this structure's build. No surprise then, that Pyramid Rover had been quietly sent up the shaft to drill its peephole *four days* before the so-called live broadcast date. Nor should it be surprising to the reader that the date selected for this secret preview, Friday the thirteenth, also held great significance for the organizers. But marking historical events might be the tip of the iceberg. It looks as if the September 17 end-of-time date was essential to the hidden Martian storyline running as the subtext to the Giza saga.

CATASTROPHE CORNER

Endings are also beginnings, and as such, the end of times need not signify doom and gloom; it can also equate with the entirely natural event of the precession of the equinoxes, leading to the end of the current age of Pisces and the beginning of the age of Aquarius. History shows that all cultures are sensitive to this great cycle, and consequently, authorities are always uneasy about such big moments. However, in any real understanding, there is no actual "moment"; the shift from one to the other is very, very gradual, and although this transition from one zodiacal house to another has been allocated a period of 2,160 years, the twelve principal ecliptic constellations do not fit neatly into such strict parameters. Some appear to the Earth-based observer to occupy a neat thirty-degrees house; others do not. The equal house divisions have their uses, but astronomers disdain the mapping and inferences that astrologers bring to the table. Those with an agenda have taken advantage of both systems since time immemorial.

Using astronomical events beyond our control as a way of engineering human perceptions still goes on, and this was demonstrated during the Jupiter Schumacher-Levy 9 comet event of 1992–1994. This event is so anomalous that it deserves further analysis, especially since the timing of it very neatly frames Gantenbrink's 1993 discovery. The Schumacher-Levy 9 comet, first spotted in March 1992, had broken up by July 1992, and was under intense scrutiny from excited astronomers worldwide: it was expected that the pieces of the comet would collide with Jupiter in 1994. And they did; the twenty-one broken pieces finally submitted to Jupiter's gravitational pull during the week of July 16 to 22. These are the very precise dates of the twenty-fifth anniversary of Apollo 11. The role of Jupiter as a shield for our solar system's debris was highlighted by this event, and the ensuing science data gleaned from this event is proving to be very instructive, and not only to climatologists. At the time unknown to the public, other indications that this event was a true anomaly were circulated within the science and Intel community. Not themselves in charge of the actual process out by Jupiter, if these Earth-bound experts spared a thought that this event had something to do with ETI, they weren't sharing that information either, instead suggesting that this event should be used for social engineering purposes.[5]

So it's interesting that in 2012 Cayce's ARE, the PSI arm of these Giza gazers, revised that quasi–religious-political Giza Pyramid time line. As it turns out, many of those interested in the Cayce school of thought, while talking about the necessity for harmony and developing one's own inner being, are often at odds with each other and are also depressingly keen on dramatic and stressful end-of-time scenarios.[6] Writing in his book *2038: The Great Pyramid Timeline Prophecy,* John Van Auken, a director of the ARE, proposed that the end date encoded in the Great Pyramid should really be extended. Having no passageways to justify this exercise, he decided to go vertically up through the King's Chamber to the apex of the roof above its relieving chambers. This took him to the year 2038.

Surely, this has nothing to do with the fact that Van Auken was publishing in 2012, the year that the Mayan Baktun cycle of 5,125 years was coming to an end; or with the fact that the King's Chamber and its five chambers are visually resonant with both the Djed pillar of Egyptian myth and the imagery on the tomb lid of the Palenque Mayan king Lord Pacal (often interpreted as "Lord Pacal in his spaceship"); or with the fact that NASA expects a hefty asteroid called Apophis (Apep, the Serpent of Chaos) to make another near-Earth return in 2038; or with the fact that 2038 is the beginning of the launch sequence for NASA's human missions to Mars—does it?

Such questions lead us straight back to the beginning-ending timings foisted on the Great Pyramid: as the nineteenth century start date was based on the beginning of the world (as understood by the religiously minded) and as such, is utterly erroneous in any practical terms; in the twentieth century, choice of an end date of 2038 is equally insignificant.

Yet the desire to mark that date infers that someone is still buying into this concept or using it for their own purposes. These are not necessarily for reasons entirely to do with practical Earth matters, for it is during this change from the Great Month of Pisces to that of Aquarius that we humans have begun stepping off the planet and started looking to Mars. Before we did that, we split the atom and landed ourselves with a legacy that is hard to bear let alone manage properly. Now, we want to go to Mars, and the nuclear issue is again on the table, as if all the disastrous events of the preceding years of nuclear mismanagement, both civilian and military, had never occurred.

Astronomers might be reticent about the effects of a Piscean/Aquarian changeover on humanity; geologists are more certain of their facts.

By 2016, geologists were suggesting that we should consider the naming of a new epoch. It is their view that the use of nuclear power, whether used for war and weaponry, for experiments with the environment, or for space travel, combined with the effects of accidents and the dangers of waste by-products of our nuclear power stations, have all combined to trigger a new geological period. These are lengthy periods of time: the Holocene epoch started 11,700 years ago, at the end of the last major glacial ice age, according to sources at the University of Berkeley. This date is much favored by those who associate it with the subsequent rising of sea levels around the planet and then make the link to a lost continent sinking under the waves. Moreover, this is where the Mars explorers join with those looking for Atlantis. Geologists are proposing that the Holocene be declared to have ended and that the new age be called Anthropocene for reasons entirely associated with our handling of nuclear fission.

As the marker for the start of this new Anthropocene geological period, geologists are considering adopting the date of July 16, 1945, the day of the first nuclear bomb test at the Trinity Test Site in New Mexico. So was the timing of the Apollo 11 launch of *three* astronauts therefore also in memoriam? Does that make the Jupiter event an extraterrestrial anthropo-event? No doubt wishing to avoid the many issues associated with the use of nuclear bombs against Japan, some geologists prefer the alternative marker date of January 1, 1950. This date is already in use relative to radiocarbon dating, where BP means "before the present year 1950." When radiocarbon dating became technically feasible, the amount of multinational aboveground nuclear bomb weapons testing had so drastically affected the proportion of carbon isotopes in the atmosphere that dating items after January 1, 1950, was considered unreliable.

This choice of 1950 is less than honest because it completely glosses over the nuclear testing in the 1940s and its disastrous consequences, and the bombs unleashed against Japan were also detonated aboveground. Furthermore, if underground detonations are considered acceptable, then what about their consequences? They potentially trigger other events such as earthquakes and the forcing of the otherwise natural tectonic plate motion, thereby promoting volcanic upheavals and tsunamis, not to mention, in all cases, the radiation risks and consequences for those involved in the experimental testing.

Would these geologists consider a middle way? While hypothesizing about the start time for such an epoch, why not split the difference and select the average of these two years, 1945 and 1950? It would take a brave academic to do that, but it would be perfectly appropriate, as it turns out to be 1947.5 or, in clear text: the middle of 1947.

And sure enough, in June 1947, as you will recall, the United States experienced a UFO wave during which pilot Kenneth Arnold saw "flying saucers" over Washington State, and at the end of June and in the early days of July, the focus of attention switched to an alleged UFO crash site near Corona, New Mexico, which became known worldwide as simply "Roswell." At that time, it would appear that an expression of ETI came into close encounters with the military near to the site of that first bomb test and even nearer to the base that was home to the USAF's 509th Composite Group, which dropped the bombs on Hiroshima and Nagasaki, Japan, but after the war was stationed in Roswell. A full analysis of Roswell can be found in *Dark Moon*,[7] but with respect to the interactions of ETI and the nuclear set, there are many stories emanating from guardians of nuclear sites in the United States and elsewhere that bear witness to the presence of UFOs over their locations. On occasions, accidents have been prevented by the disarming of missiles by forces outside the control of those on the ground.[8]

If June 1947 had major relevance for the military nuclear industry, it also served as the blueprint for the commonly held imagery associated with both spacecraft (flying saucers) and their crews. Resulting from the Roswell/Corona event, the stereotypical expression of ETI became the "little Gray."

Over the years, the Roswell/Corona event was repeatedly reintroduced to the public's consciousness, principally through official reports of investigations into an incident that officials were equally certain had never happened as previously reported! Each report had a different answer to the Roswell problem, and each of the solutions were equally implausible, but this layering of cheap wallpaper over the cracks of a wall of silence concerning the original event managed the outcome satisfactorily. Sensible people turned away from any sort of discussion. After all, if the agencies were going to try to fool us, then intelligent people should not rise to the bait. Good result, of sorts, because even today Roswell is still a problem, and no amount of "we've dealt with that, move on" is removing it from the public consciousness.

Consequently, in the context of the military hype concerning Roswell and the military's somewhat confusing attitudes toward ETI, the symbolic meaning of this ETI crash, supposedly replete with little Grays, might contain more information relevant to our space programs.

Roswell is particularly interesting because the military uses these mind game images differently. The US military (as do most military groups) dehumanizes perceived enemies by using derogatory terms to describe them. In World War II, the Japanese were referred to as "slant eyes" due to the epicanthic fold on the upper eyelid that distinguishes many Asian peoples from Westerners. It cannot be a coincidence that the potentially damaging radiation absorbed in body tissue is expressed in terms of units called *grays*. One gray (Gy) is one joule of radiation deposited per kilogram of mass. Nor can it be entirely coincidental that the so-called small and skinny extraterrestrials allegedly strewn around this June 1947 crash site were described (by those making the myths) as "little Grays with slanted eyes." Nor that this image of the "alien Gray" became linked to the alien abduction scenarios prevalent in the United States and installed in the public's mind thanks to the cover of horror writer Whitley Strieber's 1987 book *Communion: A True Story,* in which he recounted his memories of alien encounters recovered through hypnosis. Strieber worked on the authenticity of the cover's alien with artist Ted Seth Jacobs, who also painted the cover for Streiber's 1989 novel, *Majestic.* The *Communion* cover was reproduced as Jacobs had painted it originally, but that would not happen for *Majestic.* Jacobs has given an enlightening interview about his involvement with the Strieber project,[9] and as regards that *Majestic* cover what happened between conception and publication is particularly relevant here. Jacobs felt that his original cover design for *Majestic* perfectly expressed the contents of the book, concerning the opposition of the authorities to the alien phenomena. He illustrated this by painting a brutal looking soldier looming over a fragile little gray alien lying on the desert sand. When the book was actually published, the soldier had been removed and the light effects had been altered. Jacobs was an authority on the effects of light as applied to painting, so neither of these changes pleased him, to say the least. Of these two books, *Communion* is a biographical account of events perceived as real, and *Majestic* is a presentation of facts in novel form. In

Illustration: Colette Dowell.

Figure 18.2. Classic male alien Gray drawn by Colette Dowell, who has had psychic vision experiences all her life.[10]

November 2019 Strieber published a new book in which he announces the return in September 2015 of his little grays from *Communion* times. The whole series of Strieber's ETI/EDI books looks like another round of educating the public as to the presence of ETI while at the same time controlling the dose rate of information. Streiber would also be involved with the Wiltshire crop circle scene via his communications with the man on the scene, one Michael Glickman.

It is food for serious thought as to whether the Roswell myths were (a) embroidered out of whole cloth in order to justify the Cold War nuclear programs and budgets, (b) the Roswell event was a shock to the authorities themselves, or (c) as a result of their own fears and perhaps harboring some guilt concerning their own actions in 1945, the military sought to control the matter of ETI through a long term public relations exercise. Reflecting their own emotions, it was perhaps decided to create an imagery of fear around the subject of ETI. And as already mentioned, calling aliens "little grays" is using terminology assigned to ionizing radiation measurement. Then there is the fact that old nuclear ideas for human space travel—abandoned for the most part due to the challenge of working within acceptable margins of safety—are once again on the table. Not only for crewed flight to Mars but also to everywhere else, including the asteroids and, oh yes, the Lagrange points around the Moon.

Safety standards for radiation tolerance are less stringent in space than here on Earth, but as opposed to the smaller nuclear devices employed currently on space probes, if something goes wrong with a nuclear-powered human mission destined for beyond LEO, even if it's launched from LEO itself (one of NASA's pipe dreams), the consequences would be visited on every living organism on Earth. And absorbed by the material in the solar system.

TAKING THE MEDICINE

Undeterred by such considerations and faced with the impossibility of getting human beings to Mars safely (and, most importantly, in a fit state to function once there) by 2015, NASA stated that as matters stood, the nuclear fission option was probably the only way of getting to Mars in a timely fashion. Having acknowledged years ago that they know they should provide an artificial gravity environment for the crew but that they don't know how to achieve it, they are not excluding shoving their way quickly to Mars and just hoping it works out.

In the meantime, some more messaging, either to themselves or to ETI, seemed to be in order. And ancient Egyptian rituals once again emerged into twenty-first-century space travel when NASA did a repeat performance of the 1971 Apollo 15 microgravity experiment/ritual.

Here's a brief recap of original Apollo 15 scenario in which a falcon feather and a geological hammer were used to demonstrate the so-called effects of lunar gravity. As with all rituals, every action and every word is important in bringing about the desired result. So one should bear in mind when looking at these

Figure 18.3. The Book of the Dead, vignette. This ancient ritual was undertaken at the time of death in order to establish the virtues of the deceased during the lifetime. To account for its goodness, the heart of the deceased is weighed against a feather by the Moon god, Thoth. From this comes the expression "with a heavy heart" (one weighed down by sorrow). Although thinking about it, "with a leaden heart" would be the equivalent of thinking that lead might protect the human body against cosmic radiation.

experiments that messages to themselves are seeded into their own rituals at every opportunity. Paradoxically (as we have shown elsewhere), images from NASA have also demonstrated that these messages can be seen as tip-offs that something may not be as it seems. The same goes for reading any accounts of these events. In this description of the Apollo 15 ritual, note the specific mention of a *geological* hammer, and the irony of the Earth-based lunar surface practice site's location, just across the road from the USGS offices at Flagstaff, Arizona, should not be ignored. Nor the fact that this lunar practice site has not been kept as a historical memorial; quite the contrary, it is now the local trash dump. But that's enough of that; here's the description supplied by Mission Control staffer Joe Allen in the Apollo 15 preliminary science report:

> During the final minutes of the third extravehicular activity, a short demonstration experiment was conducted. A heavy object (a 1.32-kg aluminum geological hammer) and a light object (a 0.03-kg falcon feather) were released simultaneously from approximately the same height (approximately 1.6 m) and were allowed to fall to the surface. Within the accuracy of the simultaneous release, the objects were observed to undergo the same acceleration and strike the lunar surface simultaneously, which was a result predicted by well-established theory, but a result nonetheless reassuring considering both the number of viewers that witnessed the experiment and the fact that the homeward journey was based critically on the validity of the particular theory being tested.[11]

This little scientific experiment could also be interpreted as a politico-religious ritual. Having already mentioned the associations with the USGS, the use of a falcon feather might well have been attributed to the name of the Apollo 15 lunar module (said to be named after the USAF Academy mascot), yet it is also true that the Egyptian god Horus was depicted as a man with a falcon's head. The symbol of the United States is another feathered creature, the eagle, and the flag of the then Soviet Union included a hammer, symbol of the workers.

If that little sci-fi dance inferred that when it came to lunar exploration the

Figure **18.4.** Detail from the official national flag of the Union of Soviet Socialist Republics from 1922 to 1991.

Americans and the Soviets were equal partners in the vacuum of space, the hammer held in the dominant right hand of the astronaut inferred that the Soviet Union was in fact uppermost in the pecking order.

HORZ FEATHERS

Despite this author's interest in the cultural aspects of NASA's space exploration, such was the emphasis on the effects of gravity during that 1971 demonstration that the significance of its parallels with the Egyptian ritual did not occur to me until physicist Brian Cox of CERN presented, on BBC TV in October 2014, an updated version of this gravity experiment. Even then he was halfway through explaining what he was about to do when I suddenly realized that what I was actually seeing was the Egyptian ritual acting as a visual subtext and the whole exercise had been updated so as to make it relevant to the Mars expeditions. This is what happened.

The experiment had been filmed at NASA's Plum Station in Sandusky, Ohio, a part of the Glenn Research Center.[12] Ignoring the Zero Gravity Research Facility's two drop towers, which were designed for just such an experiment, Cox was filmed in the more dramatic and photogenic Space Simulation Vacuum Chamber, the largest in the world. Built by NASA in 1969 to test space hardware in simulated LEO conditions, it measures 100 feet in diameter and 122 feet in height. It was used to test space propulsion systems for many of the Mars missions, which might explain why Cox chose it over the drop towers, because should there also be a politico-symbolic element to this bit of showbiz, then it was just the place. The dimensions of the facility's drop towers could not have allowed the poetic finale planned for this episode.

In preparation for the landing on planet Plum, a sandbox was placed on the floor of the chamber while high above a red ball (replacing the hammer, but contrary to the Apollo 15 ritual, this time it was placed on the left side) and a bunch of ostrich-like feathers (replacing the single falcon feather, but on the right side) were attached to a rig suspended from the upper dome. Interestingly, back in ancient Egypt, the ostrich feather symbolized an aspect of Thoth, that of truth, as surely the red ball would have symbolized the heart of the deceased. Albeit still at the heart of the matter, that red ball could equally well be taken to represent the red planet.

And it was at this moment of the broadcast that the penny dropped! NASA was performing the ancient Egyptian ritual of the weighing of the heart, disguised as a gravity experiment.

However, steering away from the Egyptian connotations, Cox took the trouble to inform his audience that a red *bowling ball* was being used for this experiment.

Even with this choice of object, multiple clever allusions offered themselves. This feathers-bowling ball link immediately brings to mind the Coen brothers' 1998 film *The Big Lebowski:*

> The Dude again appears with a bowling ball, this time it is bright red. He holds it with strength and pride teaching Maude how to bowl. He then finds himself floating down the bowling lane between the legs of many women. Once he reaches the end of the lane he finds he is being chased by three German Nihilists in red spandex suits carrying enormous pairs of scissors.[13]

Cox's own scriptwriters might even have been cleverly alluding to Wernher von Braun's *Das Marsprojekt.* Written in 1948, it was a fully comprehensive architecture for a human piloted mission to Mars and was translated into English in 1953.[14]

If von Braun's earlier so-called competitive race to the Moon was epitomized by the Apollo 15 ritual, on paper, this 2014 exercise looked very much like a reiteration of that same message: all things looked equal, as it were, but with the heavy item now on the left and the light item on the right, inferring the United States as the dominant partner.

THE BRANDENBURG CONCERNS

Or it might be that another aspect of the Martian research undertaken post-Viking in the 1970s was being immortalized in the extravagant visual finale of this 2014 sequence. Here's what happened at the end of the experiment: filmed in slow motion, the red bowling ball thumped into the sandbox. As it did so, the bunch of feathers—which until then had made the trip as one—separated not unlike a row of bowling pins when hit by the ball, and these six quills were very floaty and feathery indeed before settling onto the sandbox.

This image of the round red ball smashing into sand seems to be another inside joke. This time referring to a colleague very much involved with Mars research since the very beginning: nuclear physicist John E. Brandenburg. At NASA's 42nd Lunar and Planetary Science Conference in 2011, devoted to solar system evolution, Brandenburg stated that the Martian atmosphere had been terminally damaged by nuclear explosions that, through drift, affected the two locations first picked up on by *Mariner 9* and then *Viking:* a region to the north of the the Elysium pyramids and the Face on Mars, respectively.[15] No elaboration was made as to whether these accidents were due to malfunctioning spacecraft.

Among the many motivations for Brandenburg's efforts to associate Mars with nuclear destruction brought about by warring alien factions, there are two front runners: it could be a mind game associated with the perils of thinking that ETI contact is a good idea, which was also Stephen Hawking's viewpoint, or it could be an extension of what is already occurring here on Earth. The fact that Brandenburg's selected sites are both predicated on the two locations where, "pyramidal structures sufficiently anomalous to their surroundings and therefore meriting a better look," as Sagan had put it concerning Elysium, would indicate the former option. Yet, if we take a look at our own history, it could be construed that our war-like behavior relative to nuclear fission had triggered a response from ETI decades ago, which has led to the DOD and its nuclear physicists mirroring their own behavior and attributing it to a war-like scenario for ETI on Mars.

Brandenburg's theory is derided by many scientists, but he is an old hand at Mars and the Cydonian issue. Brandenburg, writing only in uppercase letters during their exchanges, was the shouty member of that post-Viking conference organized by Pozos and Hoagland. The participants in that conference were the main drivers of various attitudes toward Mars exploration and the subject of ETI, and they had generated several mechanisms to explain the Cydonian complex. Brandenburg's version was this disaster scenario.

Following his 2011 announcement, Brandeburg published a paper in 2014, "Evidence of Massive Thermonuclear Explosions in Mars Past, the Cydonian Hypothesis, and Fermi's Paradox."[16] This paper covered all the essential issues that NASA has with Mars and with the public, ETI, Cydonia, and the power problems involved in feasible human space travel.*

---

*Over time Brandenburg has summed up his hypothesis more briefly: "Analysis of recent Mars isotopic, gamma ray, and imaging data supports the hypothesis that perhaps two immense thermonuclear explosions occurred on Mars in the distant past and these explosions were targeted on sites of previously reported artifacts. Analysis rules out large unstable 'natural nuclear reactors'; instead, data is consistent with mixed fusion-fission explosions. Imagery at the radioactive centers of the explosions shows no craters, consistent with 'airbursts.' Explosions appear correlated with the sites of reported artifacts at Cydonia Mensa and Galaxias Chaos. Analysis of new images from *Odyssey, MRO,* and *Mars Express* orbiters show strong evidence of eroded archeological objects at these sites. Taken together, the data requires that the hypothesis of Mars as the site of an ancient planetary nuclear massacre must now be considered. Fermi's Paradox, the unexpected silence of the stars, may be solved at Mars. Providentially, we are forewarned of this possible aspect of the cosmos. The author therefore advocates that a human mission to Mars is mounted immediately to maximize knowledge of what occurred." (J. E. Brandenburg, "Evidence for a Large, Natural, Paleo-Nuclear Reactor on Mars," NASA, 42nd Lunar and Planetary Science Conference, 2011; J. E. Brandenburg, "Anomalous Nuclear Events on Mars in the Past," Mars Society Meeting, 2014; J. E. Brandenburg, Vincent DiPietro, and Gregory Molenaar, "The Cydonian Hypothesis," *Journal of Scientific Exploration* 5, no. 1 [1991]: 1–25.)

And if those items are intractable, better to turn them into the bogeyman—or in JPL terms, the great galactic Ghoul.[17] Which duly happened; by 2015 Brandenburg had elaborated his theory to state that these explosions were generated by extraterrestrials not living on Mars.[18]

More mind games: Who is copying whom? In Brandenburg's first book of 2011, *Life And Death on Mars,* chapter 7 is titled "The Crystal Palace of Mars." In Myers and Percy's *Two-Thirds: A History of the Galaxy,* published in 1993, chapter 7 of part 3 is called "The Crystal Palace." While *Two-Thirds* does feature Mars, that section of the book is more to do with the ETI's relationship to Earth.

THREE PLUMS

In the light of all this, it would be unsurprising if the Plum Station ritual did not have a nuclear reference buried within it. Alternatively, sticking with Egyptian symbols or gods, if we stay with the mystic rituals beloved of the US space agency and the thrice-great messenger Hermes, there are three nuclear aspects or oddities that went on around that Brian Cox show-and-don't-tell episode of October 2014:

1. This show was first aired by BBC in exactly the same month that Brandenburg's paper hit the media. The paper featured a map of Mars showing two *red balls* at the locations he considered subjected to the effects of nuclear explosions. In his 2014 paper Brandenburg dissects the Cydonian artifacts in detail and satisfies himself that there are anomalous mesas at Galaxias Chaos his preferred location for the other nuclear site to the north of Utopia Elysium. Then came the PSI-Ops component.

    If one wished to provide fodder for others to discredit the ETI connection to Cydonia and indeed to Mars in general, the professional scientist John E. Brandenburg went the right way about it. Notwithstanding any purely scientific objections to his revised theory, his choice of journal was not considered appropriate by his peers. Nor were his own actions the following year. His blatant nonconformance to the ground rules of poster presentation at the 2015 Lunar and Planetary Science Conference was remarked on by his contemporaries.[19] All of which gave his peers another opportunity to chuck baby Cydonia out with Brandenburg's bathwater. The experienced Brandenburg knew perfectly well how to conform to the rules of his profession, but if anyone in the space fraternity had ever paraphrased King Henry II and muttered, apropos of Cydonia, "Who will rid us of this troublesome Face?" then Brandenburg's actions made him a

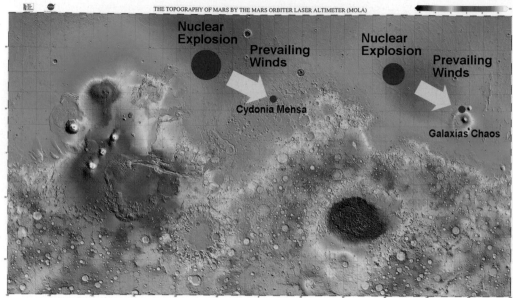

Figure 18.5. *Mars Reconnaissance Orbiter* image published in Brandenburg's 2014 paper in which red balls were used to indicate the sites of two purported nuclear explosions and arrows showed how wind drift could have contaminated the Cydonia Mensa and Galaxias Chaos locations.

good candidate for the job. Although he might not be alone in that regard.

2. In late 2015 and early 2016, Elon Musk, the billionaire founder of SpaceX, ignored all the problems associated with nuclear fallout when he proffered the opinion that nuking the skies of Mars would facilitate the terraforming of the planet for human life much faster than the current estimates.

One might wonder if Musk was not in fact providing support for Brandenburg's theory while pre-empting the issues of any nuclear incidents—accidental or otherwise—that might occur around the red planet in the future.

3. In March 2016, on the same day that progress on the Tokamak nuclear *fusion* build was announced, NASA announced in the media its intention to revive its interest in the use of nuclear *fission* for crewed missions to Mars. To justify its decision, the agency cited the historical use of nuclear fission carried on its spacecraft, but it failed to support the published Apollo record by excluding the Apollo missions from the chronology released to the media. The fact that the Plum Station vacuum chamber had originally been built to test nuclear-powered space hardware

certainly fitted with its role in that gravity 2014 ritual within its walls and Brandenburg's mapping choice of red balls to represent his nuclear explosions fitted with Brian Cox's demonstration and with Elon Musk's plans to alter the Martian atmosphere.

Back at Plum Station, once the ball and feathers had finally settled on the sandbox, Cox and those in the control room were seen expressing satisfaction, and in some cases surprise, that "everything happened at the same time, just as predicted." Did they not think the earlier 1971 Apollo ritual had been performed under vacuum conditions then?

AS ABOVE, NOT QUITE SO BELOW

With all this in mind, it's time to finish off the Hawass saga because it is associated with another space program destined for Mars but having profound connections with matters concerning human space travel, the Egyptian culture, and Hawass himself. We had left him at Giza, where he was organizing the SCA/NGS project to film the new robot in the southern shaft of the Queen's Chamber. It was going to travel up to the blocking stone that Gantenbrink had found and filmed under, around, and through it. When the film was finally screened, we saw a particularly interesting reaction from Hawass, who knew perfectly well what was going to be revealed. The *Moscow Times* recorded a very excited Hawass. "It's another sealed door . . .!" he exclaimed. Thereby completely reversing his dismissive opinion of 1993, when as you will recall, in his opinion this blocking stone wasn't a door at all!

In November 2002, two months after the SCA/NGS event, Hawass's name was inscribed on a CD destined for one of a pair of twin probes launching for Mars in the summer of 2003.[20]

Why? Was this some sort of reward for his volte-face toward the Queen's Chamber blocking stone?

Technicians at JPL say that the preparation of these twin probes was very intense, that they were ordered up in a hurry; and with production taking only three years from the designer's pad in 2000 to the launch pad in 2003, they were operating on two-thirds of the NASA mantra then in vogue: faster, better, cheaper. These probes were indeed engineered faster, were technically better, but alas, were not cheaper than previous efforts. The two Mars Exploration Rovers, or *MER-1* and *MER-2* as they were known during development and build, were considered as the harbingers of human crews exploring Mars. Although President

George W. Bush and NASA were going to wait until the first of these *MER* probes landed safely before making that sort of public statement. After their build and testing, these two probes were identified by the NASA flight-ready designations *MER-A* and *MER-B*. By launch time, *MER-A* and *MER-B* had acquired the additional names of *Spirit* and *Opportunity,* respectively.[21]

TWINS SPEAK

The MER project at JPL fits so well with the SCA/NGS blocking stone project that one might ask if anything in the intervening years since Gantenbrink's initial discovery in 1993 had inspired another silent connection between NASA and the red planet, using local myth as the messenger. The question is legitimate because the myths created by Kubrick and Clarke back in the 1960s were seemingly still alive and well and aligned with the Egyptian mythology, which could be attached to the twins *MER-A/Spirit* and *MER-B/Opportunity.*

Just prior to the 2001 SCA/NGS media tie-up, the probe *Odyssey* had launched, to arrive at Mars the month after the SCA/NGS screening. Was this *2001: A Space Probe Odyssey* referencing Kubrick's final shot of the unborn child traveling naked before creation across space, or was it to be seen as *Odyssey: The Return Home?* The link to this notion lies in the naming of these MER project twins.

The acronym MER, of itself, links to an ancient Egyptian word describing the spiraling light that was seen as a result of spinning up the mind. In other words, meditational practices produced the concept of the Mer to which was attached the concept of the Ka and the Ba. The Ka was the Egyptian concept of pure spirit. The Ba acted as the interface between the pure spirit and the physical body. In other esoteric terms this would be the All that is, the etheric Ka and the astral Ba. The combination of the Mer + Ka + Ba facilitates the ability to shift between varying energy states. It is most often illustrated by two interlocked tetrahedra on the New Age websites discussing such matters. By adopting the acronym MER and attaching the letters A and B to their probes NASA effectively encoded these Egyptian concepts into their new Mars mission. To get to *Ka,* the *Spirit* probe *MER-A* requires the additional letter *K.* To get to *Ba,* the *Opportunity* probe *MER-B*, requires the additional letter *A.* For good measure, the two additional letters which make this coding work also form the word *Ka.* In other terms, the Ka represents the invisible soul and the etheric body; the Ba represents the three-dimensional spirit and the astral body. Some people can see the Ba manifesting as light surrounding the human body. They see various colors depending on the emotions of the person and the circumstances prevalent at the time.

Figure 18.6. The Merkaba, an ancient Egyptian concept of a light that manifests as internal rotation, illustrated as within two interlocked tetrahedra.

Illustration: DSP.

Words are more than mere descriptors of things. It is pretty obvious that these Egyptian concepts illustrate the telepathic interface between states of consciousness and their manifestation in the physical world. And given the arrival of the Sri Yantra in the United States, the Barbury Castle glyph in the United Kingdom, and the triangular forms that keep occurring from out there to down here, it is now absolutely clear that NASA is also using naming techniques to encode messages from itself to Mars.

On the more physical level the tetrahedral forms chosen by NASA for landing probes on Mars, taken together with their understanding of the physics of rotating spheres (the rotation of the sphere around its notional axis produces tetrahedral pressure points within the sphere, which manifest dynamically at specific points on the surface of the sphere) reveals that they are perfectly well aware of the interlocked tetrahedral potential of the Merkaba. And to boot, that it is the combination of human consciousness with the physical that produces the required states of energy transferal. All this encoding of two probes heralding the approach of crewed travel to Mars rather indicates a wish list, as currently crewed spacecraft are not being configured to operate according to the criteria illustrated by the Merkaba. The spirit is willing but the flesh is weak. Perhaps *Opportunity* truly describes the desire to make a link with the stars and ETI in order to proceed forward.

DEFINITIONS

Relative to the layout at Giza, it's worth noting that in the Sumerian and then Akkadian languages, *Kakkabu* meant a star and was depicted as a single star. Over time, this became a horizontal row of three stars (like the three X-axis pyramids south of Menkaure Pyramid). It then changed direction and was expressed vertically (like the three Y-axis pyramids east of the Great Pyramid). Then it was written as three stars in a triangular arrangement, after which this imagery for star (or sun) became written in cuneiform.

The word for meteoric iron was preserved from the Egyptian language through the Coptic word *Ba*. It was called "the iron fallen from the sky," and Tutankhamun's mummified body had a meteoritic iron dagger sheathed in gold placed next to his right thigh (a gesture reminiscent of sacrifice and similar to the later tales of the wounded Fisher King of the Western tradition).

In so naming these two probes, NASA has not only followed the modern understanding of the Merkaba as a spacecraft but also imbued these probes with even more layers of metaphysical information than those already discussed. In ancient times, *Mer + Ka + Ba* could translate on three levels: as the literal vehicle of the gods, the communication between heavenly soul and earthly spirit, or more prosaically, the ancient record of a meteorite impact. Mesopotamian cultures considered astronomical events to be omens and wrote of them as such. These omen texts were composed in two parts: the first contained the record of the event and the second the consequences of that event for the authorities.

Hawass didn't say which of the twin probes, *MER-A/Spirit* or *MER-B/Opportunity*, his name travels with, and many critics would plump for *MER-B*, but as Hawass would eventually hand in his NGS explorer-in-residence badge in 2011—the very same year *Spirit* gave up the ghost on Mars—then *MER-A* would be the more poetic fit.

Hopefully, the twin *MER* probes indicate that NASA is thinking in terms of the more esoteric aspects of human space travel and the necessity of harmonizing the mind with the environment (their dark matter analogy) and protecting its astronauts (their dark energy analogy) through the appropriate spacecraft technology.

Another option is that these twin probes were discreetly messaging the need for help in this area by linking the Mars Exploration Rovers *Spirit* and *Opportunity* with the inspired American writer of Martian mythology Edgar Rice Burroughs. After all, *MER-A/Spirit* conveys the notion of, yes! the spirit of the era. And *MER-B /Opportunity* contains the initials ERB hidden in plain sight.

So in that spirit, and as it is how most of his fans refer to him, from henceforth, I shall often refer to Burroughs by his initials ERB.

It's time to go to Barsoom.

MINDMAPPING MARS

The planning behind the *MER* probes suggests that attention is now being directed toward the mental side of space exploration, so it's appropriate to note that a prime driver for US minds appears to be ERB. His 1917 book *A Princess of Mars* has inspired many a career in science, space exploration, and mapmaking. For those not particularly familiar with his work, it is only relevant to know that NASA is apparently very interested indeed, since the agency keeps landing its probes on or very near to the locations described in his books. Taking ERB's map of Barsoom and referencing it onto NASA's maps of Mars, one finds many interesting parallels with NASA exploration and Martian mythology as told through both *A Princess of Mars* and Myers and Percy's work from 1993.[22]

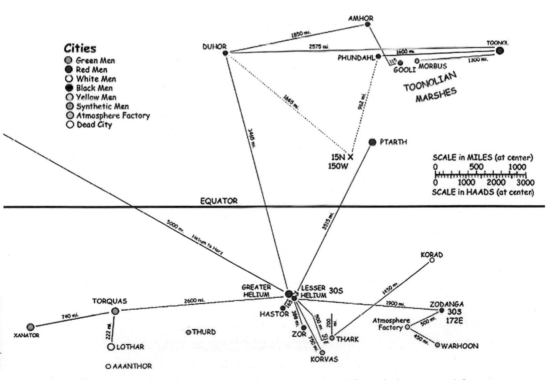

Figure 18.7. The Map of Barsoomian cities, color coded for inhabitants and function, was made by Oberon Zell, an authority on the works of Edgar Rice Burroughs.

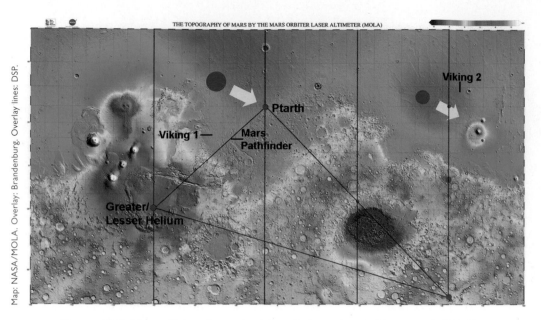

Figure 18.8. Map of Mars from the Mars Orbiter Laser Altimeter (MOLA) overlaid with the locations of Helium and Ptarth.

The Barsoom map clearly reveals that the hub around which everything else emanated for ERB was the twin cities of Greater Helium and Lesser Helium. Burroughs located these at 30° S and 90° W of the Mars meridian used by NASA today.

Looking at the location of ERB's twin cities of Greater Helium and Lesser Helium and at the location of the Cydonian Face complex that features in the *Two-Thirds* narrative, one sees that a planet-wide scale model of the Cydonian special right triangle connects these two works separated by seventy-six years of storytelling.

TWINNING

At this scale, the twin cities of Helium now find themselves as the D&M Tor analog (which is entirely appropriate, relative to the naming of this item in the *Two-Thirds* narrative, where this five-sider is called Sustanator, meaning Gateway of the Sun). The Face at the 90° angle finds Ptarth and the location of the complex at Cydonia, the subject of the Martian segment in *Two-Thirds*. The Barsoom meridian rises from the the 30° location and on this scale mapping, as calculated

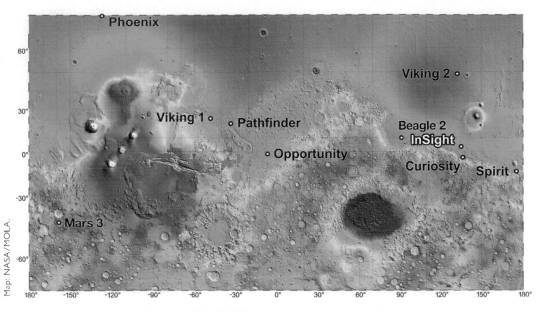

Figure **18.9.** NASA Mars landing sites to date.

by Oberon Zell from ERB's writings, has the city of Exum at the equator and the city of Horz at 40° N.

NASA put *Viking 2* down near to the Barsoom prime meridian, in the region of Brandenburg's nuclear explosion site. Coincidentally, decades later, NASA's *Pathfinder* landed along the coastline from its virtual anagram—a fictitious city called Ptarth. So *Pathfinder* could just as well have been called *Finder of Ptarth.* More accurately, this probe symbolized a return to the source, because it also contained the anagram of the Egyptian creator god, Ptah. In earlier times, Egypt was called Hwt Ka Ptah. Thus *Pathfinder* was linked to the *MER-A/Spirit,* and to the Giza Stargate, represented in space by the signature of Zahi Hawass.

However, NASA might be thinking less of mind and more of matter. The intention may be to twin these *MER* probes in order to twin the cities of Washington and Helium, thereby reinforcing the intention of mining helium in the solar system while learning how to get to Mars. (Fuel is a major issue.) Then the cities of Helium would be an analog of the underwater notional Pentagon, the actual Pentagon falling on the Barsoom cities to the southwest, the area of military training and also the nuclear dumps in the storyline from the Silbury section in *Two-Thirds.*

QUESTION TIME

Q. Did NASA consider the ERB myths when designing the probes' locations?
A. It certainly looks that way.

Q. Was Burroughs inspired by the geometry of the existing Cydonian complex when writing his stories?
A. We have no way of knowing. Burroughs wrote very rapidly, and seeing the concordances with what we do know, it's entirely likely that he was inspired by off-planet consciousness.

Q. Does that mean the smaller fractal of the special right triangle at Cydonia could infuse the atmosphere of Mars with some form of collective consciousness?
A. We are looking at self-similar models of the same geometry at different scales. As we have seen, with no conscious knowledge of Cydonia this model was laid down in Washington, DC, and according to Myers and Percy, there are three other differently scaled models here on Earth. However, all the geometric components are fairly common, so the alternative is that we humans, in laying out anything at all using this geometry, are unwittingly recalling the layout at Cydonia. And a collective consciousness that "went live" when our space probes arrived at Mars could explain both the obsession and the trepidation of NASA and indeed all our space agencies and authorities over all matters concerning Mars.

Q. Was Myers receiving inspiration from ERB's own consciousness via his writings? And when working together with Percy, were they then tapping into another, collective consciousness?
A. Myers received all the material for *Two-Thirds* intuitively, and together with Percy using both creativity and logic, they put that book together within a year. Myers being American had better awareness of ERB as an author than did David Percy who was not at all familiar with his books. At the time of writing *Two-Thirds,* the red planet's large-scale model of the anomalous Cydonian complex had not been seen by Myers and Percy. Indeed, the "Tower of Ancestors" in the center of "the Cydonian city" is described as representing one of the Helium cities, and the other would have been on the Face mesa, but it was never built. The Barsoom/Silbury city marker is off by about 69.28° of latitude. The Barsoom prime meridian is also off the Silbury mound meridian by about 56.45° of longitude. These angles have meaning as described in *Two-Thirds,* yet as already stated, in naming an area on Mars the Horz complex, Myers and Percy had not made the con-

nection with any detailed mapping of Barsoom. So one cannot accuse these two modern authors of any sort of set-up. That said, they could not have generated the complicated backstory and written their book within a single year without inspired intuitive thinking. Their work contains so much they consciously knew nothing about, matters that anticipated the discussions taking place in science today. And more importantly, it conveys useful and probably essential knowledge regarding deep-space exploration. Which should get us all thinking as to how exactly we do access our thoughts.

THE CYDONIA MODELS

Mars seems to be operating like a red rag to a bull. The very facts that the works of Burroughs and of Myers and Percy reflect back to each other; that NASA finds it important to match its probes in name and action to forms of storytelling past and present; that scientists at NASA and at SRI are involved with psychics, remote viewers, and people of exceptional talents such as Uri Geller; that Cayce's ARE is still desperately seeking the power source of Atlantis and scrabbling around the archaeological sites of our world looking for clues—all of this could be the results of consciousness working across time and with different perceptions, depending on the intentions of the participants. Any subconscious mental and spiritual affinities with Mars past and future make it easier to imagine how to do things, but at the same time, imagination, even when expressed as film, music, or art, cannot replace the very real, practical efforts involved in actually getting human beings to Mars.

The Sound of Light

Getting to Grips with the Transdimensional GPS

The mapmaker Carl Munck used the geographic coordinates of degrees, minutes, and seconds of space and time (latitude and longitude) to produce his grid point vector. He was focused on the location of the specific monument or site that he was working on. In this regard, there are two observations to be made. Although his final result visually resembles the decimal degree system used by global positioning systems and web mapping applications—even if his insistence on an accuracy of eight figures after the decimal point exceeds that of the six figures considered sufficient for the highly specialized surveying of such matters as tectonic plates—it's still important to remember that the decimal degree system and Munck's matrix system are fundamentally different in their outcome.

The decimal degree system maintains a separate identity for both latitude and longitude. The multiplication process adopted by Munck has effectively converted arc degrees and minutes to arc seconds. Munck has then merged space and time into one item through the division of the smaller number into the larger, theoretically reducing both space and time to cycles per second—of something.

THE BI-CYCLE MADE FOR TWO

Parallel to Munck's world, cycle per second units (CPS) are now defined as hertz (Hz). Used in relation to both sound and light, a hertz is a unit of frequency of change in state or cycle, in AC electrical current, a sound wave, or other cyclical waveform. And for those who want to know, 0 Hz is assigned to DC electrical current.

THE ENGINE ROOM

Bearing all this in mind, and returning to Munck's prime meridian, many of the architects, engineers, and people from other professions who have examined the Great Pyramid suggest a monument that functions as a machine. Munck, holding it as the center of the ancient-built matrix for this planet, seems to fall somewhere in between these techno types and those of the persuasion of Myers and Percy. These authors consider the Great Pyramid to be a nonfunctioning model of the computational interface between Earth, space, and the senses of self-aware beings.

The amalgamated items seen in the figure below are *metaphorical representations* of energy transfer between the Earth and the self-aware being. This particular cathedral (built between 1220 and 1258), with its two transepts, closely resembles the patriarchal cross of the Knights Templar. That cross was associated originally with the Patriarch of Jerusalem, and according to traditional symbolism, its double

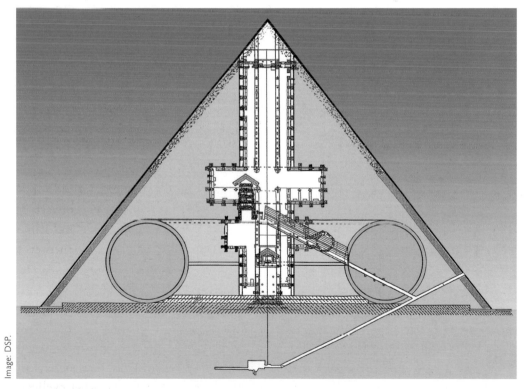

Image: DSP.

Figure 19.1. The Great Pyramid cross-section viewed from the east, incorporating a torus cross-section and the top view of the architectural floor plan of Salisbury Cathedral in Wiltshire, England.

beams incorporate several concepts of importance: The long beam (in this diagram encompassing the four chambers above the King's Chamber) represents the opening of the self to the resurrection, the sacred aspects of life. The short beam (in this diagram just above the Queen's Chamber) represents the limitation of the self when living a purely secular life. And set out in this two-dimensional illustration of four-dimensional concepts, it can be seen that the secular beam occupies a portion of the torus.

Although it is exceedingly large by our standards, Myers and Percy assert it is built to one-third scale: a functioning planetary computer of its type would therefore be some 1,440 feet in height. Although that must surely remind the reader of the discussion on gematria and the long-standing traditional relationship of the number 144 to notions of light, that piece of information also infers metaphors and requires analogies between various measurements that we on this planet, with our logical brains, tend to keep as separate.

All such multisymbolic illustrations are the equivalent of Tibetan thangkas or mandalas, designed to lead to great depths of contemplation. As such, each observer will have a different take according to her or his own knowledge, and depending on which scale is being observed: the self, the planet, the solar system, the galaxy, the universe, infinity. Everything incorporates aspects of the others. Gnosis, often translated as "knowledge," is better understood as the direct experience of the interactions of the sacred with the secular. And it is the focus of attention that defines what an individual considers important at any one time. Here it is interesting that the notional torus implied in the base of the Great Pyramid has the same diameter as the base of Silbury Hill in Avebury.[1] Also, when a 2-D side cut of Silbury Hill is inserted into this metaphorical illustration, the Queen's Chamber, with its five-stepped niche, dominates the shorter, secular transept and the median central column defined by the nave of the Cathedral.

The connection between two locations with powerful terrestrial energetic properties is also established given that Silbury Hill has a slope angle of 30°, which is virtually the latitude at which Giza finds itself today. The connection between two countries with powerful astronomical monuments is also established given that the shafts of the King's Chamber exit the Great Pyramid with a slope angle of 51.833333° and Stonehenge finds itself at latitude 51.1789°, which difference of number encodes the 80/81 ratio between the mass of the Earth and the Moon to within 0.03 percent.

If the Giza Great Pyramid marks the Earth-equinox and Stonehenge marks the solstice, then Silbury Hill, as represented by the torus, also highlights the Earth-Moon aspects, and these too are represented at Stonehenge by its four so-called station stones. Out at Cydonia, the analog and anomalous cone mound has a ditch set at an angle of 27.2° azimuth from its center; the orbital period of our Moon around our planet (99.96 percent accuracy), and this number is also the megalithic yard times ten. Here it is worth mentioning that the obsolete French word *mone* and the German word *mond* mean Moon and that the early origin of *mound* is unknown, but the roots it does have and the look of Silbury Hill in Wiltshire suggests that it too has lunar associations. Extending the Mars mound's ditch toward the northeast sees it tangent the southeast rim of the large crater. This crater's Avebury analog has the same 27.2 lunar sidereal cycle marked by the number of stones originally forming the small stone circle in the northeast quadrant. Another small stone circle marked the lunar synodic cycle of 29.5 days in the southeast quadrant and again at Stonehenge (which has twenty-nine full-height stones and one shorter stone in the outer sarsen ring). At Stonehenge, just inside the Earth ditches, the Aubrey circle of fifty-six cylindrical holes marks the eclipse cycle, while the four mounds interacting with this Aubrey circle mark the Moon's extreme positions *and* form a perfect rectangle with sides in the ratios of 12:5 and a diagonal of 13. The latitude of Stonehenge is unique for encoding this data, since any deviation would not permit the formation of a perfectly engineered rectangle. And the numbers 12 and 13 recall the twelve solar and thirteen lunar months. Astroarchaeologist and author Robin Heath has written several books on the data encoded at Stonehenge and noted that when taking in the location of the bluestones, the perfect rectangle of Stonehenge is repeated—just at a much larger scale.[2] He also addresses the issue of measure and has written a comprehensive study of Alexander Thom, who established that a yard of 2.72 feet was used in setting out the stone circles of the British Isles and Brittany. He called it the megalithic yard.[3]

ONE SMALL STEP

According to Heath, all ancient cultures applied specific ratios to the foot in order to establish the cubit measures used for measuring out their architectural works. He asserts that while different cultures used a slightly different measure for the length of the basic foot, the principle was the same across all cultures. For example, changing up the selected foot measure by 3/2 produces the culture's "common cubit." Changing that result up by 8/7 produces the culture's "royal cubit." Changing that result up by 6/5 produces the culture's "sacred cubit."

From which it is clear to me, anyway, that these cubits were used to build constructions designed to enhance the abilities and actions of those who operated within their walls. My research has led me to conclude thus far, that the common cubit was used for everyone's living spaces indoors and out, and especially the daytime quarters. The royal cubit was used for the palace and its administration buildings. The sacred cubit for all the buildings associated with the astronomer-priests and sleeping quarters.

There were many variants of the foot throughout countries of the ancient world, but suffice it to say here that experts in the history of measurement insist that whatever the size of the foot selected by the culture, the reference foot was of twelve inches.[4] That suggests to me that the notion of a culture's local prime meridian was truly based on the Sun at the zenith over the selected location. At the Great Pyramid, this occurs at the March 21 equinox. And again, its position within the Northern Hemisphere, two-thirds the distance from the North Pole, emphasizes the connection of this monument, built with the royal and sacred cubit within its structure but sitting on the natural rock, common to the land of Egypt.

Another standard across the ancient world was the means by which the foot progressed from one to four. Taking the reference foot of twelve inches as the example:

Two feet—a step of twenty-four inches
Three feet—a yard of thirty-six inches
Four feet—a pace of forty-eight inches.

(As an aside here, it's of interest that the Apollo 11 scenario of "First Man on the Moon" made a great deal of a fuss over the imaging of footprints and speaking of small steps and giant leaps, or paces.)

These progressions again might better refer to architecture because as with the variant on the foot, so there is a variant on the step, depending on the height of a human being. According to Heath, the stone circle builders had a step more in synch with the megalithic yard of 2.72 feet (32.64 inches). Which would put them nominally in a harmonic relationship with the lunar sidereal cycle. Looking at this progression of the foot and thinking of the great step at the top of the Grand Gallery leading to the King's antechamber, replete with both sacred and cubit measures, reminded me of the Sphinx and its riddle. In that version of measure, one finds children related to the pace, old age (and Google's little green men) to the yard, and adults to the step. That led to further musings about noon, the Giza plateau, and the Egyptian practice of adding five extra days to the year. This was done to compensate for the difference between their year of twelve months of thirty days each, which was about five days short of the astronomical year. In my musings, if the Great Pyramid was marking the spring equinox at noon, the Sphinx was marking the beginning/ending of the yearly circle, and human observations of the disparate cycles of the Sun and the Moon in orbit around the Earth were recorded in these five epagomenal days. As was the actual angle of 5.14°, the Moon's orbital insertion angle relative to the ecliptic. These five extra dates were attributed to five Egyptian Gods, in this order: Osiris, Horus the elder, Set, Isis, Nepthys.

TIMINGS AND DATES

Suddenly, I was back in the twentieth century, for the original dates given for these Egyptian extra days were July 14, 15, 16, 17, and 18—and here's where NASA and crop glyphs converge. The Barbury Castle tetrahedron glyph, remember, was found on July 17, which links it to Isis, and the launch of Apollo 11 on July 16 links that event to Set (and perhaps seeking). In fact the timing of the Apollo landings in many instances were designed to incorporate the rising of Orion/Osiris over the designated landing site.[5]

Finding that the birthday of Set, considered the enemy of Osiris, was the launch date for Apollo 11 is therefore remarkable. The space agency was apparently less than friendly towards the great "out there" than otherwise declared. And the sense that the US group SETI was more of a lookout than anything else also came back into my mind. As did that bit of landscaping at the Axe Majeur park, because there are more esoteric links between these two republics when noting that the French national holiday Bastille Day, on July 14, also marks the birthday of Osiris. We can now be more specific about that significant tetrahedral crop glyph that arrived in Wiltshire in the early hours of the July 17. Aset (Isis) made herself the enemy of Set, being the woman trying to repair the circle of life and continue on with it through the birth of Horus the Younger. Contemplating the myths of our ancestors and the monumental installations they left behind must be helpful in some way. Surely this is also the reason that NASA functionaries and associates are crawling all over Giza. The results of such contemplations must have a practical outcome otherwise the esoteric echelons of the space agency would not be indulging in these name games and timing rituals. I then stopped those musings and came back to the discussion of measurement, which is relevant because it is how we human beings relate to our environment on levels of which many are not consciously aware. However, subconsciously, we are all aware of our connections to both the planet we live on and the night skies above.

I then noted that even as the Great Pyramid is built with very small passageways and then very large corridors, so have human beings managed to evolve long-scale and short-scale measurements when working in a ten-based system. Some cultures call a million a billion and vice versa. Notably, the United States and the

United Kingdom diverged on this matter until relatively recently. The solution that investment bankers have found for this trading problem (getting mixed up would be a costly mistake) was interesting: the term *yard* has been adopted to define a billion of whatever currency they are trading. Which made me think about time and energy in relation to those other two ancient yard measurements we find alongside the yard of thirty-six inches.

As we know that three feet (of whatever size) make the yard, we can work out other measurements of these two ancient measures found at Avebury, Stonehenge, and Teotihuacán.

MEASUREMENTS

The megalithic yard of 2.72 feet, already discussed as memorializing the lunar orbital period and its own axial rotation period, is made up of feet that are 90.66 percent of the standard 12 inches. That gives a foot of 10.88 inches.

The Hunab found at Teotihuacán has a yard of 3.47571113 feet, which gives it a foot of 1.04919, which is 13.9 inches relative to the standard 12-inch foot. And that makes the standard 12-inch foot 86.33 percent of the Hunab. As already noted several times, this same width is found at the lintels connecting the sarsen ring at Stonehenge. Taking these two exceptional feet measures produces an average of 12.39 inches, and that's the ancient Greek foot with the long toe, while the megalithic foot is that of the Greeks at the time of the Trojan wars, and the Teotihuacán foot of 13.9 inches echoes the 13.5-inch value of the basic unit of ancient China, the Ch'ih. Thus another link between the layouts of Teotihuacán and Beijing is hiding in plain sight.

It was also a usual practice in many ancient cultures to consider a fraction, when inverted, to represent a musical tone. For example the proportion 2/3 becomes the musical tone 3/2. Remembering Heath's finding that changing up the given foot by 3/2 produced the common cubit, interestingly inverted at 2/3 that is where the solar system is to be found relative to the center of the galaxy, and as already noted, it is also where the Great Pyramid finds itself in relation to the North Pole: two-thirds of the way out from the center of the relevant system. Since Munck stipulates the twelve-inch foot as the base measure of the Great Pyramid, and its links to noon along with the Masonic partiality to pyramids and the symbolism attached to the letter *G,* which of course, is 3/2 in the diatonic musical scale, this inversion principle seemingly has legs. From that, one could surmise that this common cubit musical fifth inverted as 2/3 relates to all connections to do with the land, or environment. From that we could extrapolate that changing up from the common to the royal cubit inverted as 7/8 related to matters of the secular and daytime activities of humans, and changing up the royal cubit to the sacred cubit inverted as 5/6 concerned matters of the soul and the nighttime activities of human beings, including their dreamtimes.

Given this discussion on mind matters, let's go a step further and consider that, essentially, the notions inherent within this ancient understanding and that have been encoded via Heath's system of changing up the cubits from one to another, can be picked up by a human, and operating within a location or a monument built according to these cubits and then applying the appropriate musical tone, or a harmonic of that tone, would enhance any encoded data inherent within that place. Musically, from a fundamental tone, that common cubit would be 3/2—the perfect fifth.[6] The royal cubit, 8/7, is a flattened seventh, and the sacred cubit, 6/5, is a flattened third. In terms of the C major diatonic scale, this gives the tones of G, B♭, and E♭, respectively. The fraction of 6/5 used to get to the sacred cubit recalls the master numbers beloved of sacred geometers such as Michael Glickman. A study published in October 2020 has found that the Stonehenge sarsen ring not only contains the sound within their circumference but also that the sarsen circle stops sound from entering the circle and the tone found across the whole monument are octaves of B major. This whole subject is a work in progress and will be pursued on Aulis.com/pathway. Even if the connection of the land and the environment to the brain via music and/or measure had not been made, this step process might well have been the original reason for placing important buildings on significant natural environments such as that rock outcrop that was left in situ at Giza, ostensibly causing some inconvenience to the builders of the Great Pyramid. Clearly, the engineering of buildings and their environments takes on an interesting perspective when such matters as music theory (even at its very basic), the human brain, and the simple matter of pacing things out are taken into consideration.

The fraction 3/2 that is used to arrive at the common cubit is found in the musical circle of fifths. This is a tone circle created by proceeding in intervals of fifths clockwise (creating intervals of fourths counterclockwise, and we are back to the symbolism of the numbers 5 and 4) along with geoglyphs of Peru, the Nazca Monkey and Manos geoglyphs with their right paw/hand of four digits and their left paw/hand of five digits.[7]

The ancient Chinese originally tuned their stringed instrument, the qin, in fifths over five strings. So the tuning was done as C, G, D, A, and E. However, the strings were strung in sequential order: C, D, E, G, and A. (Much later, two other strings were added, a C octave and a D octave, and these were strung after the

A string.) These different fifths were used for rituals connected to the four seasons: The fundamental C (1/1) represented the emperor, the central authority responsible for maintaining the kingdom in harmony. The G (3/2) was used for the spring season, D (9/8) for the autumn, A (5/3) for the winter, and E (5/4) for the summer.[8] So again these harmonics are reflected within the Great Pyramid. The suspended floor of the King's Chamber is two-thirds of the way from the *flattened* top of the Great Pyramid. Then again it is one-third from the base of the pyramid, so playing with inversion, measure, fractions and percentages, and musical scaling might strip away other layers of information worth evaluating. As might the connotations of the Great Step when considered as symbolic of these basic measurement principles set out by Heath. Therefore, not all the processes referenced within the engineering of this building are necessarily three-dimensional as such. Wherever a mix of these measurements is to be found within the same monument, looking at such areas with the understanding of their ancient applications yields different insights.

Thinking of the Great Pyramid in these terms, the Egyptian royal cubit is said to be prevalent throughout the building, and given that the Egyptian sacred cubit, as noted, is found in the antechamber to the King's Chamber, the whole edifice turns into an architecture of being. We could also apply these principles to engineering projects. Translating the word *analog* as meaning "old technology" relative to digital technology, we can then note that infinitely faster computing is currently being researched and that quantum computers are now the next big step in computing. Are those *cubits* also related *Q-bits,* the abbreviation of quantum bits? Especially reflecting on the notion that the Great Pyramid is a modeling in stone of a planetary computer? And do we have here a different, transdimensional modeling of body, mind, and soul? Of course, for anyone who doesn't consider human bodies to have dimensions other than those relating to the physical body, all of the preceeding paragraph might be considered nonsense. But the fact remains that the measurements of the human body and the overall myths and traditions of all cultures support the above data.

If the Great Pyramid's catalogue of measures is designed to remind all of us of home, then in building a spacecraft to take us deep into the solar system, we could do worse than to incorporate the harmonic principles on which the structure is based. Further, if this Great Pyramid is a one-third scale model of one of three computers located within a Martian mesa designed as a Face, then one might take the information in *Two-Thirds* a step further and ascribe mental functions to each of the three large pyramids on the Giza plateau. Note that the nearest pyramid (Khafre) is 98.10 percent the height of the Great Pyramid while the farther pyramid (Menkaure) is 96.07 percent the height of Khafre. In

Two-Thirds, these percentages are linked to states of consciousness. Which makes the causeway linking the nearer Giza pyramid to the Sphinx particularly thought provoking, since it bears the name of the Khephri or Khafre. The story goes that Thutmose, an Egyptian prince of the Eighteenth Dynasty, resting between the paws of the Sphinx at noon, received in a "dream" the instructions to clear the Sphinx enclosure of sand, and in exchange for this arduous work, he would be made king. This he did, and Thutmose is credited with building the pyramid that bears the Sphinx's name.

Where does myth end and reality begin? When this dream took possession of Thutmose, it is said that he found the majestic Sphinx addressing him from his own mouth, saying, "Look at me, observe me, I am your father, Horemakhet-Kephri-Ra-Atum." For those who have read *The Only Planet of Choice,* this brings to mind a couple of conversations from those transceiving sessions with Phyllis Schlemmer. The first, in 1975, took place between the ETI TOM and Gene Roddenberry. Asked for the name of the entity who was speaking via Schlemmer, TOM replied that he was the spokesman for the Council of Nine, that in truth he was Tehuti. He added that he was also Hamarkos and Herenkar, while he was known as Thomas and also as Atum. Later, in a 1977 session, Andrija Puharich expressed particular interest in the name Hamarkos. To which inquiry TOM jumped in and said "I am the day and I am the evening and I am the mid-noon." Puharich then inquired as to how the Sphinx had come to be built and named after him (an ETI), to which TOM replied that Puharich had found the secret, but the true knowledge would be given at another time.[9] TOM also stated that the Sphinx had more to do with the founding of Egyptian culture than did the Great Pyramid. Whether physically present or existing as metaphors designed to get us thinking better and faster, there is clearly a link between the ancient and the new—out there on Mars and down here on Earth. Mind, body, spirit.

Of course, given the reduction of the Great Pyramid's coordinates to CPS by Munck, we could entertain the idea that even the coordinate system responds to this notion of transdimensional exchanges mediated by computational equipment operating at speeds unknown by present-day computer scientists. Which would mean that in reducing coordinates to one single figure (essentially arc seconds even if not considered as such), one is not only providing satellite managers with simple number strings but also locking in the essential resonance of the location in more than physical terms. Suppose our minds and our spirits are capable of operating at faster speeds than the established delta, theta, alpha, beta, and gamma hertz measurements? And that through contemplation of the data encoded in the size and geometric shape of, for example, a four-sided pyramid,

Figure 19.2. The Great Pyramid viewed from the north.

we can access other places, times, and information. *Two-Thirds* talks much of the states of change in the speeds of light *c,* and as already noted, NASA has findings that indicate that the speed of light is not a constant throughout the universe. However, it might well be that human beings can resonate with any such changes of light speed within the solar system, the galaxy, or even the universe simply by using focused meditation. Using human brains more effectively could make it possible to tap into the extra energy of these faster light speeds

A hint that this is a direction to take was given to Andrija Puharich in the 1950s when he was exploring human consciousness. During a channeled communication session by the Hindu scholar and mystic D. G. Vinod on New Year's Eve 1952, Puharich was essentially told that the methodology relating to PSI research undertaken by Puharich and his colleagues was not up to par and that the whole subject would need to be revised.[10] Vinod also emphasized and reiterated the connections between the number 7 and the output of the human being in electrical terms.*

---

*In 1974, Puharich published a full version of this speech in his book *Uri*. But by 1977, when *Briefing for the Landing on Planet Earth* was published, this speech had been judiciously edited, and the reference to Einstein and a form of the Lorentz-Einstein transformation equation was missing, along with a lot of very relevant text. One wonders why. In its full version, Vinod's words are profound and enlightening. It wasn't until 1998 that the full speech was reprinted.

At the time, no one (publicly at least) truly understood what Vinod was saying. Four decades later, when editing *Two-Thirds,* I saw that Vinod's statement had some sort of link to the means by which the light speed *c* between galaxies is calculated, which had been arrived at using data received and decoded from two crop glyphs. Myers and Percy, like everyone else, had no awareness of and therefore no understanding of that Vinod statement.[11]

Writing this book and considering the number 7 from the point of view of the human chakra system, I wondered if an access point to the local speed of light *c* mentally was via the seventh chakra, the crown chakra. Or certainly via *a crown* of some sort. Observing another "Brahman experience" given to Puharich by Vinod,[12] this speculation takes us back to the manifestation of that Sri Yantra furrowed into the ground in Oregon that consisted of five-four triangles summing to nine. I recalled the ETI TOM telling Puharich that the connection (but not the location) of the ninth chakra was at the top of the head.[13]

However, as brain waves are measured in electrical terms as CPS or Hz, can we take the information in *Two-Thirds* that says that apart from the speed of light within our solar system, there are two other distinct light speeds, interstellar and intergalactic, and do something with it? (Currently, it is considered that the solar system is within a galaxy that in turn is part of a cluster of galaxies and that this cluster is itself part of a supercluster of galaxies.) Taking this model to the astral and etheric components of our being, could we humans use our minds to connect to all other solar systems within the galaxy and possibly to all other galaxies?

This realization is no more fanciful than those RV exercises practiced by the US military—albeit on a larger scale—and it might be a better way of achieving the unspoken desire of the agencies and the military; namely, to have us all think seriously about ETI. At the very least, contemplating the ramifications of such a model should make for a good meditation, and as everything would seem to work and operate the same but at different scales, this is where we get back to the Great Pyramid, that musical tone of 3/2 (G), and its analogy to an advanced supercomputer.

Present-day computer server rooms use power frequencies as high as 400 Hz, and for technical reasons, 400 Hz systems are usually confined to a building or to vehicles such as spacecraft, submarines, and aircraft. Acoustically, 400 Hz resonates with the musical tone G4 (the fifth tone above middle C on the piano.) Relating this to the electrical current is interesting because the current used in the United Kingdom, 50 Hz, is octaves lower, at G^1.

Without going too deeply into the fascinating but arcane world of musical mathematics, it is important to note the difference between the terms *harmonic* and *overtone.* The first is preferred by physicists, the second by musicians. Sound engineers use both. There are sixteen harmonics and fifteen overtones, and this is why. In acoustics, the basic vibration of a tone is called the fundamental and is considered as the first harmonic in the numbering sequence. The term *harmonic* means "a multiple of a fundamental frequency." The sounding of the fundamental produces fifteen other frequencies, called partials by physicists and overtones by musicians. As four of these overtones are resonant with the harmonic doublings of the fundamental (the octaves), only eleven make different musical ratios with the chosen fundamental. Sequentially, the fifteen overtones are: octave, octave, perfect fifth, octave, major third, perfect fifth, minor seventh, octave, major second, major third, augmented fourth, perfect fifth, major sixth, minor seventh, and major seventh. And this overtone sequence is applicable to whatever tone is acting as the fundamental.

It is of interest here that the map of Cydonia, as observed by Hoagland and his NASA counterparts, has a west-to-east sightline that marks the partial/overtone links. The western edge of the City analogs to the fundamental, followed by the City center as the second partial, the eastern edge of the City as the fourth partial, the Face mesa as the eighth partial, and the Wall by the crater as the sixteenth partial.

Gerald Hawkins, the astronomer and author of *Stonehenge Decoded,* noticed that the crop circles published in *Circular Evidence* by Delgado and Andrews measured as diatonic intervals belonging to, as he saw it, the scale of C major

Cydonia City West		Centre	East		Face mesa						Wall by the Crater
Partials: Fundamental		2nd	4th		8th						16th
Overtones	1st	2nd 3rd	4th 5th 6th 7th		8th 9th 10th 11th 12th 13th 14th 15th						
	C	C G	C E G B♭		C D E F♯ G A B♭ B						

Chart: DSP.

Figure 19.3. The sixteen partials and their relevance to the west–east sightline at Cydonia. Here illustrated using the diatonic scale of C major. Diatonic musical scales consist of five whole tones and two semitones. As an example on the piano keyboard, C major and A minor are both diatonic. When C is the diatonic scale as illustrated here, the fundamental C^0 would be located at the City west. From there to the City center would be the first octave. From the City center to the eastern edge of the City is the second octave. From the eastern edge of the City to the Face would be the third octave. From the Face to the Wall would be the fourth octave, and the fifth octave starts at the Wall and extends eastward, but in terms of musical partials and overtones only the first 15–16 positions are relevant.*

*If the scale were A minor diatonic, the sequence of tones would be: Fundamental A, then A, E, A, E, C, G♭, A, B, C, D♯, E, F, G♭, G, A. For lateral thinkers remembering that the 24 letter Greek alphabet preceded the 26 letter English A-zed as we know it today and before that, all the ancient alphabets of the Middle East were based on the symbol of an Ox head representing the letter we now know as A.

(although the scale A minor is also diatonic, it should be noted). And as Hawkins told author Freddy Silva, sixteen out of the twenty-five circles in *Circular Evidence* had these fractions, and according to Hawkins the chances were 1:400,000 against that being accidental.[14]

The layout, apparently copied over at the Axe Majeur park near Paris, is noteworthy in that among the many associations that can be made with these doublings of the fundamental frequency seen in figure 19.3, from 2 to 4 to 8 to 16, the number 248 is referenced by Munck in relation to the Great Pyramid and by Myers and Percy with regard to a transdimensional sphere also linked to that Giza location.

Further, considering the illustration in Figure 19.3, the AC current of 50 Hz as the tone of G would be in harmonic resonance with the 3rd, 6th, and 12th overtones while making different tonal relationships with the other overtones.

If an electrical current of 50 Hz is found to have some relevance to the Great Pyramid, what about the other common frequency for alternating electric current, 60 Hz? This electrical cycle produces the tone of B^1. And this overtone will find resonance with the 15th partial. On a stand-alone basis, B^1 as the fundamental (that is, the scale of B major) is the principal scale within the acoustics of the Stonehenge sarsen ring at B^{-2}, 7.83 Hz. This scale and frequency is also the tonality of the base atmospheric electromagnetic resonant frequency. In musical terms it is the fundamental of the Schumann resonances.

THE SCHUMANN RESONANCES

Electromagnetic resonances generated by lightning discharges in the cavity between the Earth's surface and the ionosphere form a set of spectrum peaks in the extremely low frequency region of our electromagnetic spectrum.

This shell of electrons and electrically charged atoms and molecules begins some thirty miles (forty-eight kilometers) above the Earth's surface with a depth of some 620 miles (one thousand kilometers). It is most significantly distinct at heights above fifty miles (eighty kilometers). The positively charged ionosphere is codified in layers described as the D, E, F1, and F2 layers, with the D layer nearest to the negatively charged Earth's surface.

The relationship between the negatively charged Earth and the positively charged ionosphere produces the same effect as a ball condenser or capacitor. (Two electrically charged balls placed the one within the other are called ball condensers and are also known as capacitors.) Noting that, there is here perhaps a very important analogy. Electrical engineer Colin Andrews found

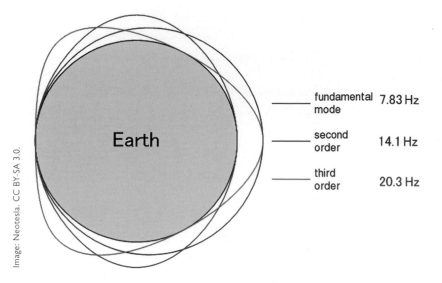

fundamental mode	7.83 Hz
second order	14.1 Hz
third order	20.3 Hz

Figure 19.4. The diagram shows the Schumann fundamental tone and the next two frequencies in the set. There are two further frequencies of 27.3 Hz and 33.8 Hz.

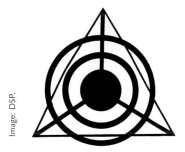

Figure 19.5. Central area of the Barbury Castle crop glyph, 1991.

magnetic anomalies within early crop glyphs that progressed from single circles to two concentric circles—dubbed "double ringers"—these would later be incorporated into crop glyphs with triangular and tetrahedral forms.

Research into the seed heads taken from these flattened areas of grain found specific variations in the subsequent growth rates relative to both the age of the crop at the time when it was laid down (flattened) and the distance from the glyph's center. The American crop circle research team BLT (named after the principal investigators John Burke, William Levengood, and Nancy Talbott) who carried out this work informed the US space agency of their findings. Whether this fact was ever acted on is unknown, but their findings surely have major implications for growing food aboard spacecraft.[15]

A LITTLE LIGHT MUSIC

When it comes to the analog of Earth attributed to the Great Pyramid, the engineers specializing in the realm of acoustic research who have investigated the monument were astonished by their findings. The specific resonances of the King's Chamber and its coffer were tested by the British acoustic engineer John Stuart Reid.

Reid registered an F♯ and a B natural. This F♯ would resonate with the 5th and 12th overtones of the Schumann resonance's B major scale, and that B would resonate with the fundamental of the Schumann resonance and its 1st, 4th, 8th, and 16th overtones. Taken along the C diatonic line of Cydonia, this augmented fourth (F♯) resonates with the 12th partial and finds the flat-topped Bastion mesa, which is the analog location of the Giza plateau.

The full technical account of Reid's research is well worth reading, as is author Alan Alford's article on this matter of resonance within the Great Pyramid.[16] Alford noted that of all the chambers in the pyramid, only the King's Chamber and its antechamber have been affected by what looks like cracking from subsidence damage. The semitone difference between the F♯ that Reid found and G (that 3/2 tone symbolic of the spring equinox mentioned earlier) can be due to many things. The offset between the center line of the pyramid and the King's Chamber and also the state of the building today, which was severely banged about inside and out by earlier seekers after treasures and further deprived of its outer casing stones by scavenging builders, are but three suggestions that might account for the slight musical tension.

While he was carrying out his experiments, Reid also experienced some healing to his back pain, and he later went on to deepen his research into the benefits of sound, how it affects matter, and how best to use the knowledge gained by such experimentation. The physicist Brian Josephson, noted for his work on superconductivity and quantum tunneling, acknowledges the importance of Reid's work, and his Cymatics website is well worth a visit. When in the King's Chamber, before Reid started experimenting he first rectified the "broken corner" of the coffer so that he could stretch a membrane tightly across it. He then scattered sand across this taut membrane and performed his acoustic experiments. These produced visual effects: when certain tones were played, Egyptian hieroglyphs formed in the sand.

The fact that sound can produce geometric forms is well attested, and examples of such cymatics can be seen online, with the patterns changing as the tone is altered. This is not quite the same thing as was found by Reid. While there were

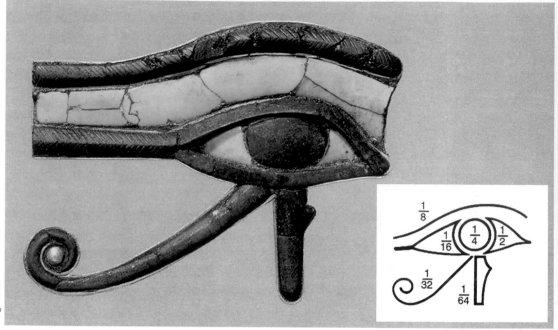

Image: DSP.

Figure **19.6**. The Eye of Horus.

some geometric responses as per cymatics, the formations he obtained were not always geometric; some were pure imagery, such as the Eye of Horus.

Note that the Eye of Horus consist of the fractions of the whole eye, which is considered as the whole, or the number 1. So we have 1/2, 1/4, 1/8, 1/16, 1/32, and 1/64. These relate to the parts of the eye as in the diagram above. Inversed we get 1/1, 2/1, 4/1, 8/1, 16/1, 32/1, and 64/1. And just like those partials, the iris of Horus has the number 4, and the fourth octave of 4/1 is located on the eastern side of the Face; it is the Lion's eye.

Egyptologists state that the eye of Horus was also used to measure quantity, and especially that of the measure of grain; one has to think of crop glyphs again, the discussion on grain size relative to the scouring of Lake Geneva by fast water movement, of the old notion of the Great Pyramid representing a grain storage facility, and that hole in the wheat fields of Echallens north of Lake Geneva. Given Reid's use of sand as his acoustic palette, it is of interest that not only did he get an image of the eye of Horus but he also got an image recalling the heel-shaped symbol used exclusively for the measurement of grain. Which takes us to Kenneth Arnold and his 1947 UFO sighting. And surely, that Star Trek insignia adopted recently by Space Command.

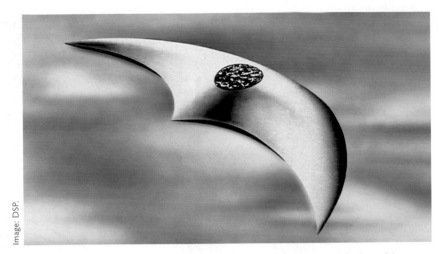

Figure **19.7.** A re-creation of the craft seen by Kenneth Arnold around Mount Rainier, Washington State, in June 1947.

Taking this further, if the Egyptians used such fractions for measuring grain, among other things, the eyeball is in fact a spherical object, and from there one can apply these fractions to portions of the whole eyeball, representing an astronomical 360°, the number of degrees in the circumference of any circle.

1/2 = 180°: The inner visible eye is now linked with a half circle or meridian—time.

1/4 = 90°: The iris and pupil are now associated with a quarter of a circle.

1/8 = 45°: The brow ridge is now associated with the diagonal of a square, or indeed the square root of two.

1/16 = 22.5°: The outer corner of the eye is now associated with the ecliptic boundary of the Sun's journey around the planet.

1/32 = 11.25°: The spiral mark is now associated with the magnetic equator.

1/64 = 05.625°: The hawk-like marking is now associated with the angle of insertion of the Moon relative to the ecliptic.

THE MOUND OF MUSIC

John Stuart Reid was not the only sound engineer to experiment in the Great Pyramid, but Thomas Danley, an American audio engineer and consultant to NASA, had a completely different experience. His profile makes interesting, if alarming, reading. He worked with the NASA hardware contractor Intersonics from 1979 to 1996, and he designed and built hardware for sounding rockets, the KC-135 reduced-gravity "vomit comet," and the Space Shuttle Program. During his time with this company, he was awarded seventeen patents for a variety of inventions, among them the Servodrive subwoofer and a variety of acoustic and electromagnetic levitation devices—the first of which was a sound source one hundred times more effective than anything Intersonic had used previously.

During the period Danley was working for NASA/Intersonics, he also found time to hop across the pond and investigate the Great Pyramid. Twice. Apparently, once having demonstrated acoustic levitation in Boris Said's 1993 documentary *The Mystery of the Sphinx,* Danley was later asked back to Egypt by the same producer to measure the acoustics of the Great Pyramid.[17] Taking his team into the pyramid, Danley specifically remarked that when investigating the resonance within the King's Chamber, the extra-low frequency of B^0/30 Hz startled everyone and made them run for the exit: 30 Hz would have exactly meshed with beta 3, the high end of the beta brain wave and the third overtone of the Schumann resonance.

That might explain the panic that ensued among these acoustic engineers, including Danley, who said, "In the pyramid, I scared the heck out of everyone. After the first TEF sweep the producer asked me to turn it up—I also went down lower—and it literally felt like it was shaking the place." It was just air moving back and forth, he added, but he moved closer to the chamber entrance anyway. "I tried not to act like it scared me, but, yeah, it did." Danley also described the acoustic reflectivity tests he applied to the same Queen's Chamber star shaft that had been probed with a robot by Gantenbrink. Danley concluded that the positive acoustic discontinuity (echo reflection) measured a sizable cavity behind the door, perhaps thirty feet deep. To date, the ongoing scan of the Great Pyramid has not corroborated this finding.

COGNITIVE DISSONANCE,
MUSICALLY SPEAKING

All of this leads to two other points. First, it might be necessary to go beyond consideration of the manifest architectural elements at any one location, as recommended by Munck, and take into account components that are very obviously *not* there. These missing links might contain as much information of importance as can be garnered from the visible remains. Second, given the extraordinary architectural puzzles all over the Giza plateau, for many people, much to the distaste of the Egyptologists, the notion of ETI lingers in the air, and the overawing presence of the Great Pyramid can induce different feelings in different people. Some have found its architectural presence "threatening." Would that effect subtly inform their opinion of the monument? Possibly. Did it affect the collective consciousness of Danley's team when they responded to the extra-low frequency 30 Hz resonance by running for the exit? Probably. And what about Danley's next find, how far would that be affected by any cultural bias when it was interpreted?

THE DEVIL IS ALWAYS IN THE DETAIL

Danley's team also found echoes of Reid's finding, in that according to Danley, the entire pyramid responded to F♯, and in his report on this matter, Danley considered it to be the augmented 4th of the diatonic scale of C major—one of the most dissonant musical intervals. That was his opinion when I first wrote this chapter, and he had noted that the Roman Catholic Church called this interval the Diabolus in Musica (the devil within the music). The church considered its edgy sound liable to incite lewd behavior in its flocks, and from the Renaissance onward, its use was banned.

Upon my return to the web archive in 2020, much has changed; data that I had not seen before, and further information concerning that F# chord are now present. Danley now writes, "Being a musician myself, I was especially interested to discover a patterned musical signature to those resonances that formed an F-sharp chord. Ancient Egyptian texts indicate that this F-sharp was the resonant harmonic center of Planet Earth. F-sharp is (coincidentally?) the tuning reference for the sacred flutes of many Native American shamans." That Danley now wants this tone to be a sacred resonance is interesting in more ways than one. When writing this chapter initially, I had followed up on my Diabolus in Musica research with the comment that other than the construction caveats expressed earlier, there is another more harmonious and rather more likely reason for the presence of this F# tone within the sonic environment of the Great Pyramid.

THE B MAJOR MASS

The relationship of the Schumann resonance to mammalian brain waves, the Earth's fundamental resonance, and the very precise location of the Great Pyramid all make it far more likely that the architecture of the pyramid has been specifically engineered to reproduce that F♯, because it is the perfect fifth of B major, the basic scale of the Schumann resonance. The most harmonic tone after the octave, the perfect fifth is expressed musically as a 3/2 ratio, and one need hardly repeat ad infinitum the association of the Great Pyramid's location with this ratio when expressed as 2/3.

The fact that the Schumann resonance is based on the B major scale is of great interest in itself. When found within this iconic monument, the harmonics related to this particular scale and that perfect fifth in particular should (and with hindsight seemingly did) alert an audio engineer such as Danley that this sonic clue is yet another indication that the Egyptian pharaoh credited with its design and build could not be the architect. Indeed, all singers and musicians will understand that the actual builders are demonstrating the evolution of Western musical theory in reverse order to that in which it was developed.

In other words: the architects of the pyramid were acoustically and sonically far, far in advance of the peoples indigenous to this planet at the time the structure was supposed to have been built. Which may be the real reason why NASA's consultant preferred initially to refer simply to the $F_\sharp$ chord. As to whether it is described as stressing, sexy, or sacred, that is apparently still dependent on the status quo.

There is another relationship relative to the difference in height of thirty feet (360 inches) between the flat-topped pyramid that we see today and the geometric build (which includes the invisible capstone at the apex). From ground level, this length divides into the built edifice fifteen times and the geometric build sixteen times, which would infer that the stones of the building as it now stands resonate with a 15th overtone, making the geometric or "invisible" original design relative to the sixteen harmonic partials. And in C major that 15th partial is of course B^1, and if using the B major scale its 15th partial would be A^1. Of course, caveat emptor! Since the outer covering has been destroyed, there is no guarantee that either the scale of B or the scale of C is at work here, but these findings do have some relevance to off-planet operations.

GOLD LIGHT, RED LIGHT—SUNLIGHT

The Munck papers are insistent that the British foot was the tool of measurement in the ancient world. (Which aligns with Heath's principle that it all starts with the foot. And following up on the earlier part of this chapter, it is better referred to as the reference foot.) These twelve inches traditionally break down into sixteen parts. Not necessarily a logical progression. But sixteen as a portion of time is two-thirds of the twenty-four hours during which the planet rotates on its axis relative to the Sun. As noted, the Sun is overhead of the Great Pyramid at noon local time on March 21 (10:00 UTC). Noon, or twelve o'clock, midday, is the time at which the ionosphere tends to dip toward the Earth. All of which has some significance relative to the ISS with its orbiting astronauts and their medical research compadres on Earth who are running a total bed rest cycle of sixteen hours awake, eight hours asleep to see how it affects the human body long term.[18]

NASA has been bound to the Egyptian culture ostensibly since the 1960s, when Farouk El Baz was in charge of much of the Apollo astronaut's training, and as already noted, the agency uses much of the Egyptian mythical pantheon in naming its programs. If the space medical researchers are taking any notice of these ancient myths, especially after the Danley audio experience, the relationship should have been made to the Schumann resonance. If they assume that the Great Pyramid offers a model of the Earth, all these measurements and the persistent myth of a missing gold-coated capstone might inform them that sunlight, or certain elements of it, would be beneficial for both the mind and the body of the astronaut in the environment of a spacecraft. Light therapy is well known and understood on Earth, and it is used, among other things, for regulating circadian rhythms. Are there any clues in this structure that might elucidate this matter? The position of the Great Pyramid as the prime meridian and therefore marking local noon infers that twelve hours out of twenty-four are indicated, while the only color in the pyramid is in the King's Chamber, and that is red. Taken together, that can indicate that 50 percent of the *infrared* component of sunlight should be operating in the spacecraft when the astronauts are operating in beta wavelength; that is, when they are awake.

NASA's space medicine scientists have also found that the tones of 5 Hz to 100 Hz are beneficial for an astronaut's neurovestibular system; 5 Hz is just into the theta brain wave, and as the tone of D sharp six octaves below middle C, it will resonate with the 5th and the 10th overtones generated by the resonance of 7.83 Hz of the Schuman resonance overtones. The top end, namely 100 Hz, is beyond the scope of human brain waves. NASA would probably do better in tuning that top end to 95 Hz. This is $F_\sharp^2$, and that will harmonize with that $F_\sharp$ found in the Great Pyramid, and it will also resonate with the 3rd and 12th overtones of the Schumann resonance.

MIND OVER MATTER

If harmonics are being tentatively explored by the space medicine researchers in terms of the brain's responses to the space environment, could it be that other experiences on the ISS have affected the astronauts in ways not yet fully appreciated? The experience of another visitor to the King's Chamber might be informative here.

Contrary to the experience of Danley and his frightened team—scurrying out of the pyramid as fast as they could—another individual had a wholly beneficial experience in the King's Chamber, one that echoed the experiences of Reid and his hieroglyphs with his own light show, or holo-image. Figure 19.8 shows a drawing made by an artist who was alone in the King's Chamber, having had special access due to his association with the team led by geologist Robert Schoch, who was then examining the geological age of the Sphinx.

It is not known if this person was creating sound by toning musical notes. However, upon seeing this image of a little Gray's head, the artist was moved and unafraid, saying that the sighting had been accompanied by a sort of telepathic transmission. He immediately "knew" that he needed to redirect his current career path, and although this was an unexpected realization, upon his return to the United States, he acted on it. As an aside, this vision experience might have another aspect to it: the intentions of this team to investigate the Sphinx with a desire to learn and document their geological findings might link to the dream/vision experience that has gone down in history as inspiring Thutmose to clean out the Sphinx enclosure.

Even if the initial research into a worldwide grid started out as a military satellite project, the Western research into RV applications has (perhaps inadvertently) brought to the fore modes of thinking that Eastern cultures and Native Americans have always remembered but that Western cultures have forgotten or dismissed as irrelevant. Namely, the ability of human beings to order their wider environment through thought.

The King's Chamber perhaps models or mirrors effects of the ionosphere's F1 layer. At nighttime this blends with the F2 layer and allows more permeability for radio transmission. Certainly the crews of the ISS and the Mir space station have found that it is much easier to be telepathic in space. Therefore, the ability to

Illustration: Frank Domingo.

Figure 19.8. Spending time alone in the King's Chamber, this image appeared to Frank Domingo, a detective with the NYPD and their senior forensic artist. Domingo was in Egypt as part of a team researching the age of the Sphinx, and he drew this portrait immediately after the event.

transmit information telepathically between crew members and between those on the ground and those in space might be given particular thought in regard to this event: the connection with human brain waves and future attempts at communication from spacecraft to the ground. The presence of the B major scale within the architecture of the pyramid and the concomitant link with human brain waves surely conveys a helpful message for us humans in future space missions.

And that thought takes us back to Switzerland.

MIND BLOCK

Amazingly, the issues raised by space exploration as set out in the *Brookings Report,* the scientist versus the mystic, and the problem of ETI can be exemplified by three of its inhabitants. In the science corner there is Albert Einstein, in the mystic's corner there is Carl Jung, and the difficult bit, ETI, is represented by the farmer Eduard "Billy" Meier and his experiences. Author and psychoanalyst Jung was born in Switzerland and lived virtually in a straight line east of the future home of Meier. And Einstein studied in Zurich, not far from Meier's home.

For most scientists, Meier's photographs and his accounts of interaction with ETI had proven difficult, but Einstein's theories were just fine. In fact, questioning Einstein's theories was, and still is, unthinkable. That the human mind is able to perceive finer states of thinking might not fit well with scientists brought up on Einstein. And even if it occurs across the board, when it comes to discussing publicly their PSI experiences on the ISS, it is a fact that the Americans are far more reticent than the Russians. This is a bit rich coming from a country that purportedly intended educating the public about ETI way back in the 1960s, but it is indicative of a certain paranoia.

Everything appears to be the wrong way around. For starters, and taking liberties with Einstein's premise in the theory of relativity, it is clear from the medical research that traveling in space (using current methods) is not going to allow a human being to stay physically younger than his counterpart on Earth.[19] Quite the contrary, experience has demonstrated that the physical body actually ages faster.[20]

LIGHT BLOCK

Thanks to probes launched from Earth and traveling into the far reaches of the solar system, Einstein's assertion that the speed of light is a constant everywhere is coming unstitched at the seams. Letting go of so much established physics is going to require a massive mind shift from scientists weaned on Einstein, but until that happens, the development of human space exploration is effectively stopped.

Regarding the question of the speed of light in any given location, whatever the result of research and studies in space or at CERN, it seems that those at NASA are going to adhere to the speed of light as a constant with grim determination. Data from probes such as *Pioneer* indicated variations well worth studying, but the data is dismissed as not sufficiently significant to warrant upsetting Einstein's premise. Which is where the ninth planet steps into the frame. Early in 2016, it was hypothesized that a large planet lurked just beyond the known limits of the solar system (and is therefore a part of it). This is a perfect mechanism for avoiding any meaningful public discussion on the variability of light speed and its relationship to any given gravitational region while accumulating information as to what exactly is happening out beyond Pluto. Spending a lot of time talking about missing dark matter and dark energy while avoiding the rather obvious phenomenon of the Swiss cylinders of missing material is another example of the head-in-the-sand approach taken by science at this time.

ROADBLOCKS

From the appearance of the Swiss cylinders of missing mass to the crop glyphs in England, highly bizarre events occurring on the surface of the planet seem to be down to one of four possible causes:

1. Offerings from ETI totally independent of and external to Earth, which would be an important conclusion worthy of full investigation
2. The collective consciousness of humans reflecting our own thoughts, hopes, and fears back to us off the ionosphere
3. A mixture of options one and two
4. The energetic imprint of our actions and intentions reflected back in visual forms so that we can begin to understand and then use the powers of our own minds in a beneficial way for our environment and our aspirations.

Regarding the second option, it may have been that unwittingly the CERN scientists were involved in the creation of these holes in Switzerland. For example, if the choice of location was tied to whomsoever would immediately benefit from its arrival and those persons' intentions coalesced into a collective consciousness that was then harnessed in order to bring about the appearance of these cylinders, then effectively, the manifestation would become a co-creative product. However, the apparent lack of a full appreciation regarding the effects of space travel on human beings and the inability to deal with getting to Mars or to the Moon infer that option two is not the case. If it were so, then the planet's authorities and space agencies would have already set about building an appropriate spacecraft.

Option three might be unwitting on the part of the Earth-bound participants (although those who fake crop circles and state that they are inspired by "upstairs" also ought to note that creation with the intention to deceive exempts the copycat from that particular brand of co-creation, however much it is proclaimed as such).

Option four brings to mind the fact that when anyone looks in a mirror, it is the silver backing that permits the light to be reflected back as an illusion of ourselves in which the right side is reflected back as the left side. Clear reflection requires insight and self-awareness.

To date there has been no viable explanation from scientists whether at CERN or elsewhere, as to how crop circles are really made—other than by these human plankers—and apart from the attempt to justify the Begnins cylinder, there has been a total silence as to how the Swiss material was swiftly and silently removed overnight or where it was taken to. Was it spun out of three-dimensional reality? That takes us straight back to option one.

It could well be that many rocket scientists have rendered themselves incapable of getting their heads around the more right-brained requirements for space travel that pertain to the unseen mental and soul components of a self-aware being. That in wishing to convince themselves as to the validity of the scientific data thus far acquired and adhering simply to the physical nuts and bolts when addressing space travel problems for both the spacecraft and its occupants, they have either failed to appreciate the validity of the diagrams or blueprints and physical oddities that have appeared on their doorstep—literally, in the case of CERN—or, contrary to their own governments stated desire to familiarize itself and us with ETI interaction, they have decided to turn a blind eye to all these anomalies. As Stanley Kubrick would put it, they are operating with their eyes wide shut.

The mapping evidence from the USGS would suggest another scenario.

Blueprints for the Red Planet

The Art of Craft Movement

VESICA PISCIS

The Mars mapping carried out by the USGS suggests that the space agencies have taken notice of what has been occurring from out there to down here, including the considerable transceived data studied by Andrija Puharich and his affiliated psychic researchers. In effect, the fundamental principles of the workings of an advanced craft appear to have been hidden in plain sight on the surface of Mars. Before discussing this in more detail, it is necessary to consider the principles adopted by the USGS and note how a geometric 2-D overlay onto a spherical 3-D object transcribes when using geometry allied with elements of history and myth to convey a message—visually and metaphorically.

Looking once again at the circles created by the USGS on the Cydonian complex as discussed in chapter 7 (figures 7.9–7.14), it was clear I was looking at a geometric vesica piscis hidden in plain sight within the USGS mapping. I also saw that the very small special right triangle was found within the slightly less accurate vesica piscis created by the USGS mapping of Cydonia Colles and Cydonia Labyrinthus. Then I found that the crater interacted with the circumference of Colles, and the Face and the D&M Tor also connected with the USGS mapping project. Notable by its absence is the key to the special right triangle: the imported and intentionally sited material that forms the cone/spiral mound and also generates a phi spiral that tangents the crater, the face and the D&M Tor.

The geometry detected at Cydonia is clearly the result of intelligent and intentional placement by extraterrestrials, unless the USGS and/or NASA claim authorship of the entire complex by means of false mapping or photographic

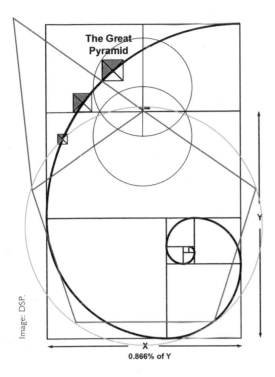

The Great Pyramid

X
0.866% of Y

Image: DSP.

Figure 20.1. The Giza big three pyramids located on the phi spiral. The yellow circle is the large Mensae circle ascribed to Cydonia by the USGS. The ideal geometric vesica piscis is derived from the USGS Cydonia modeling. The turquoise outline is the D&M Tor and its extended analog to the Pentagon's Mall gardens.

manipulation. But then, if the mapping of the USGS is intended to send a message, what about those pyramids at Giza, sitting on their own phi spiral? That too generates a five-sided figure, yet the Giza pyramids are all four-sided. I could see a link back to the Pentagon Mall gardens and the reproduction of the D&M Tor at Cydonia. Given all those seeking answers to mysteries in Egypt, it would be interesting to see what happened when taking the USGS mapping of Cydonia back to Egypt. I thought I would reverse the process and see what the interaction of the Giza phi spiral and the D&M Tor would do when placed together with a perfect Vesica Piscis.

I also reckoned that the large circle of Mensae might be, or even should be, encircling something, and from there it was just a short step to recalling Carl Munck's finding that at one time the big five-sided D&M Tor was the 0°/360° meridian for Mars. Given the Pentagon layout in Washington, DC, I realized that all those four and five symbols laid out in the ceremonial Mall gardens at the Pentagon were also referencing the association of four-sided pyramids with five-sided pentagonal forms, and so I added the Mall garden extension to the mix.

It was an even shorter step taking this image back to the USGS mapping of Cydonia. Note that the profile of the D&M Tor reveals that while the pentagonal form is inherent within the D&M Tor's five-sided form, not all its sides are derived from pentagonal geometry. The hexagonal form is also present.

I then thought that just as the USGS had overlaid a vesica piscis onto Cydonia, it would be interesting to see what the geometry on the Egyptian plateau did to Cydonia. I wanted to see how the perfect geometry at Giza interacted with the imperfect modeling of the vesica piscis established by the Colles and Labyrinthus mapping at Cydonia.

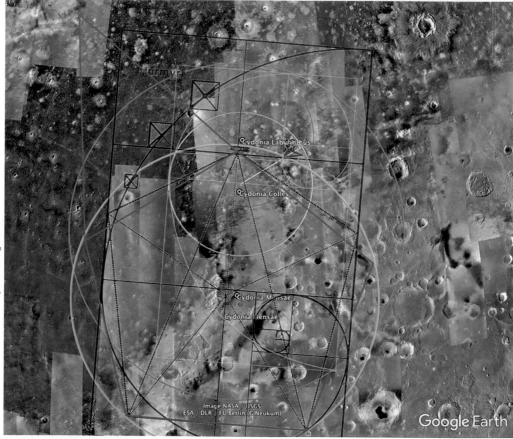

Figure 20.2. Egypt brought to Cydonia. Note how Labyrinthus interacts with the Khafre pyramid (nearest to the Great Pyramid) and with the Cydonia complex and its phi rectangle (teal). New Mensae interacts with Menkaure (farthest from the Great Pyramid) and tangents the Sphinx, which here lies parallel and looking *eastward* toward the Cydonia complex and the imported cone/spiral mound's location. The Pentagon Mall garden extension (blue) cuts the Sphinx's tail and tangents the southwest corner of the Great Pyramid. The imperfect vesica piscis generated by the USGS produces the square root of three on the turquoise horizontal line. The red spot marks the *actual* USGS New Mensae coordinate of 34° N, 13.12° W.

But there is another layer of hidden geometry operating within this USGS geometry. Creating a perfect vesica piscis from these two different circles produces interesting results. Generating a vesica piscis from the diameter of Colles produces a square root of three that runs along the body of the Sphinx, and when extended eastward, it interacts with the Cydonia anomalous complex, crossing the complex at the midpoint of the special right triangle's own square root of three. That is exactly where the flat-topped mesa (analog to the Giza plateau) tangents this line.

In Washington the analog geometry is at the Naval Heritage Center halfway up Pennsylvania Avenue. This hidden geometry certainly fulfills that peculiar reference to future sailors inscribed on the memorial. It also reconciles the profession of Neil Armstrong as both a sailor and an astronaut while explaining the choice of the Atlantic (rather than the Pacific, the lunar return ocean) for the burial of Armstrong. Not only was the sailor returning home at the end of his odyssey, he was returning home—to Atlantis.

When it comes to generating a vesica piscis from Labyrinthus, the square root of three that is found seemingly does nothing special until one remembers the frantic searching under the surface at Giza for Cayce's Hall of Records. Factoring in the Greek myth of the Cretan Minotaur hidden in the labyrinth, remember that the traditional storyline of this myth hides the fact that the Minotaur can also be taken as a combination of the Taurus and Gemini constellations. NASA's favorite constellation Orion sits on the horizon between Taurus and Gemini while the Pleiades, especially popular in the Americas, is to be found in the constellation of Taurus and of course NASA's second phase of crewed missions was called Gemini. If everything is like those twins, as above so below, a vesica piscis generated from the Labyrinthus Circle is not as insignificant as would at first appear. This is a subject that will be pursued on Aulis.com/pathway.

Whether blue-sky thinking or advised understanding, the USGS mapping seems to be entirely inspired by the mathematical principles of geometry known generally as sacred. First, let's clarify the discussion on the application of geometry and the use of metaphors in ETI messaging and, ultimately, space engineering projects as seen here in this mapping.

Everywhere that self-aware beings have created an environment for themselves in which to operate, the proportions generated by this so-called sacred geometry are to be found.[1] So it is worth a quick recap on just how much mathematics and sacred geometry bring to the table. Two-dimensional geometric designs of the equilateral triangle, the square and its double produce the square roots of two, three, and five. And all these can also be created by the interlocking of two circles through their centers. The human body finds connections with phi 1.618: from

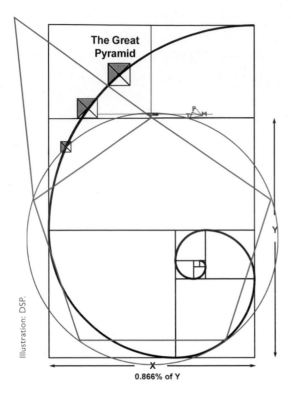

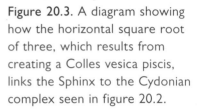

Figure 20.3. A diagram showing how the horizontal square root of three, which results from creating a Colles vesica piscis, links the Sphinx to the Cydonian complex seen in figure 20.2.

the sole of the foot to the navel is 1 and from the navel to the head is 0.618. The geometry dependent on the phi ratio rectangle and its attendant spiral is understood to pervade all forms and all living things.[2] Indeed, all of this geometry is to be found wherever self-aware beings have created something, whether in the world of art, music, or architecture.[3] It informs both philosophical and scientific thinking; it is of us humans, and we are *of it*. Myers and Percy called the spiral generated by the phi rectangle transdimensional, which is a good description of energy transfer from one state to another. On a very basic level, it simply *feels* better to live or operate within places and structures that respond to these ratios and forms that are often (in nature) sensed but not actually seen, and even that sensing is not necessarily expressed. It can be a subconscious knowing when a place feels right or doesn't. Which is why we all have personal preferences for specific locations, which is unsurprising given that the human body contains many phi proportions.[4] We are all, as is often said, star people. All these geometric forms have appeared in the fields of southern England in the form of crop glyphs, and all of these diverse forms even in two dimensions as we perceive them in the field, can subtly inform us as to how to live in harmony with the unseen energy that contributes to our personal well-being and that of our wider environment. Because *all geometric forms create energy,* deliberately designing and placing specific geometrical forms on any particular location amplifies the natural energy present at that location through the interaction on the surface of the planet with the energies upwelling from the Earth and those received from the elements above. Subconsciously, we are all connecting with the natural energy of the planet as we move about it, and live with it.

It is the combination of a self-aware being's intentions combined with the actual design that can potentially transform a location into a powerful tool. Even if one does have a form of energy at one's disposal, that power can only be properly sustained if the intention of the individual working within the confines of such a geometry is in harmony with their fellow beings and with that particular environment. Take for example, the positioning of the great Gothic cathedrals. They were initially sited to capture the perceived power of the location and thereby boost the authority of the status quo; in this case, the religious leaders. But it came about that the intentions of the clergy did not always match the harmonics of the land. And with that, the behavior of the leaders, from the top down, fell out of harmony with both their congregations and the basic tenets of their faith. Today one sees these great architectural wonders turned into virtual tourist attractions, with many smaller religious buildings being converted into nightclubs, cafés, or houses.

From that small example, much could be said about the current concerns over climate change. The lack of conscious awareness as to just how much the environment resonates with the thoughts and actions of all the self-aware beings on this planet, combined with selfishness, or rather an inability to see the wider picture, means that any natural evolution at very big scales is perceived as a threat, and fear then replaces harmony. If we are to succeed as a planetary civilization long-term, we really need to learn or remember how it can be when we interact harmoniously with both the environment and each other.

And so it is with getting to Mars.

TOWARD AN ADVANCED SPACECRAFT CONCEPT

Most of us are not going to travel to Mars, but the means by which we can do that safely can be translated onto other fractals, that of living properly with this planet and its other inhabitants, and living properly within our selves. If designing geometric symbols is a first step to acquiring the required harmonics, it is unsurprising that we see all the signs of sacred geometry mapped onto the surface of Mars by the USGS. And as the geometry just discussed is the invisible underpinning of all the cycles that sustain life, it is therefore the geometric fundamentals that lead to the design of a craft that will support and sustain human beings on long space journeys. Clearly, *some* of the above is understood. As is the fact that this geometry operates on other levels than the purely physical; when taken out of geological context, something of this intentional human interface with aspects of the human mind and the spirit is evident when looking closer at the Latin names used to describe the mapping of Cydonia. Those three circular areas—Mensae, Colles, and Labyrinthus—are all allied with the notions of mind, body, and spirit. Mensae is the plural for "mind," with this largest area suggesting the collective consciousness of all three components. Colles refers to knobs, mounds, humps, and heights, suggesting a physical, planetary component (which reinforces the fact that the diameter of this circle recalls the Earth year in the ten-base kilometer). Labyrinthus is self-evident and as already discussed immediately brings to mind the Cretan Minotaur myth. Notwithstanding Crete's links with the classical naming of this Martian region, the USGS naming also brings to mind the ancient labyrinthine Apis bull rooms found *underground* in northern Egypt and the symbolism of bull's horns as the three-dimensional representation of the spirit. And the diameter of this Labyrinthus circle recalls the lunar year, which reinforces the notion of the barycenter of the Moon, which is found within the Earth.

The distance of the locus point of Mensae from Cydonia is four hundred miles; that's the ratio of the diameter of the Sun to the Moon (400:1), therefore this large circle makes a good fit with the notion of power, both mental and solar. Eyes wide open anyone? Why would the USGS be evoking solar, lunar, and, indeed, earthly powers on Mars? Unless the mappers are talking to themselves—or perhaps encoding their intentions for the benefit of any observer. The vesica piscis laid out by the USGS unites Colles the physical and Labyrinthus the spirit—again bringing to mind the Martian probes *MER-A* and *MER-B*. But as we have seen in figure 20.2, the Labyrinthus circle is not precisely of the same dimensions as Colles. The difference in diameter is a mere nine kilometers, taking us right back to all those PSI communications and *2001: A Space Odyssey*—the gestation of the astronaut both literally and metaphorically as a star child. And perhaps, a cry for help.

I find the USGS mapping to be a subtle signal to ETI, with the undefined phi rectangle a reference to the "missing" monolith and the admission that in order to take the next step along the path to Mars, human thinking needs an upgrade. In the case of purely linear thinkers, that would be true. When it comes to interacting with ETI, producing the film called *Stargate* in 1994 and setting it against pyramids used as a means to facilitate the transfer of an incoming alien race disguised as half-human half-animals in order to hold in thrall the indigenous population and use them as slaves (pace Zitchin), is entirely linked to the remote viewers programs, Cydonia, and how the authorities would like us to think about ETI: with fear. Or is that scenario a reflection of their own fears?

The real problem is that the scientific discoveries already made, when taken together with the data supplied by all the anomalous events discussed in the preceding chapters, are of themselves enough to get us thinking outside the box, but because the data are supplied in the form of metaphor, analogy, and anomalous energetic interfaces with the planet, these are incompatible with present scientific modalities, which require only tangible repeatable experiments. Frightened of their own creative imaginations, linear thinkers are holding themselves in a prison of their own making. Sitting on the threshold of their golden cage whose door is actually wide open, should they dare to fly, they cannot see that what to them is nonsense, is a data base replete with information on many levels, and thus all these anomalies have not been taken seriously by the relevant aerospace agencies. But since it was taken seriously by the authors of *Two-Thirds,* I am using that data to explain how an advanced spacecraft might be designed, and I will come back to how mapping, symbols, and metaphor can be used as reminders of matters that in all probability were once known but forgotten along the way. Ancient memories that could be locked within our noncoding human DNA.

Even if space agencies and private contractors try to go to Mars in conventional chemical rockets in the short term, they are already aware that this is not going to be viable or beneficial for crewed flights. In the longer term, a far more robust and advanced solution needs to be found. One that that will provide all the essential requirements of a sophisticated craft fit for such a purpose. In order to protect the human beings on such a flight, the design for the drive system and the craft itself interact with each other since they are engineered around specific harmonics.

The *Two-Thirds* authors delivered a concept for a totally new form of spacecraft, one that could replace conventional rockets with a craft that has a built-in drive system that generates propulsion without propellant. Their conclusion was that the heart of this craft was a drive system built around a rapidly spinning disk (RSD). This concept was based on the understanding that all planetary bodies draw in energy in the form of real particles from the local environment through the poles, then reconfigure and power up these energized particles, which in turn are recycled through the actions of the spinning body. Initially called a *gravitron drive,* this technology would deliver a means of travel by *simulating* the way the home planet generates, or more correctly, recycles energy.

In those early days, they did not fully understand the meaning of the term *gravitron.* While realizing that adopting this word avoided confusion with the scientists' theoretical graviton, they still thought it referred to actual "particles of gravity." Although having between them analyzed the data from numerous crop glyphs and having had direct experiences of many anomalous events (notwithstanding any light or sound anomalies, walking into the crop glyphs in the late 1980s and early 1990s felt like being plugged directly into the grid), neither author was entirely convinced that this was the correct use of the term *gravitron.* They were pretty certain that light was a factor. Talking to me about their experiences when I was editing *Two-Thirds,* Percy recalled a session of inductive reasoning he had experienced two decades earlier, in the early 1970s. During that deep meditation, he had understood that a future spacecraft could be propelled by "amplified light" energy.

Looking at the Barbury Castle glyph as Myers and Percy understood it, it represented a Combinatorial Hierarchy and while noting the traditional physicists'

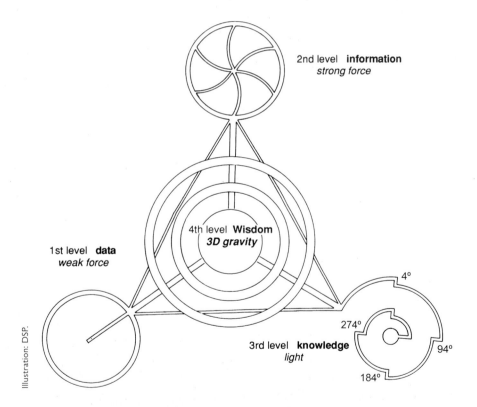

Figure 20.4. Basic illustration of the 1991 Barbury Castle crop glyph from *Two-Thirds,* activated on July 16, near Marlborough, Wiltshire, England. The ratchet-spiral on the lower right of the glyph was offset at 4° east of due north and this is the local magnetic discrepancy to true north. Note that this glyph has been annotated with aspects of consciousness set out alongside the Combinatorial Hierarchy of the four forces.

enumeration of the four fundamental forces (wherein the weak force is expressed as 10^{-6}, the strong force as 1, the electromagnetic as 1/137 and gravity as 10^{-39}) Myers and Percy considered these terms as analogs and relevant to other matters in their own energy models. In figure 20.4, for example, these four levels are represented by states of awareness.

When first elaborating on the modeling of energy relative to their spacecraft, the first two lower levels of the Hierarchy were not addressed. Later, it was understood that the weak and strong forces represented the spin of the photon (down and up, respectively) relative to the environment, while the spiral ratchet in the third level of the Hierarchy represented the spinning process and the generation of magnetic fields. This third level in the apparent Combinatorial Hierarchy they had denoted as *light,* which means, of course, photons, and the fourth level was gravity, *g.* The third and fourth levels they considered to be in an oppositional relationship to each other. Now all that had been worked out by 1993, and at that time traditional scientists were still describing gravity as a force even though no one had managed to prove that gravity consisted of any actual particles. By the twenty-first century scientist were thinking in other terms, and toying with the notion that gravity is actually an acceleration, and as such, it is not composed of particles of solid matter.

BASICS OF THE CRAFT

Back in 1993 when *Two-Thirds* was first published, the authors had also stated that solar system light speed was not a constant, even within the solar system, as it increased speed some eighty thousand miles beyond Saturn. The behavior of space probes would later appear to show that to be correct. However, at the time, although conceding that the matter of light speed was not entirely resolved, those scientists with whom they discussed the matter largely ignored their suggestion. It follows that asserting, as they did, that there were two further speeds of light in the physical universe was also ignored. A notable exception was Geoffrey Pardoe, OBE, project manager for the Blue Streak ballistic missile and an advocate for British involvement in space exploration. Taken with the suggestion that the speed of light might vary outside the solar system, Pardoe proposed that an experiment consisting of a set of lasers mounted aboard future space probes could measure the speed of light when the probe reached a point well beyond the solar system.*

---

*Early in 1995 Pardoe attended a talk given by David Percy in London titled "The Face on Mars and the Avebury Connection." Pardoe had made this suggestion during the conversation they had after that talk.

Some years later, in 2002, at a time when so-called antigravity research was all the rage and a politically sensitive topic, Percy made a full presentation of the RSD system with its gravitron drive to Hal Puthoff. After this presentation and discussion, it was learned (by back channels, as it were) that classified research activities in gravity modification might exist. However, the Myers and Percy concept was very different from the antigravity research then taking place. And as it turned out, the authors' concept of a spacecraft with a propulsion system that did not use any form of chemical propellant, did not suit politically. Although providing useful pointers, it was perhaps a bit too far ahead of its time, full of highly complex challenges even for the likes of Puthoff and his team, who were at that time researching aspects of zero-point energy and antigravity systems.

Later that same year, in November, the RSD gravitron drive concept was presented to Airbus Industrie in Toulouse, France. A number of senior engineers attended, including their flight physics senior technologist and their systems chief technologist, who remarked, "This technology will be of immense benefit to humanity." Kind words, but "as Airbus does not manufacture engines," they were unable to actually build such a drive. Had the drive itself actually been available there could have been a different outcome.[5]

These initial discussions with theoretical physicists and traditional aerospace engineers along with the apparent stalling of so-called antigravity research generally, led these two to rethink the nature of gravity as it concerned their model. The sense that a form of light, and therefore the harnessing of photons, was a potential contender as the essential element of their drive system, which had been niggling at them from the very beginning, had never really gone away. Those initial doubts that their gravitron was not actually a gravity particle re-emerged. But a particle of light—the photon—well, that's another matter entirely. Photons are in fact a bountiful resource; they are prevalent throughout the universe. A field of energy available everywhere and therefore potentially could be harnessed as a possible power source for such a craft. Light in all its forms and frequencies would have to be composed of real particles, which, though very small, must have actual mass and be energetic. Although when it came to engineering a scaled replication of that process even though stars and planets recycle this photonic energy, it was difficult to see how photons could be used by their RSD concept, given that the photon was stated to be "virtual," or massless. That was then. While Myers had retired from the fray with honors after writing *Two-Thirds,* David Percy and I continued on, intuitively knowing that we were on the right track. The intervening years saw the refinement of the system, brainstorming and research continued, and all the while we were becoming more and more convinced that everything to do with light was going to be the key to resolving the problems posed thus far for human space travelers. But the photon proved recalcitrant.

Then, in 2016, an announcement was made that supplied the missing link. What had previously been considered totally impossible actually happened: physicists in Poland had created the hologram of a single light particle. An article in the journal *Nature Photonics* stated:

> Scientists at the Faculty of Physics, University of Warsaw, successfully applied concepts of classical holography to the world of quantum phenomena. A new measurement technique enabled them to register the first ever hologram of a single light particle, thereby shedding new light on the foundations of quantum mechanics.[6]

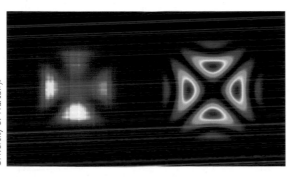

Image: FUW (Faculty of Physics, University of Warsaw).

Figure 20.5. Hologram of a single photon, reconstructed from raw measurements (left) and the theoretically predicted model (right).

This achievement clearly demonstrates that photons must have mass to be able to register such an image. As such, photons therefore must also be physical and have spin energy. Now that it is known that a photon is a physical particle with spin characteristics, it means that those early hunches were valid! Imagine a craft using photon energy for propulsion—a craft powered by a totally clean, silent, efficient technology that can take off and land vertically, anywhere. Remembering Pardoe's suggestion about using lasers to measure the speed of light and knowing that light is harnessed to produce powerful laser energy, the premise is that the photon field is a truly universal source of energy waiting to be tapped in a totally new way. And if it's used with the correct technology, we may be able to travel through space rather faster and farther than currently envisaged. This approach is based on a combination of fundamental principles and then applying the possibilities of light energy in a new way. It is about harnessing a resource readily available absolutely everywhere through the actions of a craft's internal drive system.

POWER SOURCE

Richard Feynman and John Wheeler calculated the zero-point radiation of the vacuum to be an order of magnitude greater than nuclear energy, with one teacup containing enough energy to boil all the world's oceans. and this vacuum energy has even been described as the ultimate free lunch.[7] Within this energy field are galaxies, stars, solar systems, and planets—all recycling this energy. For example the Sun receives energy from the galaxy and as a result of nuclear fusion and recycling processes, produces the solar wind and visible light and heat as well as energy in invisible wavelengths of light. All of which is sent out to the planets within its solar system. NASA has captured images of this emission and it is notably most apparent between the latitudes 30° north and south of the Sun's equator. For example on June 7, 2011, the Sun was photographed unleashing a medium-sized solar flare with a spectacular coronal mass ejection; these yellow flares are most prevalent at the tetrahedral latitudes.

It has long been understood that it is possible to model the energetic forces produced *within* a rotating sphere using geometric forms such as the platonic solids. However these are imaginary forms, used as tools to facilitate understanding: the interlocked tetrahedra do not actually exist as solid objects within the sphere they are ascribed to. Nor are they physical lines inscribed on the surface of that sphere.

Image (left): SDO, NASA Science Visualization Studio. Image (right): NASA/SDO and the AIA, EVE, and HMI science teams.Overlay lines: DSP.

Figure 20.6. The Sun both generates and recycles energy by its rotation and its position in the galaxy. In these two images the star has had a compound tetrahedra geometric form overlaid onto the Sun's surface in order to illustrate the routes the energy can take within the body of the Sun.

When a sphere recycles energy from the external environment, the interaction can best be modeled by a compound tetrahedra, which is also known as a star tetrahedron. This illustrates how the energy upwellings occur between latitudes 30° north and south of the sphere's equator. Because these are the latitudes at which these two internal tetrahedron would intersect and force extrusions of energy to the surface of the rotating sphere and beyond, these are sometimes called the star tetrahedral latitudes.

As the Earth spins on its axis, it recycles the incoming solar energy field.

Simulating the way the Earth recycles this energy through spin, the potential source of energy for the craft is drawn from this field of quantum fluctuations. However a small spacecraft is not the same as a massive spinning sphere, so that in order to compress the same principles into a drive system, the sphere of a planetary body is reduced to a *flattened,* solid disk. Thus the "poles" of the sphere become the central point of the disk. This solid disk becomes the essential heart of the drive system: manufactured from a composite of materials, resulting in a disk with extreme mass-per-volume, extra high tensile strength and cohesion, it could be encapsulated in a material such as graphene. Discovered at the University Manchester in 2004, graphene is the world's strongest material, and graphene-enhanced composite materials could be used to manufacture the disk. The rim

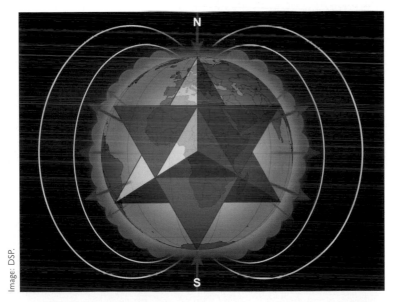

Figure 20.7. The Earth recycles spun-up photonic energy, particularly at the tetrahedral latitudes 30° north and south of the equator as indicated by the red arrows. Energy recycling output is outward, upward, in *opposition* to the inward, downward force of gravity.

needs to be thicker than the plane of the disk and should be further reinforced internally with embedded bands of metal alloy to prevent the disk from being flung apart when spun at very high speeds.

The basic premise for this revolutionary spacecraft was further distilled from the data and other information derived from the layouts of certain ancient monuments and designs that were incorporated into various crop glyphs that were activated in the late 1980s and early 1990s. The following illustrations reveal how the geometry inherent within these monuments and glyphs informed the concept for the engineering and energy transfer process. The question at this point might be: How did these two researchers manage to convert the information they were acquiring into an outline concept?

It really started in May 1990, when a virtual stranger (only familiar as being part of the community of creative businesses based in London's Covent Garden) walked up to David Percy with a photocopy of an image in his hand. Saying "You will be interested in this," he handed over a color photocopy of a photograph of new crop glyph. And without more ado, turned on his heel and left. Oddly, Percy never saw this "postman" in the area again. (It was only when putting this chapter together that I realized that the two people who had delivered the Monroe Institute brochure to me while I was standing in the Cambridgeshire Bythorn glyph had operated in precisely the same manner.) Looking the photo, David Percy could immediately see that the glyph was unusual, in that straight sides made up large parts of the formation. That, combined with the peculiar circumstances of its arrival into his hands as it were, led David Percy to consider the glyph as especially significant. Enquiries within the croppie community informed him that it was to be found in a field at Chilcomb Down in Hampshire. Contemplation on the geography of southern England and this design inspired him to combine it with the ground plan of Glastonbury Abbey, and that in turn was further combined with the layout of Stonehenge. The combination process Myers and Percy followed is set out on page 465.

Myers and Percy considered that this glyph might represent components of a spacecraft's drive system, so they produced an overlay that combined the glyph with the Abbey and the Stonehenge ground plan. First they could see how the glyph imitated the floor plan of the now ruined abbey and they married the glyph to the original ground plan of Glastonbury Abbey, and in particular the small Mary Chapel and the even smaller, *separate* St. Dunstan's Chapel to the western end of the ruins.

Myers and Percy had seen that the circle inscribed within the location of the Abbey's Chapter House was the exact same size as the earth ditch surrounding Stonehenge's architecture. Did the original builders of the Abbey make that a

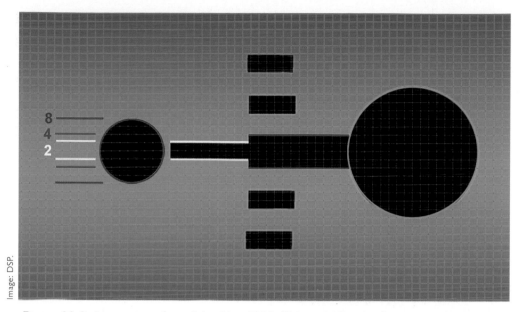

Image: DSP.

Figure 20.8. Representation of the May 1990 Chilcomb Glyph. Chilcomb, meaning the hill by the narrow valley, is the name of the field of farmed chalk downland just outside Winchester, in Hampshire that gave its name to this crop glyph.

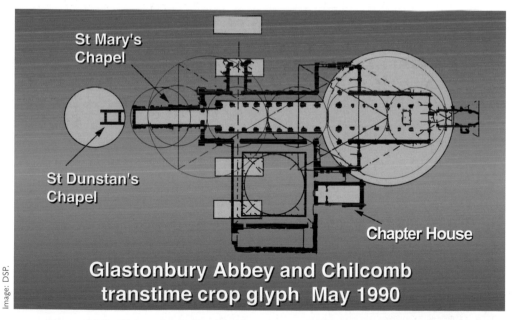

Image: DSP.

Figure 20.9. Chilcomb glyph and Glastonbury Abbey ground plan combined. The pink areas are the Glyph. The black outlines are the original Abbey floor plan and the circles are based on the diameter of an inscribed circle within the square that is the Chapter House, attached to the south side of the Abbey.

conscious choice or did they pick up such a connection subconsciously from geo-logical connections and conditions existing between these two locations? Here it might be useful to know that when Western Christianity took over from the Roman paganism, the Roman Basilica was adapted to the form of the early church. In very ancient times, the Basilica was a royal palace. Later, in Roman Imperial times it was turned into the palace of justice and an assembly hall. This consisted of a long hall with colonnades of pillars creating aisles, leading to the curved end of the building where the judge sat. This function then changed under Constantine who allowed the conversion of seven of these buildings to Christian worship. The rounded end was adapted by the church to become the apse and the site of the altar. The central nave was allocated to the clergy and the outer aisles to the congregation. Later Gothic cathedrals built their colonnades with tree-like stone supporting pillars separating the priests and monks from the congregation, but the apse and the altar remained at the rounded end of the building. Further, the word *nave* is everything to do with ships, and navigation, it being a very ancient word used for the central part of a ship's wheel, into which the axial spokes are inserted and from where these radiate out to the rim of the ship's wheel. In this amalgamation of building and wording we see that the priest-astronomer is still in charge! And when it is remembered that the ship's wheel is steered by the helms-man and that "helm" is also derived from the ancient word *haulm* meaning the stalks of all farmed plants, something interesting is happening when it comes to crop glyphs. Over time, *helm* became used for specifically describing the stalks of wheat. And then even more precisely to that of long stem wheat, grown specifically for thatching roofs. Just knowing these seemingly inconsequential facts provides much food for thought, relative to the arrival of crop glyphs in the farmed fields of plants in southern England. None of this information was known consciously by Myers and Percy as they put together their engineering schematics. How much was known within their own subconscious, and how much this had been gleaned from the greater collective consciousness and subconscious is yet another matter. And seeing these schemas and how the dimensions of their spinning disk connects with the holes in Stonehenge surely brings to mind the arrival of those holes in Switzerland—the original home of the watch industry.

The next step involved taking a ground plan view of Stonehenge and com-bining it with (a) the ground plan of Glastonbury Abbey and (b) the May 1990 Chilcomb crop glyph to produce the image in figure 20.12.

Taking the analogy of the combined stages of a specific ancient monument with the ruins of this abbey and the 1990 crop glyph, the basics emerged for what appeared to be a blueprint for a piece of technology.

Figure 20.10. Stonehenge restored. The outer ditch ring (dark band) was seen as representing the edge of a disk, the sarsen stone ring as a set of magnets, and the central trilithon sarsens as representing a set of magnets set at a *higher level* than the outer ring, pressing *down* on the disk.

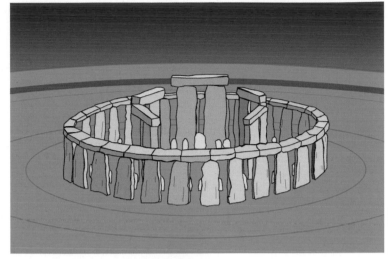

Image: DSP.

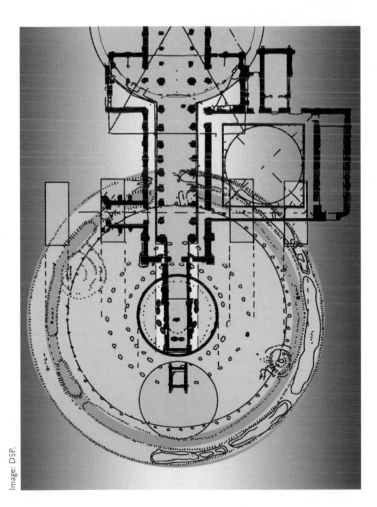

Image: DSP.

Figure 20.11. Stonehenge (green) with overlaid ground plan of Glastonbury Abbey (bold black outline) and the May 1990 Chilcomb crop glyph (pink). Here it can be seen how the glyph proportions echo those of the Abbey's Nave, how they also meet with those of Stonehenge's sarsen stone layouts, and how the three circles of holes that have been dug into the ground at Stonehenge are linked by the circle that is the same diameter as the earthworks at Stonehenge and the inscribed circle of the Chapter House.

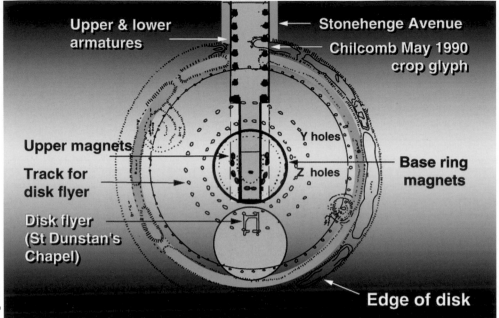

Figure 20.12. Close view of the three combined ground plans: Stonehenge (green) with overlaid ground plan of Glastonbury Abbey (bold black outline) and the May 1990 Chilcomb crop glyph (pink). The base ring magnets are the standing circle of sarsen stones, the upper magnets, the horseshoe trilithon stones. The track for the balancing disk flyer is here represented by the y and z holes (currently filled in at the monument). The area of the disk requiring extra strengthening goes from the thin dotted line surrounding the outer earthworks and the 56 Aubrey holes located on a ring just inside the earthworks here illustrated by the thin black circle just inside the earth ditch.

When it comes to putting the elements together to create such a drive, the disk at the heart of the concept needs to be supported and set spinning. For example, by electromagnetic induction (magnets implied by the arrangement of the sarsen stone ring) and then brought up to an appropriate rotation speed, even to tens of thousands of revolutions per minute (rpms) for all the engineering processes to come on line. When the disk is spun up sufficiently, the proposition is that it will release massive amounts of excited spun-up gravitrons spinning on all three axes. These highly energized, excited gravitrons will be recycled and flung off the surfaces of the disk at a distance of two-thirds from its center (which equates with thirty degrees latitude on a planetary sphere) and these are to be collected by *shielded* collection rings located above and below the disk. Then from the *collection rings,* the gravitrons will be conducted through *emission shafts* to a *distributor ring.* And from the *distributor ring,* following the paths of least resistance, the gravitrons will be delivered or vented, via magnetically shielded *connecting conduits,* to a set of nine *thrusters.* These thrusters, located around the specially shaped craft, receive the energized gravitrons, which are expelled in the same manner as propellant in chemical rocket systems.

All these processes are controlled by ultra-high-speed computers, and by deploying thrust-vector-controlled thrusters, the manipulation of the thrust output to the outer environment is enabled, thereby propelling the craft in any direction required.

Speaking to Sir John Whitmore in 1980, the ETI TOM told the former racing driver turned management consultant and the third member of the original trio working with Schlemmer and Puharich, that when the shift in consciousness that would be required from governments concerning ETI contact happens, then the planet Earth would become "a light vehicle." In 1989, at the advent of the patterned crop glyphs, he again used the term while alluding to Earth's potential when its inhabitants come into harmonic balance between the spiritual and physical aspects of existence for themselves and between themselves. However, this time he altered the term to "a light space vehicle."

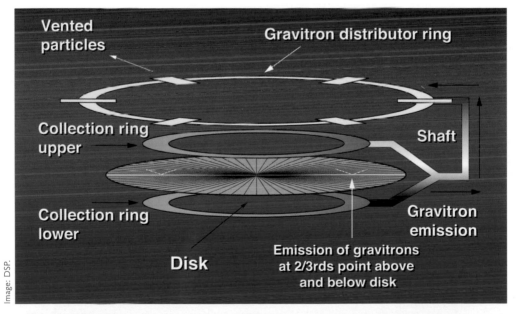

Figure 20.13. Gravitron drive basics, 1993. Although in this illustration the gravitron distributor ring (GDR) is sitting above the upper collection ring, this is only to explain the means by which these connections are made. When in a physical craft the GDR sits parallel with the disk but farther away from it.

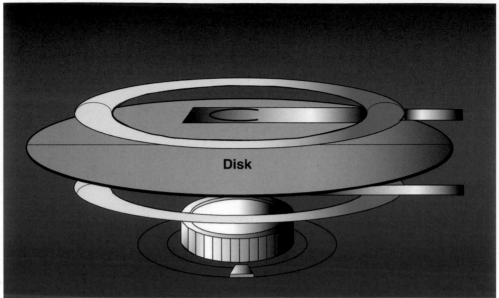

Figure 20.14. Simplified gravitron drive concept, 1993, showing the lower magnetic levitation ring and the associated stabilizing disk flyer on its track. Any change of alignment and ensuing center of gravity is immediately compensated by this flyer.

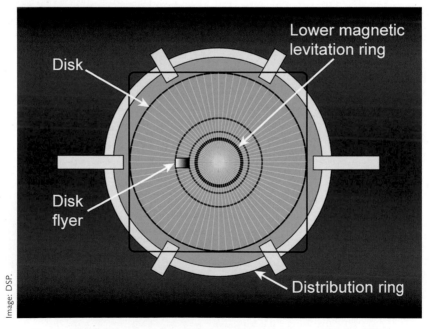

Figure 20.15. Lower level of the gravitron drive, top view, 1993. This early illustration has the distributor ring labelled as the distribution ring.

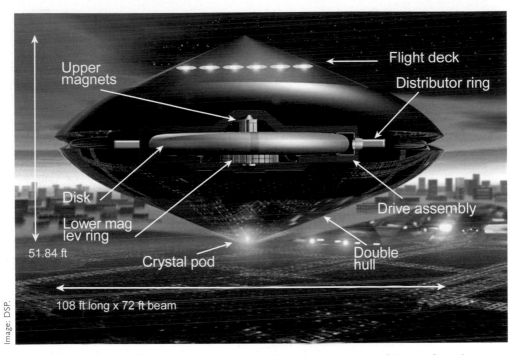

Figure 20.16. The Light Space Vehicle. Side view cutaway of the craft with superimposed internal drive assembly. Here Lower mag lev ring means the lower magnetic levitation ring.

Given that this spacecraft concept mimics the Earth's processes in order to function, it too was designed to specific harmonic ratios, which is why, when Myers and Percy first designed the craft in 1992–1993, they called it the Light Space Vehicle. And of course at that time they did not know that photons actually have mass. Looking further at how human beings can benefit from such a craft, Myers and Percy proposed that any living organism venturing beyond the confines of the Earth in a craft *should have the provision of a similar environment to the home planet to provide adequate environmental and radiation protection.* Otherwise, the organisms will be venturing naked before creation: "First they will lose heart, and then they will die," as Myers and Percy stated in 1993. At the time it was not widely known that the human muscle atrophies in space.

PROTECTION FROM SPACE RADIATION

A study published in October 2018 suggested that deep-space travel might significantly damage gastrointestinal function in astronauts. This study also raised concerns about the high risk of tumor development in the stomach and colon. Traveling long distances in space could destroy astronauts' guts, according to the major NASA-funded study. The *Independent* reported in October 2018 that this research raises substantial red flags about the possibility of humans taking journeys to places such as Mars.[8]

Myers and Percy realized that the Dharmic Bracelet crop glyph, discovered in a field adjacent to Silbury Hill, Avebury, England, in August 1992, specifically emphasized the absolute necessity to mimic the fundamental conditions of the home planet to ensure human survival in deep space. The glyph was designed to represent what can happen to self-aware beings when they do *not* live in an environment with the correct amount of flow of planetary energy or correct induced flow of energy. Consequently, in order to protect astronauts aboard craft

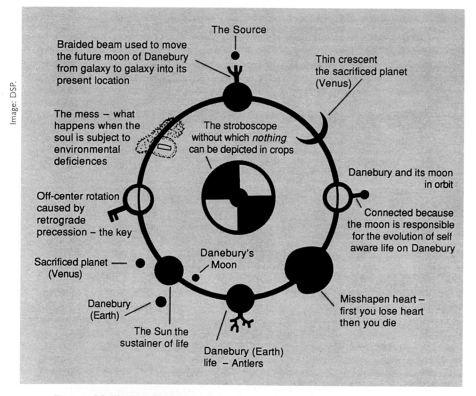

Figure 20.17. The Dharmic Bracelet or Dharmic Wheel crop glyph, Silbury Hill, August 17, 1992. Danebury is the Earth. The eight depictions are intended to be read as pairs opposite each other on the ring. The Mess was a grassy area around a rectangular metal cattle trough. The grass had been laid flat, and the water in the cattle trough was at a lower level than earlier in the night, as was evidenced by the watermark.

venturing into deep space, an appropriate *miniature* magnetosphere should be part of the craft's fundamental design. To be effective, such a shield against radiation must be sufficient to resist penetration by harmful solar and/or cosmic particles. Therefore, an active deflector shielding system should be designed into the craft, in addition to any passive shielding. Such an approach would function in a way similar to that of the Earth's magnetosphere.

The *Two-Thirds* text makes it absolutely clear that traveling anywhere in the solar system and beyond requires the replication of the environment and certain fundamental operations of the home planet. Earth has a magnetosphere that shields the planet from virtually all forms of radiation. This includes solar wind particles as well as solar (energetic) particle events and includes galactic cosmic rays (SPEs and GCRs as these are commonly known). A magnetosphere is the region of space surrounding an object in space in which charged particles are controlled by that object's magnetic field. Figure 20.19 on page 474 is a model of the Earth with its radiation belts overlaid with a spacecraft. Such a craft would in turn require its own zones of similar protection against space radiation.

The entire craft and all its working parts can be modeled by a geometric vesica piscis. This geometry is then modeled at different scales. Also involved in the blueprint are the platonic solids other than the tetrahedron: the cube, the octahedron, the icosahedron, and the dodecahedron. Since the full understanding of these relationships and the harmonics involved requires many overlays and more pages than this book can provide, these geometries will be pursued on the Aulis.com/pathway.

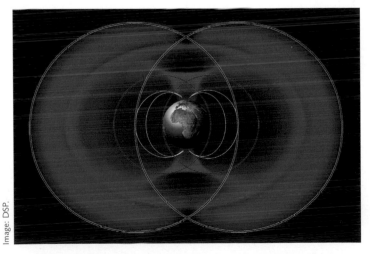

Image: DSP.

Figure 20.18. The vesica piscis–based magnetosphere around the Earth with its associated Van Allen belts.

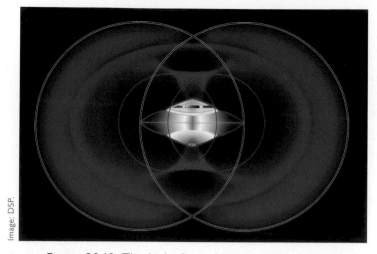

Image: DSP.

Figure 20.19. The Light Space Vehicle within its own vesica piscis field. Had the USGS mappers read *Two-Thirds* when they laid out their vesica piscis at Cydonia?

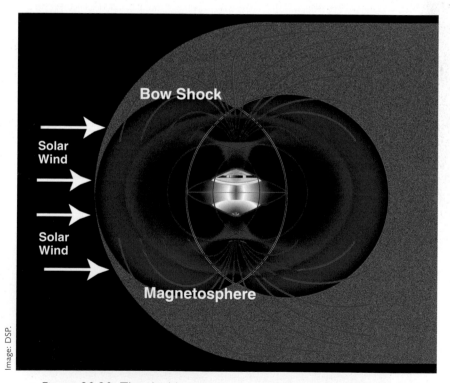

Image: DSP.

Figure 20.20. The shield generated by the Light Space Vehicle acts as its protection field. The components of the craft's internal drive system create the different vesica piscis.

THE SPACECRAFT PROPOSITION

The illustrations below show what an actual drive system might look like. An electromagnetic method of levitation would need to support the disk by way of the lower magnetic ring. A horseshoe arrangement of magnets would press down on the disk from above. Combined with the lower magnetic-levitation ring, this design would result in secure, friction-free levitation. Because the disk is suspended by magnetic bearings, nothing touches the disk as it spins.

The results of experimental research will determine the most appropriate way to develop this technology. It's not just about the means of propulsion, it's also the design of the entire craft.

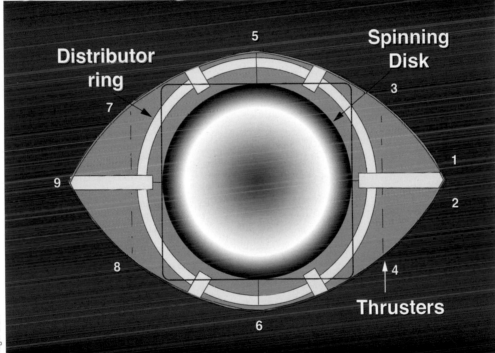

Figure 20.21. Simplified top view of the drive in situ in a craft, 1993. The numbers around the outside indicate the location of the thrusters that cannot be seen from this top view.

If converted into practical experimental research models, elements of the theoretical concept might look like the drive components shown in the following figures.

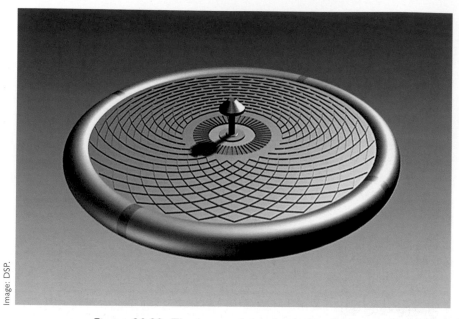

Image: DSP.

Figure 20.22. The heart of the drive: the disk.

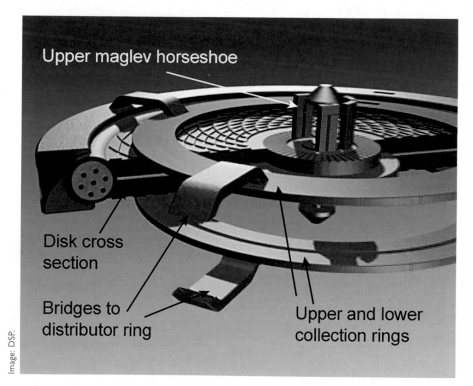

Upper maglev horseshoe

Disk cross section

Bridges to distributor ring

Upper and lower collection rings

Image: DSP.

Figure 20.23. Cutaway view of the collection rings located above and below the disk. Again maglev refers to the magnetic levitation ring.

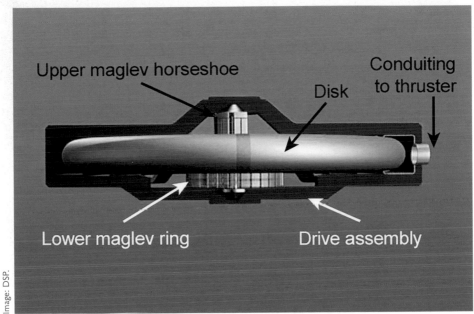

Figure 20.24. Cutaway section of the disk mounted
in the drive assembly.

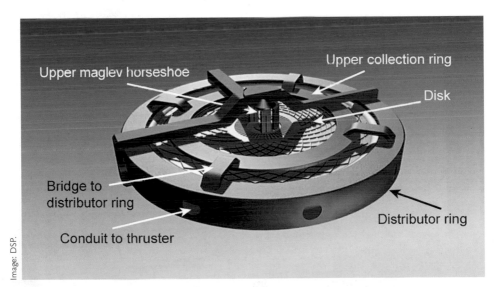

Figure 20.25. The disk set into the collection ring/distributor
ring assembly with ports at the side that connect the shielded
conduiting to the thruster locations around the craft.

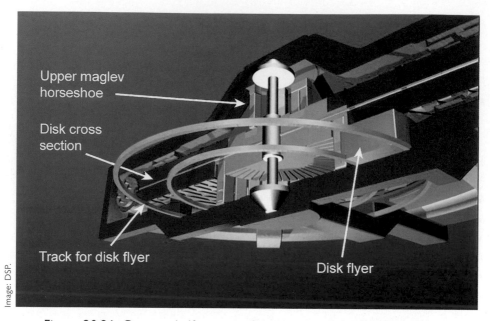

Image: DSP.

Figure 20.26. Cutaway half section of the full assembly depicting the lower maglev ring, the track for the flyer, and the disk flyer itself.

Embarking on a long-distance crewed deep-space mission in an inadequate craft could be likened to passengers crowding into an unsuitable, underpowered rubber boat with the intention of crossing an expanse of turbulent waters to distant land. They may or may not make it.

METAPHOR

Returning from the world of engineering to the process of messaging and communication, it is important to remember that many symbols and images are offered as metaphors, that is, as educational tools covering the fundamentals to those interested in gaining a more complete understanding of the way things work. As already suggested, somewhere in their noncoding DNA, humans could have everything required to enable the decoding of these metaphors, but often cultural upbringing or day-to-day life gets in the way.

Using geometric models to create relationships, pass on messages, and represent energy fields is not normally taken into account when mapping or, indeed, engineering planets or spacecraft. Yet, the USGS, by mapping Cydonia with interlocking circles and creating a vesica piscis, does all of these things. One could consider it a message to ETI because it is essentially another version of the gold-plated copper disks sent out on the *Voyager* spacecraft in 1977.[9]

The same principle applies here. The USGS layout on the Martian surface is achieved with mathematics but reveals secular and sacred geometry. Using 10s and 12s (kilometers and miles), it encodes cultural matters, classical mythology, mind, body, and spirit notions, Earth data, and biblical references. It then follows that the mixing of miles, kilometers, and feet can take on other meanings than that of simple distance and scale measurement as generally understood. It is possible to consider the mile, a 12-base system, as accessing periods of time: the distant past, the far future, or the present. The 12-base system aligns with the six-sided hexagon and interlocked tetrahedra. Kilometers are 10-base systems aligned with the five-sided and pentagonal shapes, with their internal phi-ratio stars, and with humans counting by tens with their fingers.[10] So there is a 6/5 relationship as well as a 12/10 relationship. And those are the first two blocks on that master number series, which is quite appropriate for the builders of stone models representing solar and lunar energy fields, as found in the layout of Stonehenge. For a more complicated example of this master number series, I could mention the ninth position of this series, which has $5 \times 9 = 45$ and $6 \times 9 = 54$, which together sum to 99. In its multiple progression from 9 to 18, 27, 36, 45, 54, 63, 72, 81, and 90, the 9× table is the only one that reaches 54 and then mirrors itself until the tenth position of this multiplication table. The eleventh position reiterates that number 99 the sum of 45 and 54, while the twelfth position is the number 108.

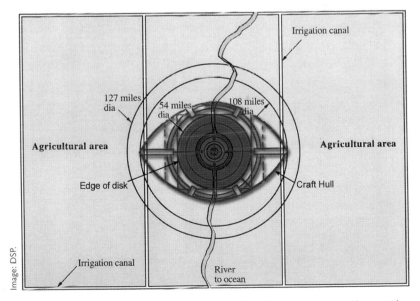

Figure 20.27. Plan of Atlantis based on Plato's writings together with the top view of a Light Space Vehicle with its 54-foot-diameter disk.

All this has subtle connections with the operation of the gravitron drive. It is built with a 54-foot diameter solid disk (in a 108-foot craft). Those who can look through the reflecting mirror will find more information behind the glass, and even more when the 54-foot-diameter disk is matched to a 54-mile circle.

The two digits, 5 and 4, sum to 9, and the ninth block of the master number series is 45/54, and the tenth block refers to the electrical frequencies used, namely 50 Hz and 60 Hz, which sums to 110 Hz. Paul Devereaux's acoustic research in the Kennet Long Barrow at Avebury comes into focus here relative to the RSD technology. Devereaux had attributed this tonality to the male voice, thinking it likely that the local priest/shaman toned within the barrow to harmonize with the Earth. (110 Hz is the tone A^2, and that would harmonize with the 13th overtone of the C major sequence, as described in chapter 19). Clearly, these researchers—Devereaux, Myers, and Percy—were getting good information, and their writings can be assessed at many different levels.

The Kennet Long Barrow faces due east, and according to those researching British energy fields, it is the recipient of the yin energy of the Mary energy line. The yang energy of the Michael line bypasses it completely, going from Avebury over to the Sanctuary, but that's another story.[11] Is there a potential union here of yin/earth and yang/human energy, and referring back to the RSD, is that a clue as to the activation of the start-up energy required for magnetic induction?

The diameter of the Stonehenge earthworks (the outer ditch ring) relative to the Avebury circle outer rim is 3.66:1. Using the USGS school of encoding time cycles onto planets, that ratio acknowledges the Earth's leap year of 366 days, which occurs once every four years. Which in turn leads to another measurement made available by those metaphorical educational purposes: the foot. If humans used their fingers to count quantities, they used the foot for other measures. Pacing out a circle from a central pole and evaluating angles relative to the position of the Sun led to the linking of time with space.[12]

When a story tells you that a six-foot RSD is the smallest diameter effective for R&D purposes, then it's useful to reverse the process and think of how technology translates to a bio-organism. A six-foot-tall person would fit nine times into a drive system of fifty-four feet and overall eighteen times into the length of the keel of the primary craft. Which is a useful hint, because the total RSD concept requires its crew to contribute to the operating system. Once again, harmonics and ratios play a part. The human head is on average around one-seventh of the physical body, leading to hints concerning the chakra system and the crown chakra in particular. Given that according to ETI there are actually nine chakras related to the human being, the other two being above the head,[13] I hypothesize that these two additional chakras relate to hyperdimensional energetic zones—and are associated with the unseen astral and etheric "bodies." There are many symbols within all cultures of self-aware beings that infer our connection to other realities than simply the physical, and apart from those used in sacred geometry and in math, various sounds, shapes, and colors also provide threads of continuity. As does human DNA. This too is a subject that requires moving images to fully do it justice, and it will be further explored online.

DOLPHINS

When it comes to biology, it is said that the mammals that most closely mirror self-aware beings are the cetaceans. The dolphin and the whale have long figured in the myths and legends of human beings, and it is no surprise therefore that these creatures were portrayed in the crop glyphs. What was surprising was the fact that three of them turned up over time, forming a triangular relationship across the Wiltshire landscape.[14] One of them was positioned near to the village of Lockeridge. What was not immediately discerned was that its length matched that of the nearby Kennet Long Barrow. Nor the fact that, scaled down, the dolphin also matched the viable part of the barrow where the shaman would be sitting (toning at 110 Hz, as Devereaux suggests), and the fact that this barrow extends beyond the inner recesses to form a sightline to Silbury Hill hints at even more energetic combinations. Geography, physics, and metaphorical illustrations combine to provide more hard, useful information pertinent to communications and spacecraft modeling.

The form that at first glance looked like a dolphin or a whale became the shorthand name for the event: this is how the 1991 Lockeridge Dolphin was born. It was named after the Wiltshire village nearest to the field in which it was found and, thinking only of the mammal it resembled, the fact that the glyph might refer to matters such as time cycles and harmonic ratios was missed by most. Upon being professionally surveyed, it revealed numbers close to the sidereal (27.3 days) and synodic (29.5 days) orbital periods of the Moon around the Earth, the mass ratio of the Earth to the Moon (80:81), and the axial period of rotation relative to the Earth and the Sun (24:25). Myers and Percy noted this data and, having had some experience of these cetacean glyphs, realized it was an important metaphor relative to the spacecraft's fundamental technology. Remembering that John C. Lilly, NASA, and the SETI group were involved

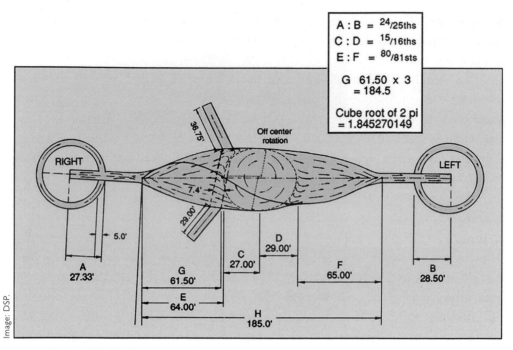

Figure 20.28. Representation of the 1991 Dolphin crop glyph, Lockeridge, Wiltshire, England. The Dolphin contained the same lunar references in the radii of its central swirled circle as there were stones originally in the two smaller circles at Avebury (twenty-seven stones in the northern so-called lunar circle and twenty-nine in the so-called solar southern circle), and its lengths A:B, C:D, and E:F, could also be inverted and expressed musical harmonics.

with research on dolphin interaction with humans, whether these same agencies ever made the connection between dolphins, dolphin crop glyphs, and safe space travel for humans is not known. As Phyllis Schlemmer told me, "We have got a huge amount of data on dolphins from TOM." She didn't offer to share it with me, and there wasn't all that much about dolphins in the material already published by her group, so I asked her why she hadn't published it. "After all," I said, "it's a subject of immense interest to all those involved in personal development and ETI." She told me that the information was all in secure storage—and changed the subject.*

I had the immediate certainty that this data had been passed over to the

---

*Thinking about the relevant passages concerning dolphins from ETI communications in both *Two-Thirds* and *The Only Planet of Choice* and taking that to the work of Lilly reveals many layers of meaning not immediately grasped at first reading.

relevant parties and subsequently locked away from public access. With this dolphin glyph arriving at the aptly named Lockeridge, that seemed a fair assessment.

Thankfully, Myers and Percy had earlier made the connection, albeit in surprising ways. The year before and to the west of the Lockeridge dolphin, another cetacean had turned up, and it was so big that many observers called it a whale. Its "fins" were not set off-center, as were those of the Lockeridge dolphin. This, they were told by their ETI contact, needed to be placed inside the ground plan of a cathedral. Fully appreciating that these buildings enclose sacred spaces is one notion, and that these enclosed spaces define areas of special benefit for human beings and bio-organisms is another, as is the biblical story of Jonah and the whale. Interestingly, the cathedral they "selected" (they were working with both intuitive and rational thinking when translating their information into illustrations and text) was Winchester in Hampshire. Analyzing these glyphs meant much more than they knew at the time. Chosen for its use in illustrating the principles of energy transconversion from one state to another, this cathedral was also located near to a site renowned for UFO activity and crop glyph formations. And it was at this location, Chilcomb Down, that the glyph depicted in figure 20.9 was activated in May 1990, as discussed earlier. Winchester Cathedral, although these two definitely did not know it, is unique in that its bells are tuned to C major diatonic—just like the frequencies found in crop glyphs by Gerald Hawkins. And although the three men were not in communication on the matter, as it turns out, Hawkins's musical theory as it related to the geometry of the early crop glyphs would later be seen to pick up on specific components of Myers and Percy's spacecraft.[15]

When Myers and Percy combined yet another cathedral, Salisbury Cathedral, with the Great Pyramid (figure 19.1, page 425) by overlaying the 2-D cathedral plan vertically onto the pyramid, this exercise involved taking a cross-section of the pyramid and observing what the combined cathedral outline was revealing. It explained so much regarding the internal pyramid layout, including why the King's Chamber is offset from the central column. As ridiculous as it may seem to some, all the interior solid areas of this metaphorical pyramid are filled with chalk. The *exception* is the central column here represented by the cathedral, the chambers, shafts, and passageways, plus the implied torus at the base level of the pyramid. The overlay of these two designs was said to be a model for a highly advanced super-speed quantum or A.I. computer. In terms of metaphor, this juxtaposition and the description are highly informative. Remembering the functional origins of this religious building, the inversion here has its palace throne, judgment seat, or apse locked to the Earth. The inference that the pyramidal structure has chalk surrounding its cavities reminds one of the human body, its bones and its organs. A pyramid filled with chalk is also a reference to the calcium of which bones are made, and if a spacecraft does not function in harmony with its environment, as has been found, astronauts' bones will suffer badly and deteriorate, as is presently the case on the ISS. In space, with no gravity provision onboard, astronauts will lose on average 2 percent of their bone mineral density per month because bones no longer have to support the body against Earth's full gravity.

The crop glyphs that first appeared in geometric forms were near to Silbury Hill, which is largely constructed of chalk and is often surrounded by water emerging from the powerful springs in its vicinity. Recall that this hill is the analog to the cone mound on Cydonia from which the phi spiral emerges to tangent the crater, the Face, and the D&M Tor. Silbury Hill's diameter fits into the base of the Great Pyramid, and the Great Pyramid's base fits the platform at Avebury. These cathedral/dolphin/pyramid overlays bring even more to the table when one remembers that the 2-D layout of a cathedral into such a pyramid infers the mathematical *net* of a hexagon. And if one considers that the imagery of the dolphin is illustrating parallel situations for a human being or an astronaut in space if not provided with the appropriate environment for living. Badly housed captive dolphins have been known to commit suicide—simply by choosing not to surface for their next breath. Which is as much a condemnation of the practice of keeping dolphins in restrictive conditions as it is of NASA's current crewed spacecraft plans.*

Which takes us back to the USGS Mars mapping that created a vesica piscis on Mars and another aspect of those dolphin crop glyphs. Rotated through 90°, the vesica piscis is less sacred geometry and more sacred art. It is used to frame images of the Virgin and Child. Or in Egyptian terms, Isis and Horus. In astronomical terms, that can read as Virgo and Pisces, and in cultural terms, as all that Masonic zodiacal imagery in Washington, DC. For the mystically inclined, the notion of fish and fishing boats has been linked to the parable of Jesus telling the fishermen how to get more fish by casting their nets on the other side of the boat than that which they were using. Those two circles of lesser Labyrinthus and greater Colles might reflect the sides of the boat and the nets inherent within this parable, while the parable itself really describes that these hardworking fishermen, weary and

---

*NASA-funded studies suggest that long-distance space travel could do significant damage to people's brains and might age them prematurely. As of February 2019, Northwestern University in the United States announced that it has researchers developing a predictive model to help NASA anticipate conflicts and communication breakdowns among crew members and to head off problems that could make or break the mission to Mars.

in despair from having worked hard with no results, were still sufficiently courageous to accept what they were told and trust the information enough to act on it. Getting an overabundance of rewards for their efforts, 153 fish were caught. A number that Munck would surely note sums to 9—a number that was also the precise difference between those two diameters of Colles and Labyrinthus. What about those 153 fish? Dolphins are not present in the freshwater Sea of Galilee, but swimming within the waters and breathing the air of land, they do represent the interface of land and sea. By taking the symbol of a dolphin to represent energetic aspects and not the literal fish, more information becomes available relevant to the management of the human-spacecraft interface. Given the dolphin's ability to function effectively with its chosen medium and given that the medium changed over eons from water to land and back to water, then the mythos is again present when recalling the alternating water/land rings on Plato's Atlantis model and the fact that the Minoan palaces on Crete contained dolphin symbols.

Bearing in mind that what was called a dolphin was not necessarily the creature itself. Nor that it was necessarily a side view of the body but rather a front-on view of processes connected with the dolphin and its functions. The necessity of thinking about all matters discussed in terms of physical organisms and as technical blueprints demonstrating the process of energy conversion becomes ever more apparent. These glyphs can also be interpreted as portraying various aspects of a dolphin's functionality, and with the dolphin, it's all in the brain.[16]

Cetaceans are noted for their innate intelligence and their affinity to human beings. Their sensitivity to very slight changes within a human organism—changes that the human is often not aware of, including incipient pregnancy—is well documented. However, when it comes to the cetacean's echolocation abilities, there are some important hints of other physics matters, such as waves and particles. Dolphins enjoy swimming alongside the bow of a ship. The motion of the ship breaks the waves at an angle of 19.47° as it moves through the water. The effect of sound traveling through water, which consists of tiny particles, is to induce those particles to move back and forth. This acoustic vibration also creates a pressure wave characterized by compression and expansion of the medium it is operating in. Within the abdominal cavity of the dolphin there is a gas-filled bladder. Gas is more compressible than water, so the swim bladder expands and contracts according to the pressure changes that occur within the sound field. If all fish detect particle motion within their environment, then only some species also detect pressure. Flat fish don't even have a swim bladder and can only sense particles. The octopus only detects particle motion, with one exception, the pelagic octopus, colloquially known as the football octopus. Dolphins detect both particle and pressure waves, and fish that can detect pressure have better sensitivity and a wider hearing range than those that do not.

The Latin name of the swim bladder is the vesica piscis. And waves and particles lead one to think of several aspects of spacecraft engineering, including the photon double slit experiment and how one might manage the input and output of photons into the craft, along with the atmospheric requirements of the occupants.

At the time of Pythagoras, the vesica piscis was considered to be a sacred form due to its length-to-height ratio. In geometrical terms, that is the square root of three divided by two (1.732/2 = 0.866). Which is the adjacent side of that Cydonian special right triangle, and its halfway point. That is marked by the flat-topped mesa no one in any of the space agencies and their affiliated organizations wants to talk about publicly. There is one final measure I have not mentioned: the square root of five. As the diagonal of the double square, it can be related to the base of advanced computers and then some. As the ratio 1.618 + 0.618, it has affinities with the golden ratio and again then some. As the foundation of the fractal Mandelbrot set, it has affinities with computing and with crop glyphs. An exquisite representation of this fractal appeared near Cambridge in the United Kingdom and was never proven to be made by human beings.[17] Interpreted by Myers and Percy as being very relevant to the RSD technology and to human thinking, it emerged in another place altogether. The Nazca plateau—another location replete with glyphs.

I contend that much of the Nazca plateau is a demonstration of an application of the technical capability of a craft with its RSD technology. The number-nine thruster of such a craft can be used to produce immensely powerful focused braided beams of energy. It could well be that such technology was used to scalp that flat-topped mesa on Cydonia, thereby producing an artificially leveled mesa totally unlike either its immediate neighbor or anything else on the Martian complex. As already suggested, similar technology could have been used to create the flat Nazca plateau, as seen in the upper part of the photograph in figure 1.1. The same technology would also make an easy job of producing the straight lines on the surface areas of Nazca. This would include the trapezoid shapes across miles of Andean mountains and define many of the glyphs found on the plain aptly named Plain of Creation. As with crop glyphs, several of the Nazca glyphs represent species not indigenous to their location. These foreigners include the monkey (Mono) and the hummingbird, which turns out not to be a hummingbird after all! And again, alas, as with crop glyphs, the originals have had their imitators. The story told by the monkey glyph with its 5-4 fingered hands and Manos with its 4-5 fingered hand and its right-angled "shoulder" recalls all those references to the 54/45 ninth place of the master number series. Furthermore, when recalling that Mono's feet have three toes each and that he is weaving three pyramids from his boustrophedon furrows while his tail is curled in a spiral brings references to the Giza big three and the multiples of the three groups of three pyramids extant on the plateau. The three-fingered toes also recalling Colette Dowell's little alien Gray with its three fingers. Manos, that head-and-hands glyph with its right angle, recalls the right angle that marks the position of the Face on Mars relative to the spiral mound and the five-sided pyramid. Taking the hint that at Nazca the Manos right angle forms one corner of a diagonal and Mono the other, we can connect their hands with this line, which when taken as the square root of five produces the geometry of the glyph found in the United Kingdom resembling the computer-generated Mandelbrot fractal.

But Nazca, where the story of humanity is laid bare, the memories possibly once stored in human noncoded DNA revealed in plain sight, is a story for another day. With all that we have discovered, can there be any doubt that the collective consciousness of the planet has been educated into the workings of the universe, in a fun way, from the very beginning? But this a game with serious intent. The fact that NASA and its affiliates spent decades sneering at those who find UFOs and ETI subjects worth considering and found the arrival of crop glyphs less than funny is not so much laughable as sad. It has taken the agency until February 2020 to reverse this attitude, and to coin a phrase, come out of the closet. Perhaps. At the time of going to press there is no clear indication as to whether the apparent recognition of sightings ascribed to ETI will end in tears of joy or sadness.[18]

AN ENDING AND A BEGINNING

We end where we began. Human beings have always looked up at the night sky and wondered, "Is there anyone out there?" We now have the capacity to answer that question, and perhaps to make a discovery that would rank as the most profound in the history of humanity. By March 2019, the *National Geographic* was announcing that "We Are Not Alone." Seems that this too was previously being said generally in terms of fear-based scenarios, but this time something else has changed. Preempting the announcement, 2020 saw another breakthrough for science when on January 25 a modification to the Einstein theory of general relativity concerning gravity was finally accepted as a viable hypothesis to pursue.

Einstein's theoretical particles of gravity are assumed to be massless. Until this year, the massive theory had met with some resistance, as it proposes that these hypothetical particles (gravitons) that are said to mediate the gravitational force have a mass. It also posited that if that hypothesis stands, then gravity would have a weaker influence on very large distance scales. These massive theory scientists are also interested in other aspects of gravity, such as its speed, which has never directly been measured, and also whether gravity moves at different speeds through different materials.[19]

While this possibly justifies the lack of any dark energy so far found and any slight alterations in light speed found to date, since dark matter, dark energy, and the graviton remain totally hypothetical, these scientists might like to switch horses and take a closer look at the photon. What if this massive theory is partly right in that the universe *is* populated with particles that do have mass, but these are actually photons. As discussed, photons have now been shown to have mass, as seen in the holographic imagery. What if *these photons* make up this missing energy?[20]

In any case, whether it's a particle or a force, weaker gravity would cede the place to the stronger photon and permit faster light speeds in certain environments, and that would lead to the understanding that the speed of light is not a constant. And that it is the mass of a photon that can be spun up to do work. The Barbury Castle glyph explained this in 1991, when taken simply as a model of photon energy conversion from slow to fast.

Whether all this extrapolation of glyphs, symbols, and metaphor into blueprints for a new technology is a question of inspired creativity leading over logic or the basis of a workable technology going forward, only experimentation and time will tell. Offering day trips to the edge of the Earth's atmosphere is all well and good, but the future of humankind may well depend on the ability to successfully travel into deep space. For that, a total conceptual renewal of spacecraft technology is required, especially concerning the generation of even a partial g-force aboard the craft, which would be essential for the sustained well-being of future human space travelers. With such a technology, it must be possible for enterprising pioneers from Earth to be able to travel safely and set foot on Mars in a healthy condition. And all the research findings thus far negate the possibility of humans making successful journeys to destinations such as Mars in NASA's current spacecraft in the near future. In 2010, Robert Zubrin, president of Pioneer Astronautics and the Mars Society, said:

> I don't think that NASA has a direction. That is precisely the problem. The administration is proposing to spend 10 years and $100 billion to accomplish nothing. NASA needs a goal. That goal should be humans to Mars by 2020.[21]

The Martian project, as von Braun called it, so far has been long and rambling, with numerous wrong turns and many failures along the way, although there have been successes with a number of probes. The matter of getting ETI contact off the pages of science fiction and into science fact has been equally convoluted. Now the game is up. Data that seemed to be all metaphor reveals itself to be all fact. And mainstream scientific concepts, written in stone for decades, are finally being reappraised. The time is now ripe for a paradigm shift, and with a totally new approach to crewed space travel, the pathway to Mars opens up before us.

Acknowledgments

First and foremost I would like to thank the diligent editorial team at Bear & Company for all their hard work in bringing this work to fruition, with a special mention for my project editor, Kayla Toher, whose patience and kindness in seeing the preparation of the manuscript through to publication knew no bounds.

My thanks go especially to Jon Graham for his continued belief in the project and to John Hays for his enthusiasm at the London Book Fair right at the outset.

Writer Sam J was invaluable—his constructive comments following his incisive read through of an early version of the text resulted in a much-improved final manuscript.

My grateful thanks also go to my colleague David S. Percy for the gift of his unique photos. He has supported and encouraged this work throughout and as well as providing a sounding board for my evolving concepts, he took my rough drawings and rudimentary sketches and transformed them into beautifully crafted illustrations and informative maps. Any errors of transmission are mine, and mine alone.

To name all those who have contributed to this project would take an inordinate amount of space—since it would mean listing everyone I have met since the first day I stepped into a crop glyph in 1990. Whether our paths have remained the same or diverged along the way, I am more than grateful for your friendship, your time, and your thoughts. Please consider that you are included in this note of thanks: as the saying goes, "None of us is as smart as all of us."

MARY BENNETT
WILTSHIRE, NOVEMBER 2020

Notes

INTRODUCTION

1. Jet Propulsion Laboratory, "Howdy, Strangers," August 19, 2002.
2. *Guardian,* "Astronomers to Sweep Entire Sky for Signs of Extraterrestrial Life," February 15, 2020.

CHAPTER 1.
THE QUEST FOR ANSWERS

1. Carl Sagan, *Cosmos* (New York: Random House, 1980), 5, 30–31.
2. Alice Calaprice, ed., *The Ultimate Quotable Einstein* (Princeton, N.J.: Princeton University Press, 2010), 474. Section misattributed to Einstein.
3. Jacquetta Hawkes, "God in the Machine," *Antiquity* 41, no. 163 (1967): 174.
4. Hugh Newman, *Earth Grids: The Secret Patterns of Gaia's Sacred Sites* (Glastonbury, England: Wooden Books Ltd., 2008), 1.
5. Peter Tompkins, *Mysteries of the Mexican Pyramids* (London: Thames & Hudson, 1987), 226–37.
6. Scott Olsen, *The Golden Section: Nature's Greatest Secret* (Glastonbury, England: Wooden Books Ltd., 2006), 28.
7. David P. Myers and David S. Percy, *Two-Thirds: A History of Our Galaxy* (London: Aulis Publishers, 1993), 413–14.
8. Phyllis V. Schlemmer, *The Only Planet of Choice: Essential Briefings from Deep Space,* 2nd ed., edited by Mary Bennett (Bath, England: Gateway Books, 1994), 53.
9. Tompkins, *Mysteries of the Mexican Pyramids,* 247–48.
10. Myers and Percy, *Two-Thirds,* 362.
11. Schlemmer, *Only Planet of Choice,* 156–59.

12. Sagan, *Cosmos,* 129–30.

13. Tompkins, *Mysteries of the Mexican Pyramids,* 247, 251–53.

14. Myers and Percy, *Two-Thirds,* 27, 358.

15. Mark Lehner, *The Complete Pyramids* (London: Thames & Hudson, 1997), 222.

16. Stan Gooch, *The Neanderthal Legacy: Reawakening Our Genetic and Cultural Origins* (Rochester, Vt.: Inner Traditions, 2008), 129.

CHAPTER 2.
THERE'S NO BUSINESS LIKE SHOW BUSINESS

1. Phil Kouts, "Towards a Moon Base: Leaving Apollo's Legacy Behind," AULIS Online, April 2016. Also addresses Mars travel plans.

2. Jessica Orwig, "Docking with the International Space Station Is So Insanely Complicated It's a Wonder We Ever Get It Right," *Business Insider,* March 30, 2015.

3. Mike Wright, "The Disney–Von Braun Collaboration and Its Influence on Space Exploration," NASA monograph, NASA website.

4. John M. Logsdon, *Exploring the Unknown: Selected Documents in the History of the United States Civilian Space Program,* vol. 6, Space and Earth Science (NASA History Division, Washington, DC, 2004), 158; James Van Allen, "Radiation Belts around the Earth," *Scientific American* 200, no 3 (March 1959): 41–44.

5. Clive Dyer, personal interview, June 1997; Mary Bennett and David S. Percy, *Dark Moon: Apollo and the Whistle-Blowers* (London: Aulis Publishers, 1999), 97.

6. Bennett and Percy, *Dark Moon,* 101.

7. Ellen Stofan, BBC-TV *Newsnight* interview, November 2014; Mary Bennett, "Orion, the Van Allen Belts & Space Radiation Challenges," AULIS Online, October 2015.

8. Patrick L. Barry, "Radioactive Moon," NASA Science website, September 8, 2005.

9. Ernst Stuhlinger, *Wernher von Braun: Crusader For Space: A Biographical Memoir* (Malabar, Fla.: Krieger Publishing, 1994), 126; Roger D. Launius and J. D. Hunley, compilers, *An Annotated Bibliography of the Apollo Program,* NASA Monographs in Aerospace History, no. 2, July 1994, notes 32 and 33.

10. *Huntsville Times,* Walt Disney interview, 1965; Wright, "Disney–Von Braun Collaboration," notes 43 and 44.

11. Michael Benson, *Space Odyssey: Stanley Kubrick, Arthur C. Clarke, and the Making of a Masterpiece* (New York: Simon & Schuster, 2018).

12. Christopher Frayling, *The 2001 File: Harry Lange and the Design of the Landmark Science Fiction Film* (London, Reel Art Press, 2015).

13. William J. Broad, "The Bomb Chroniclers," *New York Times,* September 13, 2010.

14. Mike Glyer, "And He Built a Crooked Air Force Base," File 770 website,

January 14, 2015; Kevin Hamilton and Ned O'Gorman, *Lookout America!: The Secret Hollywood Studio at the Heart of the Cold War* (Hanover, N.H.: Dartmouth College Press, 2019).

15. Brian Cox, *Carpool,* television interview, July 24, 2009 (no longer available online).

16. John Cloud, "Imaging the World in a Barrel: CORONA and the Clandestine Convergence of the Earth Sciences," Social Studies of Science, Sage Publications, April 1, 2001; CIA, "CORONA: America's First Satellite Program," 1–54, pdf file available on the Central Intelligence Agency website.

CHAPTER 3.
VIKING WARRIORS

1. Michael Benson, *Space Odyssey: Stanley Kubrick, Arthur C. Clarke, and the Making of a Masterpiece* (New York: Simon & Schuster, 2018), 64–66; Carl Sagan and Jerome Agel, eds., *The Cosmic Connection* (London: Hodder & Stoughton, 1974), 181–84.

2. Carl Sagan, *Cosmos* (New York: Random House, 1980), 118–21.

3. Oliver Morton, *Mapping Mars: Science, Imagination, and the Birth of a World* (London: Harper Collins, Fourth Estate, 2002); David S. F. Portree, "Humans to Mars: Fifty Years of Mission Planning, 1950–2000," NASA History Office website.

4. Irene K. Fischer, *Geodesy? What's That?* (New York: iUniverse, 2005).

5. Richard C. Hoagland and Mike Bara, *Dark Mission: The Secret History of NASA* (Los Angeles: Feral House, 2007).

6. Robert Bauval and Adrian Gilbert, *The Orion Mystery: Unlocking the Secret of the Pyramids* (London: William Heinemann, 1994).

7. Sagan, *Cosmos,* 118–21.

8. Stan Gooch, *The Dream Culture of the Neanderthals* (Rochester, Vt.: Inner Traditions, 2006); John Michell and Christine Rhone, *Twelve Tribe Nations and the Science of Enchanting the Landscape* (London: Thames & Hudson, 1991).

9. David Ovason, *The Secret Zodiacs of Washington DC: Was the City of Stars Planned by Masons?* (London: Century Books, 1999).

10. Mary Bennett and David S. Percy, *Dark Moon: Apollo and the Whistle-Blowers* (London: Aulis Publishers, 1999), 462–71.

11. Graham Hancock, Robert Bauval, and John Grigsby, *The Mars Mystery: A Tale of the End of Two Worlds* (London: Michael Joseph, 1998).

12. Bennett and Percy, *Dark Moon,* 462–71.

13. Morton, *Mapping Mars.*

14. Extract from *Proposed Studies on the Implications of Peaceful Space Activities for Human Affairs (The Brookings Report).* NASA publishes some of the report's 250 pages under

different references. Verified February 2020. The Hathi Trust Digital Library holds a complete copy of the report given to the 87th Congress, 1st Session.

15. *The Brookings Report* NASA Document ID 19640053196; NASA Report/Patent Number NASA-CR-55643, vol 1; NASA Document ID 19640053194; NASA Report/Patent Number NASA-CR-55640, vol. 3.

CHAPTER 4. NEMES OR NEMESIS

1. Carl Sagan, *Cosmos* (New York: Random House, 1980), 130.

2. Sagan, *Cosmos,* 135.

3. Arthur C. Clarke, *The Snows of Olympus: A Garden on Mars* (London: Victor Gollancz, 1994).

4. Christopher Mellon, "The Military Keeps Encountering UFOs. Why Doesn't the Pentagon Care?" *Washington Post,* March 9, 2018; "Glowing Auras and 'Black Money': The Pentagon's Mysterious U.F.O. Program," *New York Times,* December 16, 2017; "On the Trail of a Secret Pentagon UFO Program," *New York Times,* December 18, 2016.

5. Eric M. Jones. "'Where Is Everybody?' An Account of Fermi's Question," U.S. Department of Energy, Office of Scientific and Technical Information website, March 1985. This is Eric Jones's complete account of the Fermi Paradox discussion.

6. David Baker, *Spaceflight and Rocketry: A Chronology* (New York: Facts on File Books, 1996).

7. Russell Targ, *Limitless Mind: A Guide to Remote Viewing and Transformation of Consciousness* (Novato, Calif.: New World Library, 2004); NASA contract 953653, 1975; Russell Targ, *Do You See What I See?: Memoirs of a Blind Biker* (Hampton Roads, Va.: Hampton Roads Publishing, 2010).

8. Mary Bennett and David S. Percy, *Dark Moon: Apollo and the Whistle-Blowers* (London: Aulis Publishers, 1999), 504.

9. Carl Sagan, *The Demon-Haunted World* (London: Headline Book Publishing, 1996).

10. Richard C. Hoagland, *The Monuments of Mars: A City on the Edge of Forever* (Berkeley, Calif.: North Atlantic Books, 1987), 4–5, 11.

11. Walter Hain, *We, from Mars: Old and New Hypotheses about the Red Planet* (Vienna, Austria: Karl Werner, 1992), 120–22; previously published under the title *Wir, vom Mars* (Cologne, Germany: Ellenberg Verlag, 1979).

12. David P. Myers and David S. Percy *Two-Thirds: A History of Our Galaxy* (London: Aulis Publishers, 1993), part 2, chapter 7.

13. Phyllis V. Schlemmer, *The Only Planet of Choice: Essential Briefings from Deep Space,* 2nd ed., edited by Mary Bennett (Bath, England: Gateway Books, 1994).

CHAPTER 5. MEASURE FOR MEASURE

1. USGS, "Astrogeology Science Center: Maps," USGS: Science for a Changing World website.
2. Mark Carlotto, *The Martian Enigmas: A Closer Look* (Berkeley, Calif.: North Atlantic Books, 1997).
3. MarsNews, "A 'New Cydonia' of Ancient Extraterrestrial Monuments Found on Mars." MarsNews website, August 18, 2009. Web archive summary posted originally by futurist Alfred Lambremont Webre.
4. Zecharia Sitchin's published works are available from Inner Traditions, Bear & Company, Rochester, Vermont.
5. Bryan Bender, "A New Moon Race Is On. Is China Already Ahead?" *Politico,* June 13, 2019.
6. Jacquelin Feldscher and Lui Zhen, "Are the U.S. and China on a War Footing in Space?" *Politico,* June 16, 2019.

CHAPTER 6. CURIOSITY KILLED THE CAT

1. NASA, "'Jake Matijevic' Contact Target for Curiosity," NASA website, September 19, 2012.
2. Clara Moskowitz, "Curiosity Rover's Pet Mars Rock 'Jake' Unlike Any Seen on Red Planet," SPACE.com website, October 11, 2012; E. M. Stolper et al., "The Petrochemistry of Jake_M: A Martian Mugearite," *Science,* September 27, 2013.
3. Alistair Munroe, "Welcome to Glenelg: Twinned with Mars," *The Scotsman,* October 2, 2012.
4. Wernher von Braun, *The Mars Project* (University of Illinois Press, 1953, 1962, 1991); German edition: *Das Marsprojekt* (Esslingen, Germany: Bechtle Verlag, 1952).
5. J. Nigro Sansonese, *The Body of Myth: Mythology, Shamanic Trance, and the Sacred Geography of the Body* (Rochester, Vt.: Inner Traditions, 1994); Homer, *Iliad* and *Odyssey.*
6. U.S. Department of Defense, "Department of Defense Establishes U.S. Space Command," Department of Defense website, August 29, 2019.
7. Iain McGilchrist, *The Master and His Emissary: The Divided Brain and the Making of the Western World* (New Haven, Conn.: Yale University Press, 2009).
8. Randolfo Rafael Pozos, *The Face on Mars: Evidence for a Lost Civilization?* (Chicago: Chicago Review Press, 1986).
9. Arthur M. Young, "About Arthur M. Young," Arthur M. Young website.
10. Lynn Picknett and Clive Prince, *The Stargate Conspiracy* (London: Little, Brown and Company, 1999).

11. For a list of books by Russell Targ, search for "Russell Targ at the ESP Research website; Sheila Ostrander and Lynn Schroeder, *Psychic Discoveries behind the Iron Curtain* (US: New York: Marlowe and Company, 1971, 1997; UK: London: Souvenir Press, 1997).

12. Mary Bennett and David S. Percy, *Dark Moon: Apollo and the Whistle-Blowers* (London: Aulis Publishers, 1999), 466.

13. Pozos, *Face on Mars,* 50.

14. Jim Channon, *First Earth Battalion Operations Manual* (CreateSpace [self-publishing service owned by Amazon], November 6, 2009; originally published in the 1970s); Jon Ronson, *The Men Who Stare at Goats* (London: Picador, 2004) 29–57; Ronald M. McRae, *Mind Wars: The True Story of Government Research into the Military Potential of Psychic Weapons* (New York: St. Martin's Press, 1984).

15. Ronson, *Men Who Stare at Goats.*

16. Ronson, *Men Who Stare at Goats.*

17. David Adams, "Gore's Climate Film Has [Nine] Scientific Errors," *Guardian,* October 12, 2007.

18. Richard C. Hoagland, *The Monuments of Mars: A City on the Edge of Forever* (Berkeley, Calif.: North Atlantic Books, 1987).

19. For more information, see the National Endowment for the Arts home page.

20. Graham Hancock, Robert Bauval, and John Grigsby, *The Mars Mystery: A Tale of the End of Two Worlds* (London: Michael Joseph, 1998).

21. Edgar Mitchell, *The Way of the Explorer* (New York: G. P. Putnam's Sons, 1996).

22. "From Mind Control to Murder? How a Deadly Fall Revealed the CIA's Darkest Secrets," *Guardian,* September 6, 2019.

CHAPTER 7.
CYDONIA LOST AND FOUND

1. Oliver Morton, *Mapping Mars: Science, Imagination, and the Birth of a World* (London: Harper Collins, Fourth Estate, 2002).

2. Walter Hain, *We, from Mars: Old and New Hypotheses about the Red Planet* (Vienna, Austria: Karl Werner, 1992); previously published under the title *Wir, vom Mars* (Cologne, Germany: Ellenberg Verlag, 1979).

3. Morton, *Mapping Mars.*

4. Patrick Moore, *Patrick Moore on Mars* (London: Cassell, 1998).

5. Google Earth Pro/Mars extract from electronic link placed on the Face on Mars black-and-white photograph.

6. USGS Astrogeology Science Center, GEOLOGIC INVESTIGATIONS SERIES I–2811 ATLAS OF MARS: QUADRANGLES MTM 40007, 40012, 40017, 45007,

45012, AND 45017 https://pubs.usgs.gov/imap/i2811/i2811.pdf. This McGill 2005 map contains within it the Tanaka 2003 map and all the geological data.

7. Mark Carlotto, *The Martian Enigmas: A Closer Look* (Berkeley, Calif.: North Atlantic Books, 1992).

8. Carlotto, *Martian Enigmas.*

9. David P. Myers and David S. Percy, *Two-Thirds: A History of Our Galaxy* (London: Aulis Publishers, 1993), preface.

CHAPTER 8. THE WASHINGTON CONNECTION

1. David Ovason, *The Secret Zodiacs of Washington DC: Was the City of Stars Planned by Masons?* (London: Century Books, 1999).

2. United States Navy Memorial, "Visitor Center," United States Navy Memorial website.

3. David Baker, *Spaceflight and Rocketry: A Chronology* (New York: Facts on File Books, 1996), year 1985.

4. United States Navy Memorial, "Visitor Center," United States Navy Memorial Website, Rear Admiral William Thompson.

5. Philippe Coppens, *The Stone Puzzle of Rosslyn Chapel* (Amsterdam: Frontier Publishing, 2004).

6. Mary Bennett and David S. Percy, *Dark Moon: Apollo and the Whistle-Blowers* (London: Aulis Publishers, 1999), 504.

7. Zecharia Sitchin, *The Earth Chronicles Expedition* (Rochester, Vt.: Bear & Company, 2007).

8. David P. Myers and David S. Percy, *Two-Thirds: A History of Our Galaxy* (London: Aulis Publishers, 1993), 208.

9. Ovason, *Secret Zodiacs.*

10. Gordon Strachan, *Chartres: Sacred Geometry, Sacred Space* (Edinburgh, Scotland: Floris Books, 2003), 11–12.

CHAPTER 9.
A REAL FRENCH CONNECTION

1. Steve Vogel, *The Pentagon: A History; The Untold Story of the Wartime Race to Build the Pentagon—And to Restore It Sixty Years Later* (New York: Random House, 2008).

2. Dobroslav Líbal, *Castles of Britain and Europe: Fortifications in Britain and Western Europe from the Roman Empire to the Last Century* (Prague: Aventinum Publishing House, 1992); English edition: (Leicester, Blitz Editions 1999), 231.

3. Philip Coppens, "Mitterrand's Great—Unknown—Work," Eye of the Psychic

website; article first appeared in *Les Carnets Secrets,* vol. 9 (2007), and *Atlantis Rising* (September–October 2011).

4. Claude Mollard, *La Saga de L'Axe Majeur: Dani Karavan A Cergy-Pontoise* (Paris: Beaux Arts Edition, June 2011); Agnes Sander and George Duby, *L'Axe Majeur, Cergy-Pontoise* (Paris: Beaux Arts Edition, June 2009); both titles published in French.

CHAPTER 10. SETI, CETI, OR DETI:
THAT IS THE QUESTION

1. SETI Institute, "Mission," SETI Institute website; SETI Institute, "History of the SETI Institute," SETI Institute website.

2. Guiseppe Cocconi and Philip Morrison, "Searching for Interstellar Communications," *Nature,* September 19, 1959.

3. Cocconi and Morrison, "Searching for Interstellar Communications."

4. Mary Bennett and David S. Percy, *Dark Moon: Apollo and the Whistle-Blowers* (London: Aulis Publishers, 1999).

5. Billy Meier, *Message from the Pleiades: The Contact Notes of Eduard Billy Meier,* edited by Wendell C. Stevens (Tucson, Ariz.: privately published, 1988).

6. Frank Drake and Dava Sobel, *Is Anyone Out There?* (New York: Delacorte Press, 1992); SETI Institute, "Early SETI: Project Ozma, Arecibo Message," SETI Institute website.

7. Green Bank Observatory, "History," Green Bank Observatory website.

8. Stephen J. Garber, "Searching for Good Science: The Cancellation of NASA's SETI Program," *Journal of the British Interplanetary Society* 52 (1999): 3–12.

9. SETI Institute "Drake Equation," SETI Institute website.

10. Douglas A. Vakoch, ed., "Archaeology, Anthropology, and Interstellar Communication," NASA History Series, NASA SP-2013-4413; Mark J. Carlotto, "Detecting Patterns of a Technological Intelligence in Remotely Sensed Imagery," *Journal of the British Interplanetary Society* 60, no. 1 (January 2007): 28–39.

11. Vakoch, "Archaeology, Anthropology, and Interstellar Communication."

12. Vakoch, "Archaeology, Anthropology, and Interstellar Communication."

13. James Fletcher, NASA administrator, 1975, text of speech, 1975, https://history.nasa.gov; go to 1975.

14. Fletcher, text of speech

15. Vakoch, "Archaeology, Anthropology, and Interstellar Communication."

16. John C. Lilly, *Man and Dolphin* (Garden City, N.Y.: Doubleday & Co, 1961); Christopher Riley, "The Dolphin Who Loved Me: The NASA-Funded Project That Went Wrong," *Observer,* June 8, 2014.

17. Drake and Sobel, *Is Anyone Out There?*

18. Drake and Sobel, *Is Anyone Out There?*

19. Garber, "Searching for Good Science."

20. SETI Institute, "Bernard M. Oliver (1916–1995)," SETI Institute website; "PROJECT CYCLOPS: A Design Study of a System for Detecting Extraterrestrial Intelligent Life," NASA STI Repository website.

CHAPTER 11. SWITZERLAND:
THE ANOMALOUS YEARS (1972–1990)

1. Michael Glickman, "The Bishops Cannings Basket," Michael Glickman on Crop Circles website, January 23, 2014; Geoffrey Gibbs and Sally James Gregory, "Fined— For Running Rings Round Crop Circles," *Guardian,* November 6, 2000; *Gazette & Herald* (Trowbridge, UK), "Secrets of Crop Circles," May 2, 2002.

2. CSIRO, "Australian Square Kilometre Array Pathfinder," CSIRO website.

3. Stuart Holroyd, *Briefing for the Landing on Planet Earth* (London: W. H. Allen/ Virgin Books, 1977).

4. Jim Dilettoso, "The Billy Meier Case: The Simple 'Farmer' Who Talked with the Star People," TJR Research website, circa 2000.

5. Dilettoso, "Billy Meier Case."

6. Werner Anderhub and Hans Peter Roth, *Crop Circles: Exploring the Designs and Mysteries* (New York: Sterling Publishing, 2002); also published in German in 2000 (Aarau, Switzerland: AT Verlag).

7. John Stuart Reid, "What Is Cymatics?" Cymascope: Sound Made Visible website.

8. *The Economist,* "Lake Monsters," November 3, 2012.

9. "Contournement de Genève Suisse [The construction tunnel of Confignon]," pdf file available online

10. Anderhub and Roth, *Crop Circles;* Walter A. Fuchs, "Mysterious Holes in Switzerland," pdf file available online.

11. *Astronomy Now,* "'Stealth Dark Matter' Theory May Explain Universe's Missing Mass," September 2015.

12. Ethan Siegel, "Could the Large Hadron Collider Make an Earth-Killing Black Hole?" *Forbes,* March 11, 2016.

13. Scott Henderson, initial research with additional new findings, "Apollo Space Suits: Shenanigans and Shortcomings," AULIS Online, September 2019.

14. CERN, "The Birth of the Web," CERN website.

15. Archie Roy, introduction to *Circular Evidence: A Detailed Investigation of the Flattened Swirled Crops Phenomenon,* by Pat Delgado and Colin Andrews (London: Bloomsbury, 1989).

16. Old Crop Circles website, "Historic Old Crop Circles." Charlton is the eighteenth entry in the UK circles section.

17. Island Lagoon Tracking Station: 1959–1972, "Island Lagoon Station, Woomera," Honeysuckle Creek Tracking Station website.

18. Delgado and Andrews, *Circular Evidence.*

19. Garvit Rawat, *NASA Squirming and the New Moon Order,* Kindle 2020 ISBN 978-1-7771665-0-2.

20. Stephen J. Garber, "Searching for Good Science: The Cancellation of NASA's SETI Program," *Journal of the British Interplanetary Society* 52 (1999): 3–12.

21. Michael Brooks, *13 Things That Don't Make Sense: The Most Intriguing Scientific Mysteries of Our Times* (London: Profile Books, 2009).

CHAPTER 12. OBSERVERS, SURVEYORS, DODMEN, AND DREAMERS

1. Christopher Priest, *The Prestige* (London: Simon & Schuster, 1995).

2. Richard C. Hoagland, *The Monuments of Mars: A City on the Edge of Forever* (Berkeley, Calif.: North Atlantic Books, 1987).

3. Priest, *Prestige.*

4. M. D. Johnston, P. B. Esposito, V. Alwar, S. W. Demcak, E. J. Graat, and R. A. Mase, "Mars Global Surveyor Aerobraking at Mars," Jet Propulsion Laboratory, California Institute of Technology, NASA: Mars Exploration Program website, 1998.

5. Johnston et al., "Mars Global Surveyor."

6. Priest, *Prestige.*

7. Jim Dilettoso, "The Billy Meier Case: The Simple 'Farmer' Who Talked with the Star People," TJR Research website, circa 2000; Mark Carlotto, *The Martian Enigmas: A Closer Look* (Berkeley, Calif.: North Atlantic Books, 1992).

8. Carlotto, *Martian Enigmas.*

9. Steve Vogel, *The Pentagon: A History; The Untold Story of the Wartime Race to Build the Pentagon—And to Restore It Sixty Years Later* (New York: Random House, 2008).

10. Michael Glickman, *Crop Circles: The Bones of God* (Berkeley, Calif.: Frog Books, 2009).

11. Elizabeth Suckow and Chris Jedrey, "Hidden Headquarters," NASA HQ History Division and the Office of Headquarters Operations website, March 24, 2009.

CHAPTER 13. THE NAME GAME AND THE A-WORD

1. Mary Bennett, "Mojave to Mars . . . via the Way of the Dead," AULIS Online.

2. Stanton T. Friedman, *Flying Saucers and Science: A Scientist Investigates the Mysteries of*

UFOs: Interstellar Travel, Crashes, and Government Cover-ups (Pompton Plains, N.J.: Career Press/New Page Books, 2008); Jim Marrs, *Alien Agenda: The Untold Story of the Extraterrestrials among Us* (New York: HarperCollins Publishers, 1997); Mary Bennett and David S. Percy, *Dark Moon: Apollo and the Whistle-Blowers* (London: Aulis Publishers, 1999), part 2, chapter 6.

3. Andrei Bulatov and Alexander Boyko, "The April Odyssey and the November Boat: The 1970 Event from the Russian Perspective," AULIS Online, November 2016.

4. Mary Bennett and David S. Percy, "The Odyssey of the Lost Apollo CM: A Detailed Analysis of the April 1970 Event," AULIS Online, November 2016.

5. NASA's voting rules can be found on their History of the STS (Space Shuttle Transport System) website.

6. Phyllis V. Schlemmer, *The Only Planet of Choice: Essential Briefings from Deep Space,* 2nd ed., edited by Mary Bennett (Bath, England: Gateway Books, 1994).

7. *Project Scanate: Exploratory Research in Remote Viewing.* Approved for release 2002/11/13 and available online.

8. Jim Marrs, *PSI Spies* (Franklin Lakes, N.J.: Career Press/New Page Books, 2007).

9. Andrija Puharich, *Uri: The Authorised Biography of Uri Geller, the World's Most Famous Psychic* (London: W. II. Allen, 1974).

10. Marrs, *PSI Spies.*

11. Courtney Brown, review of *Mind Trek: Exploring Consciousness, Time, and Space through Remote Viewing,* by Joseph McMoneagle, courtneybrown.com.

12. Joseph McMoneagle, *Mind Trek: Exploring Consciousness, Time, and Space through Remote Viewing* (Charlottesville, Va.: Hampton Roads Publishing, 1993, 1997).

13. John Cloud, "Imaging the World in a Barrel: CORONA and the Clandestine Convergence of the Earth Sciences," *Social Studies of Science,* Sage Publications, April 1, 2001.

14. Cloud, "Imaging the World in a Barrel."

15. Cloud, "Imaging the World in a Barrel."

CHAPTER 14. MATRIX MAN

1. John Cloud, "Imaging the World in a Barrel: CORONA and the Clandestine Convergence of the Earth Sciences," *Social Studies of Science,* Sage Publications, April 1, 2001.

2. Cloud, "Imaging the World in a Barrel."

3. Joseph McMoneagle, *Mind Trek: Exploring Consciousness, Time, and Space through Remote Viewing* (Charlottesville, Va.: Hampton Roads Publishing, 1993, 1997).

4. Carl Munck, *The Code* (1996).

5. David Kahn, *The Codebreakers* (New York: Scribner, 1967; rev. ed., 1996).

6. Mary Bennett and David S. Percy, *Dark Moon: Apollo and the Whistle-Blowers* (London: Aulis Publishers, 1999); For information on Roswell: Kevin C. Ruffner, ed., *Corona: America's First Satellite Program,* CIA Cold War Records Series (Washington, D.C., History Staff Center for the Study of Intelligence Central Intelligence Agency, 1995.)

7. Association for Research and Enlightenment, "Our Ancient Egyptian Heritage," Edgar Cayce's AR.E website.

8. Mark Lehner, *The Complete Pyramids* (London: Thames & Hudson, 1997).

9. Munck, *The Code.*

10. Reconstructed extract from dialogue in Steven Spielberg's movie *Close Encounters of the Third Kind* (Culver City, Calif.: Columbia Pictures, 1997).

11. David Wood, *Genesis: The First Book of Revelations* (Kent: Baton Press, 1985); Henry Lincoln, *Key to the Sacred Pattern: The Untold Story of Rennes-le-Château* (Oxfordshire: Windrush Press, 1997).

12. Anatoly I. Kandiew in a Carl Munck Newsletter, circa 1993.

13. Kandiew in a Carl Munck Newsletter.

CHAPTER 15.
WHAT A WAY TO RUN A RAILROAD

1. John Cloud, "Imaging the World in a Barrel: CORONA and the Clandestine Convergence of the Earth Sciences," *Social Studies of Science,* Sage Publications, April 1, 2001.

2. Cloud, "Imaging the World in a Barrel."

3. Cloud, "Imaging the World in a Barrel"; Irene K.Fischer, *Geodesy? What's That?* (New York: iUniverse, 2005).

4. Astrogeology Science Center, "Maps," USGS, Science for a Changing World website.

CHAPTER 16. CLOSER ENCOUNTERS

1. Reconstructed extract from dialogue in *Close Encounters of the Third Kind* (Culver City: Columbia Pictures, 1997).

2. Taken from the top center of Devil's Tower using Google Earth Pro, 2018.

3. Stuart Holroyd, *Briefing for the Landing on Planet Earth* (London: W. H. Allen, 1977).

4. Phyllis V. Schlemmer, *The Only Planet of Choice: Essential Briefings from Deep Space,* 2nd ed., edited by Mary Bennett (Bath, England: Gateway Books, 1994), 44, 142, 160.

5. Eltjo H. Haselhoff, "Opinions and Comments on Levengood WC, Talbott NP (1999)

Dispersion of Energies in Worldwide Crop Formations. Physiol Plant 105: 615–624," *Physiologia Plantarum* 111 (2001): 123–25.

6. Tijana Radeska, "Walt Disney Produced Propaganda Films for the U.S. Government during WWII," Vintage News website, August 2, 2016; Diana Brown, "Disney World and Its Tangled Web with CIA Ops," How Stuff Works website, January 5, 2018; Kevin Hamilton and Ned O'Gorman, *Lookout America!: The Secret Hollywood Studio at the Heart of the Cold War* (Hanover, N.H.: Dartmouth College Press, 2019).

7. J. Deardorff, B. Haisch, B. Maccabee, and H. E. Puthoff, "Inflation Theory Implications for Extraterrestrial Visitation," *Journal of the British Interplanetary Society* 58 (2005): 43–50.

8. James W. Deardorff, "Possible Extraterrestrial Strategy for Earth," *Quarterly Journal of the Royal Astronomical Society* 27 (1986): 94–101.

9. National Research Council, *Forging the Future of Space Science: The Next 50 Years: An International Public Seminar Series Organized by the Space Studies Board: Selected Lectures* (Washington, D.C.: National Academies Press, 2010), 147–48.

10. Jacques Vallée, *Dimensions: A Casebook of Alien Contact* (London: Souvenir Press, 1988).

11. Curtis Fuller, Proceedings of the First International UFO Congress (New York: Warner Books, 1980), 156–65.

CHAPTER 17. GIZA AND CYDONIA

1. Maria Carmelo Betro, *Hieroglyphics: The Writings of Ancient Egypt* (New York: Abbeville Press, 1996), 57.

2. Mark Lehner, *The Complete Pyramids* (London: Thames & Hudson, 1997).

3. Association for Research and Enlightenment, "Ancient Wisdom and Civilizations in the Cayce Readings," Edgar Cayce's A.R.E. website.

4. Robert Temple with Olivia Temple, *The Sphinx Mystery: The Forgotten Origins of the Sanctuary of Anubis* (Rochester, Vt.: Inner Traditions, 2009).

5. Lehner, *Complete Pyramids.*

6. David P. Myers and David S. Percy, *Two-Thirds: A History of Our Galaxy* (London: Aulis Publishers, 1993), see the Great Pyramid figures in the appendix.

7. Hugh Newman, *Earth Grids: The Secret Patterns of Gaia's Sacred Sites* (Glastonbury, England: Wooden Books, 2008).

8. Myers and Percy, *Two-Thirds.*

9. Robert Schoch with Robert Aquinas McNally, *Voices of the Rocks: A Scientist Looks at Catastrophes and Ancient Civilizations* (New York: Harmony Books, Penguin Random House, 1999); Robert M. Schoch and Robert Bauval, *Origins of the Sphinx: Celestial Guardian of Pre-Pharaonic Civilization* (Rochester, Vt.: Inner Traditions, 2017).

10. The "Scan Pyramids" Project online, October 19, 2015.

11. Lucy Pringle, *Crop Circles: The Greatest Mystery of Modern Times* (London: Thorsons, 1999).

CHAPTER 18. GIZA GAZING

1. Rudolf Gantenbrink, "The Upuaut Project," The Upuaut Project official website.

2. Robert Bauval, *The Orion Mystery: The Revolutionary Discovery That Rewrites History* (London: Heinemann, 1994).

3. See Dr. Zahi Hawass's home page for an insight into Hawass and his viewpoint of all matters Egyptian.

4. Charles Piazzi Smythe, *Our Inheritance in the Great Pyramid* (London: Wm. Isbister, 1874).

5. Mary Bennett and David S. Percy, *Dark Moon: Apollo and the Whistle-Blowers* (London: Aulis Publishers, 1999), 464–65.

6. William Hutton, "Locations of the Records of the Atlantean Civilization and Its Firestone: Including Speculations on Where and How They Will Be Found," Biblioteca Pleyades website, October 2001, updated January 2, 2004; John Van Auken, *2038: The Great Pyramid Timeline Prophecy* (Virginia Beach, Va.: 4th Dimension Press, 2012).

7. Bennett and Percy, *Dark Moon,* 210–36.

8. Adam Janos, "Why Have There Been So Many UFO Sightings near Nuclear Facilities?" History Channel website, June 21, 2019.

9. Ryan Sprague, "The Artist behind the 'Communion' Book Cover," Rogue Planet website April 23, 2019. Clicking on the link "full interview" in this article will take you to the October 6, 1999, archived article "Ted Seth Jacobs: An Interview with the Artist" by Will of Beyond Communion.

10. Colette Dowell, Circular Times (Dowell's home page).

11. David R. Williams, "The Apollo 15 Hammer-Feather Drop," NASA Space Science Data Coordinated Archive website.

12. Brian Cox, BBC TV, *Human Universe,* episode 4, first shown October 28, 2014; a script transcript, TVO50 Website.

13. Michael Nicholson, *The Dude Painting,* Saatchi Art website; "The Big Lebowski" plot summary, Internet Movie Database.

14. Wernher von Braun, *The Mars Project* (Champaign: University of Illinois Press, 1953, 1962, 1991); German edition: *Das Marsprojekt* (Esslingen, Germany: Bechtle Verlag, 1952).

15. J. E. Brandenburg, "Evidence for a Large, Natural, Paleo-Nuclear Reactor on Mars," 42nd Lunar and Planetary Science Conference, 2011; pdf file available online.

16. J. E. Brandenburg, "Evidence of Massive Thermonuclear Explosions in Mars Past, The Cydonian Hypothesis, and Fermi's Paradox," 2014 Annual Fall Meeting of the APS Prairie Section, Monmouth, Illinois, November 21–22, 2014. The full text of his paper is available at archive.org.

17. Bennett and Percy, *Dark Moon*, 96–97.

18. J. E. Brandenburg, "Evidence for Large, Anomalous Nuclear Explosions," "Program of Technical Sessions," The Forty-Sixth Lunar And Planetary Science Conference, The Woodlands, Texas, March 16–20 2015.

19. "Science Conferences (#LPSC2015), Ivory Gates, and Who Gets In," Exposing PseudoAstronomy website, March 26, 2015.

20. Dr. Zahi Hawass home page.

21. NASA, "Mars Exploration Rovers," Mars Exploration Rovers website.

22. Oberon Zell, "A New Map of Barsoom," ERBZine website.

CHAPTER 19. THE SOUND OF LIGHT

1. Steve Marshall, *Exploring Avebury: The Essential Guide* (Cheltenham, England: History Press, 2016).

2. Robin Heath, *Sun Moon and Stonehenge: Proof of High Culture in Ancient Britain* (England: Bluestone Press, 1998); Robin Heath, "A New Year Message from the Past," Sky and Landscape: Megalithic Research by Robin Heath website, December 29, 2018.

3. Robin Heath, *Alexander Thom: Cracking the Stone Age Code* (Cardigan, Wales: Bluestone Press, 2007).

4. John Neal, *All Done With Mirrors: Opus 2* (London: Secret Academy, 2000).

5. Richard Hoagland, *Dark Mission: The Secret History of NASA* (Port Townsend, Wash.: Feral House, 2009).

6. Jonathan Goldman, *Healing Sounds: The Power of Harmonics* (Rochester, Vt.: Healing Arts Press, 1992).

7. Eric Taylor, *The AB Guide to Music Theory,* part 1, 2nd ed. (London: Associated Board of the Royal Schools of Music, 1990).

8. Cris Foster, *Musical Mathematics: On the Art and Science of Acoustic Instruments* (San Francisco, Calif.: Chronicle Press, 2010), 484–504; R. H. van Gulik, *The Lore of the Chinese Lute: An Essay in the Ideology of the Ch'in,* 3rd ed. (Bangkok: Orchid Press, 2011).

9. Phyllis V. Schlemmer, *The Only Planet of Choice: Essential Briefings from Deep Space,* 2nd ed., edited by Mary Bennett (Bath, England: Gateway Books, 1994), 5, 156.

10. Andrija Puharich, *Uri: The Authorised Biography of Uri Geller, the World's Most Famous Psychic* (London: W. H. Allen, 1974); H. G. M. Hermans, *Memories of a*

Maverick: Andrija Puharich M.D., LL.D (Holland: Pi Publications, 1998), 56–57; Stuart Holroyd, *Briefing for the Landing on Planet Earth* (London: W. H. Allen, 1977, Corgi Edition, 1979), 46–47; Rick J. Carlson, ed., *The Frontiers of Science and Medicine,* The May Lectures (London: Wildwood House, 1975): 156–62.

11. David P. Myers and David S. Percy, *Two-Thirds: A History of Our Galaxy* (London: Aulis Publishers, 1993) part 4, chapter 4.

12. Puharich, *Uri,* 16–17.

13. Schlemmer, *Only Planet of Choice,* 284.

14. Freddy Silva, *Secrets in the Fields: The Science and Mysticism of Crop Circles* (Charlottesville, Va.: Hampton Roads Publishing, 2002), 118–23; Freddy Silva, "The Biophysics of Crop Circles," Crop Circle Secrets website, 1998 and 2000.

15. John Burke and Kaj Halberg, *Seeds of Knowledge Stones of Plenty* (San Francisco/Tulsa, Okla.: Council Oaks Books, 2005).

16. John Reid, *Egyptian Sonics: A Preliminary Investigation Concerning the Hypothesis That the Ancient Egyptians Had Developed a Sonic Science by the Fourth Dynasty* (Cumbria, UK: Sonic Age Limited, 2001); CymaScope: Sound Made Visible website, "Cymatics Experiment in the Great Pyramid"; Alan Alford, "Pyramid of Secrets—The Singing Pyramid and the Myth of Creation," Robert Bauval official website, article first appeared in the *Daily Mail,* June 21, 2003.

17. Thomas Danley, "The Great Pyramid: Early Reflections & Ancient Echoes," Live Sound International website, July/August 2000.

18. NASA, "Bed Rest FAQs," NASA Analog Missions website.

19. Could This Happen? Website, "Double Take: NASA's Twin Study," May 22, 2014.

20. P. D. Hodkinson, R. A. Anderton, B. N. Posselt, and K. J. Fong, "An Overview of Space Medicine," *British Journal of Anaesthesia* 119, issue suppl. 1 (December 2017): i143–i153; Compare and contrast: NASA website, "Space Medicine," August 12, 2004.

CHAPTER 20.
BLUEPRINTS FOR THE RED PLANET

1. John Martineau, ed., *Megaliths: Studies in Stone* (Glastonbury, England: Wooden Books Ltd., 2010).

2. Gyorgy Doczi, *The Power of Limits: Proportional Harmonics in Nature, Art, and Architecture* (Boulder, Colo.: Shambhala Publications, 1981).

3. John Martineau, ed., *Quadrivium: The Four Classical Liberal Arts of Number, Geometry, Music, and Cosmology* (Glastonbury, England: Wooden Books Ltd., 2010).

4. Michael S. Schneider, *A Beginner's Guide to Constructing the Universe, the Mathematical*

Archetypes of Nature, Art, and Science (New York, Harper Perennial, HarperCollins, 1994); Warwick Cairns, *About the Size of It: The Common Sense Approach to Measuring Things* (London: Macmillan, 2007).

5. Nick Cook, *The Hunt for Zero Point* (London: Century Books, 2001).

6. Radosław Chrapkiewicz, Michał Jachura, Konrad Banaszek, and Wojciech Wasilewski, "Hologram of a Single Photon," *Nature Photonics* 10 (July 18, 2016): 576–79; Photonics Views, "Hologram of a Single Photon," Photonics Views website, July 19, 2016.

7. Charles Seife, "Quantum Mechanics: The Subtle Pull of Emptiness," *Science* 275, no. 5297 (January 10, 1997): 158; Conrado Salas Cano, "The Universe's Storehouse of Energy," Check the Evidence website, April 11, 2008.

8. Karen Teber, "Animal Study Suggests Deep Space Travel May Significantly Damage GI Function in Astronauts." Georgetown University Medical Center News, October 1, 2018; Andrew Griffin, "Travelling to Mars and Deep Space Could Kill Astronauts by Destroying Their Guts, Finds NASA-Funded Study," Independent, October 1, 2018.

9. Carl Sagan, F. D. Drake, Ann Druyan, Timothy Ferris, Jon Lomberg, and Linda Salzman Sagan, *Murmurs of Earth: The Voyager Interstellar Record* (New York: Random House; London: Hodder & Stoughton, 1978).

10. George Ifrah, *The Universal History of Numbers: From Prehistory to the Invention of the Computer* (London: Harvill Press, 1998); first published in French as *L'Histoire Universelle des Chiffres*, 1994.

11. John Michell, *New View over Atlantis* (London: Thames & Hudson, new edition, 1986); Hamish Miller and Paul Broadhurst, *The Sun and the Serpent* (Cornwall, UK: Mythos, 1990).

12. Ifrah, *Universal History of Numbers.*

13. Phyllis V. Schlemmer, *The Only Planet of Choice: Essential Briefings from Deep Space*, 2nd ed., edited by Mary Bennett (Bath, England: Gateway Books, 1994), 284.

14. Michael Glickman, *Crop Circles* (Glastonbury, England: Wooden Books, 2005), 16.

15. Freddy Silva, *Secrets in the Fields: The Science and Mysticism of Crop Circles* (Hampton Roads, Va.: Hampton Roads publishing, 2002), 193–200; Ivan Peterson, "Geometric Harvest," *Science News,* February 1, 1992, 76; available only to subscribers, but a scan of original article can be found at the Crop Circle Research website.

16. Schlemmer, *Only Planet of Choice,* 140.

17. David P. Myers and David S. Percy, *Two-Thirds: A History of Our Galaxy* (London: Aulis Publishers, 1993), part 3, chapter 13.

18. Michael Horn, "Michael Horn to Present NASA Discoveries Confirming Billy Meier UFO Case Is Real," Cision PR Newswire website, May 21, 2019.

19. Claudia de Rham, "Has Physicist's Gravity Theory Solved 'Impossible' Dark Energy Riddle?" *Guardian,* January 25, 2020.

20. Ervin Laszlo, *Science and the Akashic Field: An Integral Theory of Everything* (Rochester, Vt.: Inner Traditions, 2004).

21. Robert Zubrin, "Voices: Experts and Analysts Weigh In On NASA's New Direction," Space website, October 1, 2010; Robert Zubrin, "The Mars Decision," *The New Atlantis,* no. 60 (Fall 2019): 46–60.

Index

About the Authors

Mary Bennett studied music at Dartington College of Arts and then spent much of her life as a researcher, writer, and translator. Always maintaining a balance between the professional world and the psychic abilities she has had since childhood, she also held executive positions at major brand corporations, including les *must* de Cartier, in Paris, France. This involved travel to North and South America, the Middle East, England, Switzerland, Belgium, Holland, and Spain. In the late 1970s, the internal realization that she had books to write became such a driving force that she felt impelled to prepare her departure from the corporate world. With the aim of acquiring a wider knowledge of art, architecture, and sculpture, she joined the Parisian art gallery Artcurial.

In 1983, she completed the transition, becoming an independent translator in Paris. After working as French liaison for the production of the British movie *The Ebony Tower*, filmed in France, she visited India and Kashmir before relocating to England and beginning a full-time writing career, working freelance in Manchester for various publishers and Granada TV while continuing to investigate the interaction of consciousness with alternative energy systems.

A further move to Hampshire, in southern England, in 1989 meshed with the emergence of crop circles into public consciousness. Bennett was in the right place at the right time. The numerous PSI and EDI events that have always been part of her life increased in quantity and quality. One of these had influenced her decision to move south and another stimulated the organization of an experiment in Salisbury designed to explore the effect of sound on local energy lines and the relationship of crop circles to sound. Following the success of the Salisbury experiment, she set up a conference on crop glyphs and ETI connections, at which David S. Percy was invited to give a talk on the Face on Mars and the Avebury connection. After that event, Bennett was asked to edit *Two-Thirds: A History of Our Galaxy* (Aulis Publishers, 1993). This was followed by a request from Sir John Whitmore to edit the second edition of *The Only Planet of Choice* (Gateway Books, 1994), a bestseller that has been recognized as unique in its category by the trade and the public alike. She then researched and co-wrote *Dark Moon: Apollo and the Whistle-Blowers* (Aulis Publishers, 1999).

A freelance consultant and translator to publishers, media organizations, and private individuals, Bennett practices Qigong, Taiji, and Zen meditation in order to maintain and improve the psychic abilities that have also provided valuable insights, as well as checks and balances, on the astonishing information unearthed during her research for *Alien Intelligence and the Pathway to Mars,* her latest book.

Bennett lives in Wiltshire, England, and is the co-editor of Aulis Online, a platform covering subjects ranging from the Apollo Moon record and the future of human space travel through to the origins of humans and the Mars/Earth connections.

Photo: Frances Pinter

David S. Percy is a fellow of the Royal Society of Arts and a long-standing associate of the Royal Photographic Society. An award-winning cinematographer and film and television producer, he is a well-established professional communicator in the world of commerce. His expertise has been sought by multinational corporations as well as not-for-profit organizations. An early pioneer in computer graphics and animation, Percy is also a still photographer, graphic designer, and author.

Traveling the world researching ancient sites and making documentaries, Percy has filmed in Nagaland at the foothills of the Himalayas, from open-sided helicopters, out of the rear of Hercules aircraft flying at ten thousand feet, and from camera-adapted ambulances in the Middle East.

Percy produced the world's first annual report on video for the international Emhart Corporation in 1982. He was director of photography on one of the first British 35mm anamorphic widescreen short movies with Dolby sound for theatrical release, and his first movie, *The Anna Contract,* ran continuously in London's West End for three months in the 1970s. Further cinema credits include director of photography for *Discomania* and *Knights Electric,* often referred to as an early precursor to the music video.

Percy was a trailblazer in computer animation and video-wall design in the 1980s, and he and his team produced one of the first fully interactive multimedia resources on laser disc.

He created the missile-tracking visual effects sequences for video supervisor Ira Curtis-Coleman featured in the 1985 John Landis film *Spies Like Us.* That same year he photographed a major theatrical movie featuring the new Tornado aircraft for the British Ministry of Defence: *The Third Dimension.*

In 1997, Percy produced the film *The Face on Mars: The Avebury Connection,* a follow-up to a previous production, *The Terrestrial Connection,* filmed at the United Nations Headquarters in New York in 1992. In 2000, Percy continued his investigation into the Apollo record and directed the film *What Happened on the Moon?*

Percy lives in London and is the co-editor of Aulis online and the head of Aulis Publishing.